TOPICS IN THE THEORY OF CHEMICAL AND PHYSICAL SYSTEMS

Progress in Theoretical Chemistry and Physics

VOLUME 16

The titles published in this series are listed at the end of this volume.

Topics in the Theory of Chemical and Physical Systems

Proceedings of the 10th European Workshop on Quantum Systems in Chemistry and Physics held at Carthage, Tunisia, in September 2005

Edited by

SOUAD LAHMAR

Laboratoire de Spectroscopie Atomique et Moléculaire et Applications, Université de Tunis El-Manar, Tunisia

JEAN MARUANI

Laboratoire de Chimie Physique, CNRS and UPMC, Paris, France

STEPHEN WILSON

Rutherford Appleton Laboratory, Oxfordshire, U.K.

and

GERARDO DELGADO-BARRIO

Instituto de Matemáticas y Física Fundamental, CSIC, Madrid, Spain

Springer

A C.I.P. Catalogue record for this book is available from the Library of Congress.

ISBN-13 978-94-017-8123-7
ISBN-13 978-1-4020-5460-0 (eBook)

Published by Springer,
P.O. Box 17, 3300 AA Dordrecht, The Netherlands.

www.springer.com

Printed on acid-free paper

PROGRESS IN THEORETICAL CHEMISTRY AND PHYSICS

A series reporting advances in theoretical molecular and material sciences, including theoretical, mathematical and computational chemistry, physical chemistry and chemical physics

Aim and Scope

Science progresses by a symbiotic interaction between theory and experiment: theory is used to interpret experimental results and may suggest new experiments; experiment helps to test theoretical predictions and may lead to improved theories. Theoretical Chemistry (including Physical Chemistry and Chemical Physics) provides the conceptual and technical background and apparatus for the rationalisation of phenomena in the chemical sciences. It is, therefore, a wide ranging subject, reflecting the diversity of molecular and related species and processes arising in chemical systems. The book series *Progress in Theoretical Chemistry and Physics* aims to report advances in methods and applications in this extended domain. It will comprise monographs as well as collections of papers on particular themes, which may arise from proceedings of symposia or invited papers on specific topics as well as from initiatives from authors or translations.

The basic theories of physics – classical mechanics and electromagnetism, relativity theory, quantum mechanics, statistical mechanics, quantum electrodynamics – support the theoretical apparatus which is used in molecular sciences. Quantum mechanics plays a particular role in theoretical chemistry, providing the basis for the spectroscopic models employed in the determination of structural information from spectral patterns. Indeed, Quantum Chemistry often appears synonymous with Theoretical Chemistry: it will, therefore, constitute a major part of this book series. However, the scope of the series will also include other areas of theoretical chemistry, such as mathematical chemistry (which involves the use of algebra and topology in the analysis of molecular structures and reactions); molecular mechanics, molecular dynamics and chemical thermodynamics, which play an important role in rationalizing the geometric and electronic structures of molecular assemblies and polymers, clusters and crystals; surface, interface, solvent and solid-state effects; excited-state dynamics, reactive collisions, and chemical reactions.

Recent decades have seen the emergence of a novel approach to scientific research, based on the exploitation of fast electronic digital computers. Computation provides a method of investigation which transcends the traditional division between theory and experiment. Computer-assisted simulation and design may afford a solution to complex problems which would otherwise be intractable to theoretical analysis, and may also provide a viable alternative to difficult or costly laboratory experiments. Though stemming from Theoretical Chemistry, Computational Chemistry is a field of research in its own right, which can help to test theoretical predictions and may also suggest improved theories.

The field of theoretical molecular sciences ranges from fundamental physical questions relevant to the molecular concept, through the statics and dynamics of isolated molecules, aggregates and materials, molecular properties and interactions, and the role of molecules in the biological sciences. Therefore, it involves the physical basis for geometric and electronic structure, states of aggregation, physical and chemical transformation, thermodynamic and kinetic properties, as well as unusual properties such as extreme flexibility or strong relativistic or quantum-field effects, extreme conditions such as intense radiation fields or interaction with the continuum, and the specificity of biochemical reactions.

Theoretical chemistry has an applied branch – a part of molecular engineering, which involves the investigation of structure–property relationships aiming at the design, synthesis and application of molecules and materials endowed with specific functions, now in demand in such areas as molecular electronics, drug design or genetic engineering. Relevant properties include conductivity (normal, semi- and supra-), magnetism (ferro- or ferri-), optoelectronic effects (involving nonlinear response), photochromism and photoreactivity, radiation and thermal resistance, molecular recognition and information processing, and biological and pharmaceutical activities; as well as properties favouring self-assembling mechanisms, and combination properties needed in multifunctional systems.

Progress in Theoretical Chemistry and Physics is made at different rates in these various fields of research. The aim of this book series is to provide timely and in-depth coverage of selected topics and broad-ranging yet detailed analysis of contemporary theories and their applications. The series will be of primary interest to those whose research is directly concerned with the development and application of theoretical approaches in the chemical sciences. It will provide up-to-date reports on theoretical methods for the chemist, thermodynamician or spectroscopist, the atomic, molecular or cluster physicist, and the biochemist or molecular biologist who wishes to employ techniques developed in theoretical, mathematical or computational chemistry in his research programme. It is also intended to provide the graduate student with a readily accessible documentation on various branches of theoretical chemistry, physical chemistry and chemical physics.

CONTENTS

PREFACE

This volume contains a representative selection of papers presented at the *Tenth European Workshop on Quantum Systems in Chemistry and Physics* (QSCP-X), held at *Beït al-Hikma*, seat of the *Académie Tunisienne des Sciences, des Arts et des Lettres*, in Carthage, Tunisia, September 1-7, 2005. About 90 scientists from 18 countries, half of them from North Africa, attended the meeting, which focussed on the evolution of current issues and problems in methods and applications.

This workshop continued the series that was established by Roy McWeeny near Pisa (Italy), in April 1996, then continued on a yearly basis: Oxford (1997), Granada (1998), Paris (1999), Uppsala (2000), Sofia (2001), Bratislava (2002), Athens (2003), Grenoble (2004) .... The purpose of QSCP workshops is to bring together chemists and physicists with a common field of interest – the quantum mechanical theory of the many-body problem – and foster collaboration at the fundamental level of innovative theory and conceptual development. Quantum mechanics provides a theoretical foundation for our understanding of the structure and properties of atoms, molecules and the solid state in terms of their component particles, electrons and nuclei. The study of 'Quantum Systems in Chemistry and Physics' therefore underpins many of the emerging fields in science and technology: nanostructures, smart materials, drug design, and so on.

The tenth workshop was the first in the series held outside Europe. Participants gathered on the coast of North Africa, in one of the most influential cities of the ancient world, Carthage. Founded by Phoenicians from Tyre in the ninth century BC, it challenged the power of Rome. The situation of *Beït al-Hikma* on the Gulf of Tunis provided an excellent venue for scientists from different scientific and cultural backgrounds. They came from Western and Eastern Europe and North and South America as well as from Algeria, Morocco and, of course, Tunisia. Participants from overseas discovered a young and vibrant local scientific community engaged in theoretical molecular physics and chemistry.

The Carthage QSCP workshop was divided into 5 morning and 3 afternoon plenary sessions, during which a total of 36 lectures of about 30 minutes each, including discussion, were delivered by leading experts. There were also 2 evening sessions where 63 posters were presented, each being first described in a 3-minute oral presentation.

The fifteen papers collected in this volume are gathered into three sections, each addressing different aspects of the study of quantum systems in chemistry and physics. They are:

Part I: Advanced Methodologies
Part II: Interactions and Clusters
Part III: Excited States and Condensed Matter

We are pleased to acknowledge the support given to the Carthage workshop by the *University of Tunis,* the *Société Tunisienne d'Optique* and the *Académie Tunisienne des Sciences, des Arts et des Lettres.* The efforts of all members of the Local Organizing Committee were very much appreciated, especially the team of LSAMA (Laboratoire de Spectroscopie Atomique, Moléculaire et Applications), led by Pr Zohra Ben Lakhdar, including Drs Hassen Ghalila, Zoubeida Dhaouadi and Nejmeddine Jaidane. The material and logistic help of the *Tunisian Academy* and of the *Hotel Amilcar* is also gratefully acknowledged.

We are grateful to the participants not only for the high standard of the talks and posters presented at the workshop, which is reflected in this proceedings volume, but also for the friendly and constructive atmosphere throughout the formal and informal sessions. The QSCP workshops continue to provide a unique forum for the presentation and discussion of new ideas and developments.

As usual, since the 2001 workshop in Sofia, an impressive ceremony took place at the banquet dinner, held at the *Hotel Amilcar* in Carthage. The *Promising Scientist Prize* of the *Centre de Mécanique Ondulatoire Appliquée* (PSP of CMOA) was shared between the two selected nominees: Dr Majdi Hochlaf and Dr Richard Taïeb: http://www.lcpmr.upmc.fr/prize.html.

We hope that this volume has captured some of the excitement and enthusiasm that participants showed during the QSCP-X workshop, and that it will convey to a wider audience some of the concepts and innovations considered at *Beït al-Hikma.*

Souad Lahmar

Jean Maruani

Stephen Wilson

Gerardo Delgado-Barrio

OBITUARY - RAYMOND DAUDEL (1920-2006)

Raymond Daudel was born in Paris, France, on February 2, 1920, the only child in a middle-class French family. As early as eight he was fascinated by scientific experiments, and at ten he was deeply impressed by a visit with his father to a Museum of Chinese Arts in Paris. From these early experiences he kept, throughout his life, a common interest in the arts and the sciences.

In 1942, Raymond Daudel received a first-class degree in engineering from the prestigious *Ecole Supérieure de Physique et Chimie Industrielles de la Ville de Paris* (ESPCI). Then he became an assistant of Irène Joliot-Curie (the daughter of Pierre and Marie Curie and the wife of Frédéric Joliot, all Nobel Laureates), who was at that time a Professor of Chemistry at the *Sorbonne*, and of Antoine Lacassagne, a Professor of Medicine at the *Collège de France* (known for discovering carcinogenic effects of female hormones). His two supervisors were co-directors of the *Institut du Radium* (now part of the *Institut Curie*). While helping them in their research on the applications of radio-elements to the treatment of malignant tumours, Raymond Daudel prepared a *Thèse de Doctorat ès-Sciences* on chemical separation of radio-elements formed by neutron bombardment, which he presented in 1944.

After following the lectures of Louis de Broglie (who received the Nobel Prize for his discovery of matter waves: $\lambda = h / p$), Raymond Daudel realized that wave mechanics was becoming an essential tool in the understanding of the structure and dynamics of the large molecules from which living beings were built. In 1944, he founded the *Centre de Chimie Théorique de France* (CCTF), with the backing of Irène Joliot-Curie, Antoine Lacassagne, Louis de Broglie, and famous chemists, in order to foster scientific research on the applications of wave mechanics in chemistry and medicine.

In 1954, CCTF became the *Institut de Mécanique Ondulatoire Appliquée à la Chimie et à la Radioactivité*, under the sponsorship of the *Centre National de la Recherche Scientifique* (CNRS); and, in 1957, this body changed its name to that of *Centre de Mécanique Ondulatoire Appliquée* (or CMOA) with Louis de Broglie as the President of the Board of Directors. In 1962 the *CMOA du CNRS* was transferred to a location closer to the large computers of the *Institut Blaise-Pascal*.

CMOA then involved about 40 academics, half of them originating from 10 different countries, especially the USA. There Daudel, Lefebvre and Moser had become known through one of the first textbooks in Quantum Chemistry that was widely used in American universities. CMOA was at that time structured in four main teams:

- That of Raymond Daudel proper, mostly oriented towards (bio) chemical reactivity, with Odilon Chalvet, Simone Odiot, Federico Peradejordi, Georges and Nadine Bessis and a few others, from France and abroad.

- That of Carl Moser, rather oriented towards elaborated methods for the computation of smaller systems. In the late 1960's Moser was to found the *Centre Européen de Calcul Atomique et Moléculaire* (or CECAM), while his co-worker Hélène Lefebvre-Brion joined the newly founded *Laboratoire de Photophysique Moléculaire* (PPM) at Orsay.

- That of Roland Lefebvre, especially involved in the interpretation of magnetic resonance spectra in condensed phases with, e.g., Philémon Kottis and Jean Maruani, but also in open-shell methodology, with Yves Smeyers at Madrid, and in molecular dynamics, with M. Garcia Sucre at Caracas. Roland Lefebvre was to become a cofounder of the PPM, with Sydney Leach and a few others.

- That of Savo Bratos, especially involved in the interpretation of infrared spectra in condensed phases. In the late 1960's, Bratos founded the *Laboratoire de Physique Théorique des Liquides* (PTL). Then, in 1984, former co-worker Marcel Allavena founded the *Laboratoire de Dynamique des Interactions Moléculaires* (DIM). This latter attracted other researchers from Pullmans' and Salem's groups and became the *Laboratoire de Chimie Théorique* (LCT).

In the mid-1970's, other teams were created within the framework of former CMOA: Earl Evleth (an organic chemist from UCSB); Jean Maruani (symmetries and properties of non-rigid molecules); Pierre Becker (molecular structure by X-ray and neutron diffraction); Nicole Gupta (band structure in metal alloys); as well as other groups from the existing teams.

During its 30 years of existence, the *CMOA du CNRS* developed a broad activity in scientific research, education, and animation. Over a thousand papers and twenty volumes were published, and 80 doctorate theses presented, between the late 1950's and mid 1980's. Numerous workshops, congresses and Summer schools were organized under the auspices of the CMOA. The *International Academy of Quantum Molecular Sciences* (IAQMS), which has held triannual congresses since 1973, was founded in 1967 by Raymond Daudel, together with the Pullmans and scientists from Sweden (Löwdin), England (Coulson, Pople), the USA (Parr, Roothaan), and other places. The *World Association of Theoretical Organic Chemists* (WATOC), which also organizes a congress - alternating with those of IAQMS - every three years, was founded in 1982 on the same pattern.

Hundreds of scientists from all over the world have paid visits to CMOA, and dozens have worked there, for periods ranging from a few days to a few years. One remembers, for instance: Atkins, Bader, Bagus, Barojas, Beveridge, Christov,

Cizek, Coope, Csizmadia, Dannenberg, Goodman, Heilbronner, Jaffe, Karplus, Kaufman, Lorquet, Ludeña, Löwdin, Lund, Lunell, Matsen, Mezey, McConnell, McDowell, McWeeny, Mulliken, Nesbet, Pople, Pyykkö, Richards, Smeyers, ….

By 1984, most of the former members of CMOA had joined other laboratories, except a few (S. Besnaïnou, G. Giorgi, J. Maruani) who were invited by Pr Christiane Bonnelle to follow Pr Raymond Daudel in her *Laboratoire de Chimie Physique* (LCP), close to the *Institut Curie.* A strong impulse in theoretical and computer-oriented methods was then given to this mostly experimental laboratory. But the name of CMOA was retained by Jean Maruani to found an international, non-profit organization devoted to the promotion of scientific exchange and the organization of scientific meetings.

Raymond Daudel was a man of the cities: he disliked the countryside and wild life and was inclined to sedentarity and meditation. On June 18, 1944, he had married his student, Pascaline Salzedo, who initiated him into mountain climbing and stimulated his interest in exotic journeys and artworks. She was a rather tiny woman but with a strong will, who gave him a steady help throughout his career, although she was also working as a scientist at the *Institut Curie.* When she was hit by cancer, an illness on which she had worked for years, she insisted on going regularly from her room at *Hôpital Curie* to her office at *Institut Curie.* After Pascaline died, in 1976, Raymond Daudel went through a depressive period in which he lost nearly all interest in his scientific activities. Another blow came when his long-time rival, Bernard Pullman, was elected as a full member of the *Académie des Sciences* in succession to Daudel's mentor, Louis de Broglie, thus putting an end to his hope of getting into that prestigious body.

At that time Senator-Mayor Francis Palmero, who had known Pr Raymond Daudel since the first IAQMS congress, held at Menton in 1973, was a political friend of the painter Nicole Lemaire D'Aggagio, who was a municipal counsellor at the nearby city of Antibes-Juan les Pins. He asked her to involve Pr Raymond Daudel in some common project. Mrs Lemaire was then President of the National Commission of Fine Arts at the Women's Professional Union and, as such, had been invited, in 1972, to a meeting in the Soviet Union. She was received at the Kremlin and, in the midst of the Cold War, she advocated the creation of an international and interdisciplinary academy to foster peace through cooperation, between scientists and artists from Western and Eastern Europe. After trying to involve various academics, she came to Pr Daudel. It was as if a heavenly voice had told him: "If you can't join them, beat them".

The *European Academy of Sciences, Arts and Humanities* was founded, in 1979, at the very same address (60 rue Monsieur-le-Prince in Paris) where Pierre de Fermat had created, three centuries earlier, the informal group that was to become the French *Académie des Sciences.* The founding members of the *European Academy* were Armand Lanoux from *Académie Goncourt*, René Huygues from *Académie Française*, Jean Bernard from *Académie de Médecine*, Louis Leprince-Ringuet from *Académie des Sciences*, and such foreign scientists as Ilya Prigogine

(Belgium), Per-Olov Löwdin (Sweden), and Camille Sandorfy (Canada) - now all deceased. Today the *European Academy*, under the new presidency of Pr de Thé, from the *Pasteur Institute*, involves about 300 full members (all academician in their own country), including 70 Nobel Laureates, and 700 corresponding members, from all over the world. It has the status of a non-governmental organization and acts as a consultant for such international bodies as UNESCO and WHO.

I have a personal debt towards the *European Academy*. It was through this body that I was invited, in 2002, by the President of the *Tunisian Academy* to give a talk at one of the *European Academy* meetings, held at Carthage. This allowed me to visit again, for the first time since forty years, the country where I was born. There I approached some Tunisian colleagues, whom I had never met before, and convinced them to organize the tenth QSCP workshop at Carthage.

In recent years, Pr Daudel was mainly involved in establishing a world network on research against retroviruses, together with Luc Montagnier (whose team discovered the virus of AIDS), and also in promoting the teaching of global issues and cultural diversity to engineering students, in the context of rising concern for sustainable development.

Raymond Daudel passed away in Paris, France, on June 20, 2006, at the age of 86. He leaves two sons, both married and having children: Olivier (working in computer science in Paris) and Sylvain (involved in education for management in Singapore). A short ceremony took place in his honour during the eleventh QSCP workshop held at *Kochubey Palace*, St Petersburg, Russia, on August 23, 2006, and a larger one was organized by various bodies, on November 30, at the *Royal Chapel* of Versailles.

Raymond Daudel was *not* a religious man, in the sense that he did not belong to any creed - even though he respected even the weirdest creeds. He was, I would say, rather close to the philosophy of the Stoics. But he *was* religious in the original sense of *religare*: to *link*. To link men of different backgrounds and cultures; and also to link the visible world, which can be accessed by Science, and the invisible realm, which - he thought - could be best approached through Art.

Jean Maruani

Honorary Director
of Research at CNRS
President of CMOA

(Approved by Olivier and Sylvain Daudel
and by Nicole Lemaire D'Aggagio)

Part I
Advanced Methodologies

LITERATE MANY-BODY PERTURBATION THEORY PROGRAMMING: THIRD-ORDER "RING" DIAGRAMS

STEPHEN WILSON

Rutherford Appleton Laboratory, Chilton, Oxfordshire OX11 0QX, UK

Abstract The application of literate programming methods in many-body perturbation theory is illustrated by considering the computation of the third-order "ring" diagram in the correlation energy expansion for a closed-shell, singlet system. An *a posteriori* application of literate programming techniques to a major component of the first published computer program for performing many-body perturbation theory calculations within the algebraic approximation is given.

1. INTRODUCTION

Science, by it very nature, is a body of *public* knowledge, to which each researcher makes his personal contribution, and which is corrected and clarified by mutual criticism. It is a corporate activity in which each of us builds upon the work of our predecessors, in competitive collaboration with our contemporaries. The nature of the communication system is thus vital to science; it lies at the very heart of the 'scientific method'.

This quotation is taken from a volume entitled *The Force of Knowledge: The Scientific Dimension of Society* (p. 90) published in 1976 [1] by the late Professor J. Ziman, FRS. Ziman continues by pointing out that

The actual form of communication are varied, and have changed in emphasis over the centuries.

In his volume *Public Knowledge: An Essay Concerning the Social Dimension of Science* [2], Ziman writes

... physicist and engineers will sometimes make available to their colleagues the tapes of instructions for computer programs that they have devised for some particular purpose, such as for the solution of some difficult equation, or for the reduction and analysis of certain types of data. These tapes are collected in 'libraries' at computer centers, and are used directly to generate further knowledge. It seems to me that these are not just tools of research; they embody information, and play just the same role as would mathematical formulae published in books, or tables of physical data in scientific papers.

S. Lahmar et al. (eds.), Topics in the Theory of Chemical and Physical Systems, 3–33.

He continues

> A computer tape is ... not unlike an algebraic formula, but this is not an essential limitation, except at the stage when it is being prepared by a human mind. It can be transformed, in the computer, to something much more complex, which is unintelligible to the bare human intellect, and can only be read, so to speak, by another computer.

Although Ziman made these comments in the late 1960s, even today it remains the case that many scientific computer programs are not placed in the public domain. On the other hand, results obtained by executing these codes for specific applications are widely published in the literature. Objectively, this cannot be regarded as a *bone fide* application of the scientific method.

In a recent paper entitled *Theory and computation in the study of molecular structure* [3], Quiney and I have advocated the use of literate programming methods, first introduced by Knuth [4], but now little used [5], as a means of placing quantum chemistry computer code in the public domain along side the associated theoretical apparatus. Such publication not only places the work in the body of scientific knowledge but also serves to establish authorship.

In two other recent publications, Quiney and I have given examples of the application of literate programming methods in quantum chemistry. In a paper entitled *Literate programming in quantum chemistry: A simple example* [6], we describe an application to the calculation of an approximation to the ground state of the helium atom. The paper, we submit, demonstrates the pedagogical advantages of literate programming. In a second paper entitled *Literate programming in quantum chemistry: A collaborative approach to code development for molecular electronic structure theory* [7], we describe the use of literate programming methods in collaborative code development. We are also preparing a volume with the title *Literate programming in quantum chemistry: an introduction* [8].

In the present paper, the application of literate programming methods in many-body perturbation theory will be illustrated by considering the computation of the third-order "ring" diagram in the correlation energy expansion for a closed-shell, singlet system. The FORTRAN77 code for calculating this component of the correlation energy was first published by the author [9] in 1978. The present work, therefore, represents an *a posteriori* application of literate programming techniques to a major component of the first published computer program for performing many-body perturbation theory calculations within the algebraic approximation, an approximation which is the essential ingredient of molecular applications. Our central purpose here is to demonstrate how literate programming methods can be used to place the computer programs developed in quantum chemistry in the public domain.

The code presented here was published as part of a program package by D.M. Silver and the present author [9–11] which calculated the components of the correlation energy corresponding to all second and third-order diagrammatic terms. The whole package was originally published in *Computer Physics Communications* in 1978.

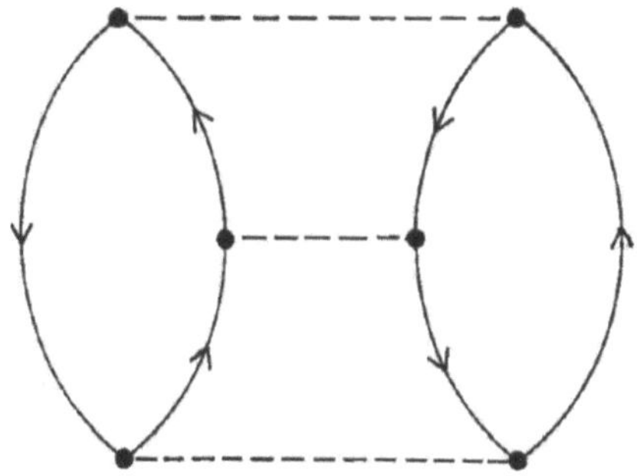

Figure 1. Brandow diagram for the third-order "ring" energy.

The particular energy component that will be studied here is the third-order "ring" energy also referred to as the third-order hole-particle energy, $E_3(hp)$ (for details see, for example, [12] or [13]). The corresponding Brandow diagram (a diagram with antisymmetrized vertices) is shown in Figure 1. The associated Goldstone diagrams are shown in Figure 2.

The algebraic expression corresponding to the Brandow diagram shown in Figure 1 takes the form

$$E_3(hp) = -\sum_{ijk}\sum_{abc} \frac{\langle ij|\, O\, |ab\rangle \langle ak|\, O\, |ic\rangle \langle bc|\, O\, |jk\rangle}{D_{ijab} D_{jkbc}}$$

where $\langle pq|\, O\, |rs\rangle$ denotes a two-electron integral with "antisymmetrized" interaction

$$O = \frac{1 - P_{12}}{r_{12}}$$

and D_{pqrs} is a denominator factor which depends on the choice of zero-order Hamiltonian for the perturbation expansion. The indices i, j and k label hole states whilst the indices a, b and c label particle states.

2. A LITERATE MANY-BODY PERTURBATION THEORY PROGRAM

2.1. Background

Literate programming was introduced by Knuth in 1984 [4]. He suggested that

> the time [was] ripe for significantly better documentation of programs, and that we [could] best achieve this by considering programs to be works of literature.

This requires a radical shift of emphasis in the writing of computer programs. Knuth writes

> instead of imagining that our main task is to instruct a computer what to do

we should

> concentrate rather on explaining to human beings what we want a computer to do

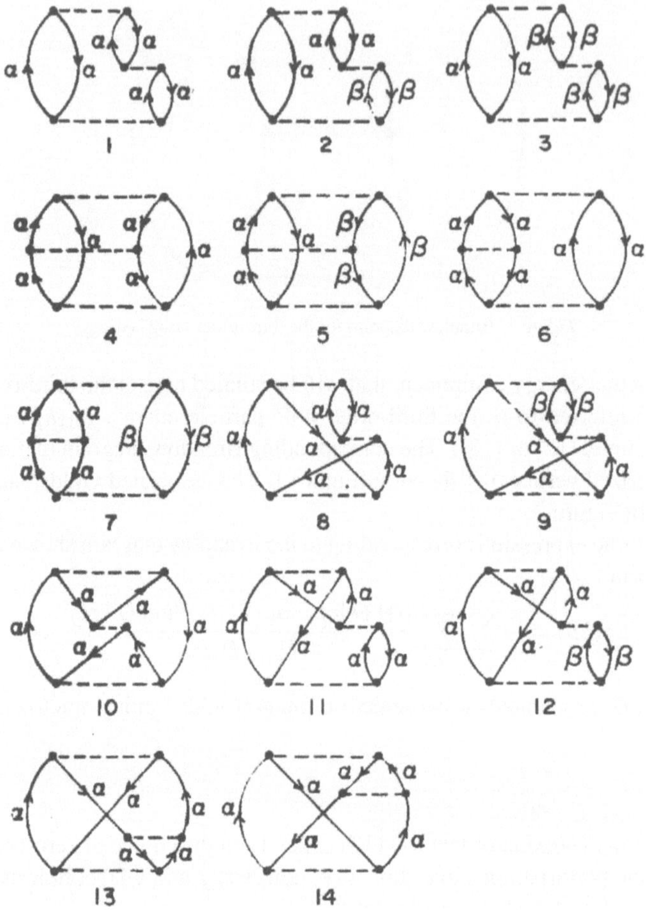

Figure 2. All Goldstone diagrams for the third-order "ring" energy, including those for which the spin labels differ. Taken from S. Wilson, *Comput. Phys. Commun.* **14**, 91 (1978).

Thus the task facing a literate programmer extends beyond that of a computer programmer. The literate programmer must strive not only to create correct and efficient code, but also a description of the theoretical concepts that lie behind the code.

Unfortunately, it appears that literate programming techniques are today little used [5]. But, in recent work [3], we have emphasized their value as a means of placing quantum chemistry computer code in the public domain along side the associated theoretical apparatus. Such publication not only places the work in the body of scientific knowledge, where it can be [14]

> fully and freely available for open criticism and constructive use,

but also serves to establish authorship.

2.2. A literate program for third-order many-body perturbation theory "ring" diagram components

The following literate program has been constructed by taking the original FORTRAN77 listing of the "ring" diagram code, removing all "comment" cards, and then adding LaTeX text, together with Figures, to provide a clear description of the code and its functionality.

2.2.1. *Controlling routine:* `sw`

The controlling routine for the computation of $E_3(hp)$ is called `sw` and begins with the following call, declarations and assignments:

```
subroutine sw  (io,ktf,ktr,kts,iprnt)
implicit real*8 (a-h,o-z)
dimension gijk(2,1100),ktf(10),ktr(10),
  dint(35,35)
nocctm=10
norbtm=35
nvirtm=25
```

In the arguments of `sw`, `io` is the print data set, `ktf(10)` data sets for labelled lists of integrals, `ktr(10)` work data sets, `kts` is an internal output data set, and `iprnt` is a print control parameter such that

`iprnt=1`: print out of intermediate results
`iprnt=0`: no print out of intermediate results

The three variables `nocctm` ($n_{occ}^{\max}$), `norbtm` ($n_{orb}^{\max}$) and `nvirtm` ($n_{virt}^{\max}$) define the maximum number of occupied orbitals, orbitals and virtual orbitals, respectively, that can be consider with the array dimension settings indicated. The array `gijk` is used to store the energy components corresponding to a given i, j, k combination

$$g_{ijk} = \sum_{abc} \frac{\langle ij|\, O \,|ab\rangle \langle ak|\, O \,|ic\rangle \langle bc|\, O \,|jk\rangle}{D_{ijab} D_{jkbc}}$$

and is of dimension $2 \times n_g^{\max}$, where

$$n_g^{\max} = 2n_{occ}^{\max} \left(\left(n_{occ}^{\max}\right)^2 + n_{occ}^{\max} \right)$$

The controlling routine `sw` calls three subroutines `inpt`:

```
call inpt(io,ktf,ktr,nocctm,norbtm,nvirtm,dint,
          iprnt)
```

which initials certain arrays and controls the subroutines `symt` and `ordr`. `symt` handles degenerate symmetry species for linear systems, whilst `ordr` arrange the two-electron integrals of the type $\langle ij|\, O \,|ab\rangle$ in separate data sets in a manner that will be described in more detail below; `ring`:

```
call ring(gijk,io,ktf(5),ktr,dint)
```

which is the routine where the required correlation energy components are computed; `gwrt`:

```
call gwrt(io,kts,gijk,iprnt)
```

which controls the printing of intermediate data where it is required. Control is then returned to the controlling routine:

```
return
end
```

and this completes execution of the controlling routine `sw` for the "ring" diagram energy components. Control is passed back to the calling routine, which is not considered explicitly here, where the "ring" diagram energy is added to other components of the correlation energy.

2.2.2. *Initialization of arrays:* `inpt`

The subroutine `inpt` initializes certain arrays and also controls the subroutine `symt`, which handles degenerate symmetry species for linear systems, and `ordr`, which arranges the two-electron integrals of the type $\langle ij|\, O\, |ab\rangle$. All of the quantities appearing in the argument list for `inpt` have been defined above:

```
subroutine inpt(io,ktf,ktr,nocctm,norbtm,nvirtm,
dint,$iprnt)
```

The declarations and common blocks are as follows:

```
implicit real*8 (a-h,o-z)
dimension ktf(10),ktr(10),label(20),dint(35,35)
logical lsym1(550),lsym2(550)
common/ptind/ ind(60),norb,nnorb,nocc,nnocc,
  nvirt,nnvirt
common/ptsym/ lsym1,lsym2,invlab(60)
common/ptres/ etwo(2),etotal(2),ediag,eorb(60)
```

The three common blocks store: (i) various indices (`ptind`), (ii) symmetry data (ptsym), (iii) results (`ptres`)

Execution begins with the reading of a `label`, `nocc` (n_{occ}) – the number of occupied orbitals, `norb` (n_{orb}) – the number of orbitals, `eorb` (ϵ_p) – the orbital energies, `invlab` – a symmetry label, and `dint` – the denominator shift integrals. These quantities are read from the data set `ktf(10)`. `invlab` is stored in the common block `ptsym` along with the logical arrays `lsym1` and `lsym2`.

```
kt=ktf(10)
read(kt) label,nocc,norb,eorb,invlab
read (kt) dint
rewind kt
```

If intermediate print out is switched on, print the title "`Third-order hole-particle ring diagram`":

```
      if(iprnt.ne.0) write (io,6000)
 6000 format(39h1Third-order hole-particle ring
              diagram/)
```

The number of virtual orbitals, n_{virt}, is determined and the following conditions checked

$$n_{occ} \leq n_{occ}^{\max}; \quad n_{orb} \leq n_{orb}^{\max}; \quad n_{virt} \leq n_{virt}^{\max}$$

If any of these inequalities are not satisfied then an error message is written and execution terminated.

```
      nvirt=norb-nocc
      if(nocc.le.nocctm.and.norb.le.norbtm.and.
     $nvirt.le.nvirtm) go to 3
      write(io,6007) nocctm,nvirtm,norbtm
 6007 format(38h0Dimension problem in ring
        program sw  //10x
     $36hNumber of doubly occupied orbitals =,i2,5x,
     $31hNumber of unoccupied orbitals =,i3,5x,
     $20hNumber of orbitals =,i3)
      stop
    3 continue
```

Various indices stored in the common block `ptind` are set up. `ind(i)` contains values of

$$\frac{1}{2}\left(i\left(i-1\right)\right)$$

This index is required to determine the position of the elements of a symmetric matrix whose elements are stored in a one-dimensional array. `nnocc` ($\bar{n}_{occ}$), `nnorb` ($\bar{n}_{orb}$) and `nnvirt` ($\bar{n}_{virt}$) are given by

$$\begin{aligned}\bar{n}_{occ} &= \frac{1}{2}\left(n_{occ}\left(n_{occ}+1\right)\right);\\ \bar{n}_{orb} &= \frac{1}{2}\left(n_{orb}\left(n_{orb}+1\right)\right);\\ \bar{n}_{virt} &= \frac{1}{2}\left(n_{virt}\left(n_{virt}+1\right)\right)\end{aligned}$$

```
      j=0
      do 1 i=1,norb
      ind(i)=j
      j=j+i
    1 continue
      nnorb=(norb*(norb+1))/2
```

```
      nnocc=ind(nocc+1)
      nnvirt=ind(nvirt+1)
```

Arrays in which the calculated energy components will be stored are set to zero. `etotal` will contain the total energy whilst `etwo` will contain the two-body component. `iz=1` corresponds to the Hartree–Fock model zero-order Hamiltonian, that is the Møller–Plesset expansion whereas `iz=2` identifies the shifted denominator scheme which uses the Epstein–Nesbet zero-order Hamiltonian. `ediag` will be used to store the diagonal component. These energies are stored in the common block `ptres` together with the orbital energy (`eorb(60)`).

```
      do 5 iz=1,2
      etotal(iz)=0.0d+00
      etwo(iz)=0.0d+00
    5 continue
      ediag=0.0d+00
```

Control this then passed to the subroutine `symt` and then subroutine `ordr` before it is returned to the controlling routine `sw`.

```
      call symt
      call ordr(ktf(6),ktr)
      return
      end
```

2.2.3. Handling of degenerate symmetry species for linear molecules: `symt`

Symmetry can be exploited to improve the efficiency of *ab initio* quantum chemical programs. This program recognizes degenerate symmetry species for linear molecules.

```
      subroutine symt
      implicit real*8 (a-h,o-z)
      logical l1,l2,l3,l4,l5,l6,l7,l8,l9,lik,
        lsym1(550),
     $lsym2(550)
      common/ptind/ ind(60),norb,nnorb,nocc,nnocc,
        nvirt,nnvirt
      common/ptsym/ lsym1,lsym2,invlab(60)
```

Element i of the array `invlab` identifies the symmetry of the orbital i. This index is used for handling degenerate symmetry species that occur for linear molecules. All elements of `invlab` are set equal to 1 if the molecule is not linear or if real spherical harmonics are not used. For linear molecules, the elements of `invlab` are assigned as follows:

$$\sigma \rightarrow 1$$
$$\pi \rightarrow 2$$
$$\bar{\pi} \rightarrow 3$$

If a π orbital is occupied then the corresponding $\bar{\pi}$ orbital must also be occupied and must follow the π orbital in consecutive order.

The logical arrays `lsym1` and `lsym2` are initialized by setting all elements to `.false.`.

```
      do 1 i=1,550
      lsym1(i)=.false.
      lsym2(i)=.false.
    1 continue
```

By considering all sets of indices

$$i,\ j,\ k \text{ with } i \geq k$$

all cases which will have a non-zero contribution because of symmetry considerations can be identified. Set `isym`, `jsym` and `ksym` to `invlab(i)`, `invlab(j)` and `invlab(k)`, respectively.

```
      icount=0
      do 10 i=1,nocc
      isym=invlab(i)
      do 10 k=1,i
      lik=(i.ne.(k+1))
      ksym=invlab(k)
      do 10 j=1,nocc
      jsym=invlab(j)
```

Set the logical variables `l1`, `l2`, ..., `l9` according to values of `isym`, `jsym` and `ksym`.

```
      icount=icount+1
      l1=(isym.eq.1)
      l2=(isym.eq.2)
      l3=(isym.eq.3)
      l4=(jsym.eq.1)
      l5=(jsym.eq.2)
      l6=(jsym.eq.3)
      l7=(ksym.eq.1)
      l8=(ksym.eq.2)
      l9=(ksym.eq.3)
```

Now set `lsym1(icount)` according to the symmetry of the orbital triple labelled by the indices i, j and k.

```
      if(l1.and.l6.and.l7) lsym1(icount)=.true.
      if(l3.and.l4.and.l7) lsym1(icount)=.true.
      if(l1.and.l4.and.l9) lsym1(icount)=.true.
      if(l3.and.l6.and.l7) lsym1(icount)=.true.
      if(l3.and.l4.and.l9) lsym1(icount)=.true.
```

```
      if(l1.and.l6.and.l9) lsym1(icount)=.true.
      if(l3.and.l5.and.l7) lsym1(icount)=.true.
      if(l3.and.l4.and.l8.and.lik) lsym1(icount)
     =.true.
      if(l1.and.l6.and.l8) lsym1(icount)=.true.
      if(l3.and.l6.and.l9) lsym1(icount)=.true.
      if(l3.and.l6.and.l8.and.lik) lsym1(icount)
     =.true.
      if(l3.and.l5.and.l9) lsym1(icount)=.true.
      if(l3.and.l5.and.l8.and.lik) lsym1(icount)
     =.true.
```

Similarly, set `lsym2(icount)` according to the symmetry of the orbital triple labelled by the indices i, j and k.

```
      if(l1.and.l5.and.l7) lsym2(icount)=.true.
      if(l2.and.l4.and.l7) lsym2(icount)=.true.
      if(l1.and.l4.and.l8) lsym2(icount)=.true.
      if(l2.and.l5.and.l7) lsym2(icount)=.true.
      if(l2.and.l4.and.l8) lsym2(icount)=.true.
      if(l1.and.l5.and.l8) lsym2(icount)=.true.
      if(l2.and.l6.and.l7) lsym2(icount)=.true.
      if(l2.and.l4.and.l9) lsym2(icount)=.true.
      if(l1.and.l5.and.l9) lsym2(icount)=.true.
      if(l2.and.l5.and.l8) lsym2(icount)=.true.
      if(l2.and.l5.and.l9) lsym2(icount)=.true.
      if(l2.and.l6.and.l8) lsym2(icount)=.true.
      if(l2.and.l6.and.l9) lsym2(icount)=.true.
   10 continue
      return
      end
```

2.2.4. Processing of the two-electron integrals of the type $\langle ij|\, O\, |ab\rangle$: `ordr`

The subroutine `ordr` reads the integrals $\langle ij|\, O\, |ab\rangle$ from the data set `jt` and creates n_{occ} data sets `ktr(i)`, `i=` $1, n_{occ}$, where data set `ktr(i)` contains $\langle ij|\, O\, |ab\rangle$ for a given `i` arranged in blocks labelled by `j`. The disposition of the integrals $\langle ij|\, O\, |ab\rangle$ in the data sets `ktr(i)`, `i=` $1, n_{occ}$ is illustrated in Figure 3.

	`ktr(i)`:$i=1$	`ktr(i)`:$i=2$	...	`ktr(i)`:$i=n_{occ}$
$j=1$	$j=1$, all a,b	$j=1$, all a,b	...	$j=1$, all a,b
$j=2$	$j=2$, all a,b	$j=2$, all a,b	...	$j=2$, all a,b
...	...	...		...
$j=n_{occ}$	$j=n_{occ}$, all a,b	$j=n_{occ}$, all a,b	...	$j=n_{occ}$, all a,b

Figure 3. Organization of the integrals $\langle ij|\, O\, |ab\rangle$.

The argument list consists of `jt` and `ktr`:

```
      subroutine ordr(jt,ktr)
```

The declarations, common block and equivalence statements are

```
      implicit real*8 (a-h,o-z)
      dimension a(625),ktr(10),indx(625),lblds(20),
        na(1)
      common/ptind/ ind(60),norb,nnorb,nocc,nnocc,
        nvirt,nnvirt
      equivalence (na(1),a(1))
```

Here `a(625)` is an array used to store values of the integral $\langle ij|\, O\, |ab\rangle$ for all a and b for a given i and j. It has the dimensions n_{virt}^2. The array `indx(625)` is used to store the corresponding integral labels. The various indices in the common block `ptind` were set up by the subroutine `ordr`. Note the equivalence of the arrays `na` and `a`, which is used for efficient reading and writing of integral lists.

```
      read (jt) lblds
      ies=1
      is=0
      js=0
      do 7 i=1,nocc
      do 7 j=1,i
      n=0
      if(ies.eq.0) go to 20
   10 continue
      if(is-i) 12,11,20
   11 continue
      if(js-j) 12,13,20
   12 continue
      read (jt,end=19) is,js,ns
      go to 10
   13 continue
      n=ns
      go to 20
   19 continue
      ies=0
   20 continue
      if(n.eq.0) go to 5
```

Now read the integral labels into the array `indx` and the integrals into the array `a`, which is related to the array `na` by an equivalence

```
      m=n+n
      call rfst(indx,n,jt)
      call rfst(na,m,jt)
    5 continue
```

Write the labels and integrals to the data set `ktr(i)`

```
      kt=ktr(i)
      write(kt) i,j,n
      if(n.eq.0) go to 6
      call wfst(indx,n,kt)
      call wfst(na,m,kt)
    6 continue
```

If $i \neq j$ write the labels and integrals to the data set `ktr(j)`

```
      if(i.eq.j) goto 7
      kt=ktr(j)
      write(kt) i,j,n
      if(n.eq.0) go to 7
      call wfst(indx,n,kt)
      call wfst(na,m,kt)
    7 continue
```

Execution of the subroutine `ordr` concludes by rewinding the data sets `jt` and `ktr(i)`, `i=` $1, n_{occ}$.

```
    3 continue
      rewind jt
      do 4 i=1,nocc
      kt=ktr(i)
    4 rewind kt
      return
      end
```

2.2.5. *Evaluation of the third-order "ring" diagram energy components:* `ring`

`ring` is the subroutine in which the third-order "ring" diagram energy components are actually evaluated. The computation proceeds in two main steps.

In the first step, the integrals $\langle ij| O |ab\rangle$ and $\langle bc| O |jk\rangle$ together with the corresponding denominator factors D_{ijab} and D_{jkbc} are combined to form an intermediate f_{ijkac} by summing over the index b

$$f_{ijkac} = \sum_b \frac{\langle ij| O |ab\rangle \langle bc| O |jk\rangle}{D_{ijab} D_{jkbc}}$$

The summation shown here is over spin orbitals, but in practice it is carried out over spatial orbitals. The different spin cases which can arise are summarized in Figure 4. The difference spin cases are distinguished by the index μ.

The second step involves the contraction of the integrals of the type $\langle ak| O |ic\rangle$ by summation over the indices a and c to give

$$g_{ijk} = \sum_{ac} f_{ijkac} \langle ak| O |ic\rangle .$$

μ	$\langle IJ\vert g\vert AB\rangle$	$\langle IJ\vert g\vert BA\rangle$	$\langle JK\vert g\vert BC\rangle$	$\langle JK\vert g\vert CB\rangle$	F^{μ}_{IJKAC}
1	$(\alpha\alpha\alpha\alpha)$	$(\alpha\alpha\alpha\alpha)$	$(\alpha\alpha\alpha\alpha)$	$(\alpha\alpha\alpha\alpha)$	$(\alpha\alpha\alpha\alpha\alpha)$
2	$(\alpha\alpha\alpha\alpha)$	$(\alpha\alpha\alpha\alpha)$	$(\alpha\beta\alpha\beta)$	$(-)$	$(\alpha\alpha\beta\alpha\beta)$
3	$(\alpha\beta\alpha\beta)$	$(-)$	$(\beta\alpha\beta\alpha)$	$(-)$	$(\alpha\beta\alpha\alpha\alpha)$
4	$(\alpha\beta\alpha\beta)$	$(-)$	$(\beta\beta\beta\beta)$	$(\beta\beta\beta\beta)$	$(\alpha\beta\beta\alpha\beta)$
5	$(-)$	$(\alpha\beta\alpha\beta)$	$(-)$	$(\beta\alpha\beta\alpha)$	$(\alpha\beta\alpha\beta\beta)$

$$g = \frac{1}{r_{12}}$$

Figure 4. Spin cases which arise in the calculation of the intermediate F^{μ}_{IJKAC}.

μ	$\langle IC\vert O\vert AK\rangle$	F^{μ}_{IJKAC}	G^{μ}_{IJK}
1	$(\alpha\alpha\alpha\alpha)$	$(\alpha\alpha\alpha\alpha\alpha)$	$(\alpha\alpha\alpha)$
2	$(\alpha\alpha\alpha\alpha)$ $(\alpha\beta\beta\alpha)$	$(\alpha\beta\alpha\alpha\alpha)$ $(\alpha\beta\alpha\beta\beta)$	$(\alpha\beta\alpha)$
3	$(\alpha\beta\alpha\beta)$	$(\alpha\alpha\beta\alpha\beta)$	$(\alpha\alpha\beta)$
4	$(\alpha\beta\alpha\beta)$	$(\alpha\beta\beta\alpha\beta)$	$(\alpha\beta\beta)$

Figure 5. Spin cases which arise in the calculation of the intermediate G^{μ}_{IJK}.

In practice, the summation is carried out over spatial orbitals. The different spin cases are again distinguished by the index μ. The spin case which arise are defined in Figure 5.

The `ring` subroutine calling arguments are as follows:

```
subroutine ring(gijk,io,jt,ktr,dint)
```

where

`gijk`: energy components for a given i, j, k
`io`: printed output
`jt`: data sets containing $\langle ic|\, O\, |ak\rangle$ integrals
`ktr`: data sets containing $\langle ij|\, O\, |ab\rangle$ integrals
`dint`: denominator integrals

The declarations, data, common block and equivalence statements are:

```
      implicit real*8 (a-h,o-z)
      dimension g1(2),g2(2),g3(2),g4(2),
     $f1(2,2),f2(2,2),f3(2,2),f4(2,2),f5(2,2),
     $d1(2),d2(2),d3(2),d4(2),d5(2),
     $vijab(625),vjkbc(625),gijk(2,1),
     $ijab(625),jkbc(625),ktr(10),lblds(20),
        dint(35,35),
     $itypea(625),itypec(625),locab(325),locbc(325),
```

```
$rint(950),icak(950),
$nint(1),nijab(1),njkbc(1),iar(950),icr(950),
   ityper(950)
 logical j1,j2,j3,j4,j5,l1,l2,l3,l4,l5,l6,l7,l8,
   l9,l10,
$lsym1(550),lsym2(550)
 data t1/0.0d+00/,t2/0.0d+00/,t3/0.0d+00/,
$t4/0.0d+00/,t5/0.0d+00/
 common/ptind/ ind(60),norb,nnorb,nocc,nnocc,
   nvirt,nnvirt
 common/ptsym/ lsym1,lsym2,invlab(60)
 common/ptres/ etwo(2),etotal(2),ediag,eorb(60)
 equivalence (nijab(1),vijab(1)),(njkbc(1),
   vjkbc(1))
1,            (nint(1),rint(1))
 limit=nvirt*nvirt+nnvirt
 nd=2
 read(jt) lblds
 isym0=0
 indx=0
```

Start a loop over i and assign the data set `it` containing the $\langle ij|\ O\ |ab\rangle$ for a given i.

```
do 1 i=1,nocc
it=ktr(i)
```

and then start a loop over k, assign `isym`, test for the case $i = k$, and assign the data set `kt` containing the $\langle jk|\ O\ |bc\rangle$ for a given k.

```
do 2 k=1,i
isym=isym0
l9=(i.eq.k)
kt=ktr(k)
```

The integrals of the type $\langle ic|\ O\ |ak\rangle$ are now read from the data set `jt`. First the labels i (`ir`) and k (`kr`) are read together with the number of integrals in the block n_r (`nr`).

```
read(jt,end=900) ir,kr,nr
```

If an error is encountered control passes to statement `900`. Check that the indices `ir` and `kr` correspond to the required block of integrals:

```
if(ir.ne.i.and.kr.ne.k) goto 901
```

If they do not then handle the error by passing control to statement `901`. Also check that the number of integrals in the block, `nr`, is not greater than the program limitations imposed by array dimensions.

```
      if(nr.gt.limit) goto 903
```

If this limit is exceeded, handle this error by passing control to statement `903`. If the number of integrals in the (i,k) block is 0 then the processing of this block can be skipped.

```
      if(nr.eq.0) go to 2
```

The integral labels and the integrals $\langle ic|\ O\ |ak\rangle$ themselves are read from the data set `jt`.

```
      nnr=nr+nr
      call  rfst(icak,nr,jt)
      call  rfst(nint,nnr,jt)
```

The integral labels for the (i,k) block of integrals are now unpacked. In this code, `iar(iz)` contains the index a, `icr(iz)` the label c, and `ityper(iz)` contains an index which defines the integral type.

```
      do 3 iz=1,nr
      label=icak(iz)
      laba=label/1200
      iar(iz)=laba-(laba/60)*60-nocc
      labc=label/20
      icr(iz)=labc-(labc/60)*60-nocc
      ityper(iz)=label-(label/20)*20
    3 continue
```

The integrals $\langle ic|\ O\ |ak\rangle$ are now available for all (a,c) for a given (i,k). The program now loops over the third occupied orbital index j.

```
      do 7 j=1,nocc
      isym=isym+1
      read(it,end=900) mi,mj,mn
```

If an error is encountered control passes to statement `900`. Check that the indices `mi` and `mj` correspond to the required block of integrals:

```
      if((mi.ne.i.or.mj.ne.j).and.
        (mi.ne.j.or.mj.ne.i)) goto 901
```

If they do not then handle the error by passing control to statement `901`. Also check that the number of integrals in the block, `mn`, is not greater than the program limitations imposed by array dimensions.

```
      if(mn.gt.limit) goto 903
```

If they do not then handle the error by passing control to statement `903`. Also check that the number of integrals in the block, `mn`, is not greater than the program limitations imposed by array dimensions.

```
      if(mn.eq.0) go to 5
```

The integral labels and the integrals $\langle ij|\, O\, |ab\rangle$ themselves are read from the data set `it`.

```
      mnn=mn+mn
      call  rfst(ijab,mn,it)
      call  rfst(nijab,mnn,it)
    5 continue
```

If $i = k$, $\langle jk|\, O\, |bc\rangle$ the block is equivalent to the $\langle ij|\, O\, |ab\rangle$ block and so the former do not have to be read. Control therefore passes to `8`.

```
      if(l9) goto 8
```

The integrals of the type $\langle jk|\, O\, |bc\rangle$ are now read from the data set `kt`. First the labels j (`ni`) and k (`nj`) are read together with the number of integrals in the block n_n (`nn`)

```
      read(kt,end=900) ni,nj,nn
```

If an error is encountered control passes to statement `900`. Check that the indices `ni` and `nj` correspond to the required block of integrals:

```
      if((ni.ne.j.or.nj.ne.k).and.
         (ni.ne.k.or.nj.ne.j)) goto 901
```

If they do not then handle the error by passing control to statement `901`. Also check that the number of integrals in the block, `nn`, is not greater than the program limitations imposed by array dimensions.

```
      if(nn.gt.limit) goto 903
```

If they do not then handle the error by passing control to statement `903`. If the number of integrals in the (j,k) block is 0 then the processing of this block can be skipped.

```
      if(nn.eq.0) go to 7
```

The integral labels and the integrals $\langle jk|\, O\, |bc\rangle$ are read from the data set `kt`.

```
      nnn=nn+nn
      call  rfst(jkbc,nn,kt)
      call  rfst(njkbc,nnn,kt)
```

If the number of integrals in the $\langle ij|\, \mathcal{O}\, |ab\rangle$ block is 0 then processing of this block can be skipped.

```
      if(mn.eq.0) go to 7
      goto 9
```

The following code is only executed if $i = k$. The block of integrals $\langle jk|\, O\, |bc\rangle$ is obtained from the block $\langle ij|\, O\, |ab\rangle$.

```
    8 continue
      if(mn.eq.0) go to 7
      nn=mn
      do 10 iz=1,nn
      vjkbc(iz)=vijab(iz)
      jkbc(iz)=ijab(iz)
   10 continue
    9 continue
      if(lsym1(isym)) goto 7
      l6=lsym2(isym)
```

Set the elements of the arrays `g1`, `g2`, `g3` and `g4` corresponding to the intermediate G^{μ}_{IJK}, $\mu = 1, 2, 3, 4$, to 0. The index `iz=1, nd` distinguished different denominator factors.

```
      do 4 iz=1,nd
      g1(iz)=0.0d+00
      g2(iz)=0.0d+00
      g3(iz)=0.0d+00
      g4(iz)=0.0d+00
    4 continue
      l7=(i.gt.j)
      l8=(j.gt.k)
```

The arrays pointing to the integrals $\langle ij|\, O\, |ab\rangle$ and $\langle jk|\, O\, |bc\rangle$ in the integral arrays `vijab` and `vjkbc`, respectively, are now assigned. Set the elements of the arrays `locab` and `locbc` to 0.

```
      do 20 ix=1,nnvirt
      locab(ix)=0
      locbc(ix)=0
   20 continue
```

The following code assigns the values of `locab`:

```
      iz=0
   21 continue
      iz=iz+1
      label=ijab(iz)
      laba=label/1200
      laba=laba-(laba/60)*60
      laba=laba-nocc
      labb=label/20
      labb=labb-(labb/60)*60
      labb=labb-nocc
      itypea(iz)=label-(label/20)*20
      indab=ind(laba)+labb
```

```
      if(locab(indab).eq.0) locab(indab)=iz
      if(locab(indab).eq.iz-1) locab(indab)=1-iz
      if(iz.lt.mn) goto 21
```

Execution reaches this point when `iz=mn` and then continues by assigning values of `locbc`:

```
      iz=0
   22 continue
      iz=iz+1
      label=jkbc(iz)
      labb=label/1200
      labb=labb-(labb/60)*60
      labb=labb-nocc
      labc=label/20
      labc=labc-(labc/60)*60
      labc=labc-nocc
      indbc=ind(labb)+labc
      itypec(iz)=label-(label/20)*20
      if(locbc(indbc).eq.0) locbc(indbc)=iz
      if(locbc(indbc).eq.iz-1) locbc(indbc)=1-iz
      if(iz.lt.nn) goto 22
```

Execution reaches this point when `iz=nn` and all the non-zero elements of `locbc` are assigned.

The index `ir` counts the "ring" integrals. The program now processes the "ring" integrals $\langle ic|\, O\, |ak\rangle$ to first form the intermediates F^{μ}_{IJKAC} and then the G^{μ}_{IJK}.

```
      ir=0
   25 continue
      ir=ir+1
      ia=iar(ir)
      ic=icr(ir)
      l10=(ia.eq.ic)
      itype=ityper(ir)
```

Assign the three "ring" integrals to the scalars `vr1`, `vr2`, `vr3`:

```
      vr1=0.0d+00
      vr2=0.0d+00
      vr3=0.0d+00
      if(itype.ne.1.and.itype.ne.3.and.itype.ne.5.and.
     $itype.ne.7) goto 50
      vr1=rint(ir)
      if(ir.ge.nr) go to 52
      iaa=iar(ir+1)
      icc=icr(ir+1)
```

```
      if(iaa.ne.ia.or.icc.ne.ic) goto 52
      ir=ir+1
      itype=ityper(ir)
   50 continue
      if(itype.ne.2.and.itype.ne.4.and.itype.ne.6.and.
     $itype.ne.8) goto 51
      vr2=rint(ir)
      if(itype.ne.8) go to 52
      if(ir.ge.nr) go to 52
      iaa=iar(ir+1)
      icc=icr(ir+1)
      if(iaa.ne.ia.or.icc.ne.ic) goto 52
      ir=ir+1
      itype=ityper(ir)
   51 continue
      if(itype.ne.9) goto 901
      vr3=rint(ir)
   52 continue
```

Set the arrays used for storing the intermediates F^{μ}_{IJKAC} to 0. `f1`, `f2`, `f3`, `f4`, `f5` correspond to the five different spin cases: $\mu = 1, 2, 3, 4, 5$. `ix=1,2` distinguish F^{μ}_{IJKAC} and F^{μ}_{IJKCA}. `iz=1,2` allows calculations for different denominator factors to be carried out at the same time.

```
      do 42 ix=1,2
      do 42 iz=1,nd
      f1(ix,iz)=0.0d+00
      f2(ix,iz)=0.0d+00
      f3(ix,iz)=0.0d+00
      f4(ix,iz)=0.0d+00
      f5(ix,iz)=0.0d+00
   42 continue
```

Now the summation over b begins:

```
      do 40 ib=1,nvirt
      kb=ib+nocc
```

and the two cases F^{μ}_{IJKAC} and F^{μ}_{IJKCA} considered.

```
      do 41 ix=1,2
      ka=ia+nocc
      kc=ic+nocc
      l1=(ia.ge.ib)
      l2=(ib.ge.ic)
      if(l1) iab=ind(ia)+ib
      if(l2) ibc=ind(ib)+ic
```

```
      if(.not.l1) iab=ind(ib)+ia
      if(.not.l2) ibc=ind(ic)+ib
      jab=locab(iab)
      jbc=locbc(ibc)
      if(jab.eq.0.or.jbc.eq.0) goto 43
      v1=0.0d+00
      v2=0.0d+00
      v3=0.0d+00
      v4=0.0d+00
      if(jab.lt.0) goto 54
      itype=itypea(jab)
      if(itype.lt.5) v1=vijab(jab)
      if(itype.eq.5) v3=vijab(jab)
      goto 55
   54 continue
      jab=iabs(jab)
      itype=itypea(jab)
      v1=vijab(jab)
      v3=vijab(jab+1)
   55 continue
      if(jbc.lt.0) goto 56
      jtype=itypec(jbc)
      if(jtype.lt.5) v2=vjkbc(jbc)
      if(jtype.eq.5) v4=vjkbc(jbc)
      goto 57
   56 continue
      jbc=iabs(jbc)
      jtype=itypec(jbc)
      v2=vjkbc(jbc)
      v4=vjkbc(jbc+1)
   57 continue
```

The denominator factors are now required. They are stored in the arrays `d1`, `d2`, `d3`, `d4`, `d5` corresponding to the different spin cases labelled by $\mu = 1, 2, 3, 4, 5$. This program handles both the Møller–Plesset and the Epstein–Nesbet perturbation series. For the Møller–Plesset expansion, the denominators do not depend on the spin case and are given by

$$D_{ijab} = \varepsilon_i + \varepsilon_j - \varepsilon_a - \varepsilon_b$$

and

$$D_{jkbc} = \varepsilon_j + \varepsilon_k - \varepsilon_b - \varepsilon_c.$$

The product of these denominator factors are assigned in the following code:

```
      dijab=eorb(i)+eorb(j)-eorb(ka)-eorb(kb)
```

```
djkbc=eorb(j)+eorb(k)-eorb(kb)-eorb(kc)
d=dijab*djkbc
if(d.lt.1.0d-10) goto 902
d=1.0d+00/d
iz=1
d1(iz)=d
d2(iz)=d
d3(iz)=d
d4(iz)=d
d5(iz)=d
```

which includes a check for vanishing denominator factors which would cause overflow. The detection of such factor provokes an error condition via a `goto 902`.

The third-order "ring" energy component for the perturbation series corresponding to the Epstein–Nesbet zero-order Hamiltonian is given by

$$E_3(hp) = -\sum_{ijk}\sum_{abc} \Theta_{ik,ac} \frac{\langle ij|\, O\, |ab\rangle \langle ak|\, O\, |ic\rangle \langle bc|\, O\, |jk\rangle}{\left(D_{ijab} - d_{ijab}\right)\left(D_{jkbc} - d_{jkbc}\right)}$$

where

$$\Theta_{ik,ac} = 2^{-\gamma_{pq}\gamma_{rs}} \left(\gamma_{pq} + \gamma_{rs}\right)$$

in which

$$\gamma_{pq} = \begin{cases} 0, & \text{if } p = q \\ 1, & \text{if } p \neq q \end{cases}$$

and

$$\begin{aligned} d_{pqrs} = & \left(\langle pq|\frac{1}{r_{12}}|pq\rangle - \langle pq|\frac{1}{r_{12}}|qp\rangle \right) + \left(\langle rs|\frac{1}{r_{12}}|rs\rangle - \langle rs|\frac{1}{r_{12}}|sr\rangle \right) \\ & + \left(\langle pr|\frac{1}{r_{12}}|pr\rangle - \langle pr|\frac{1}{r_{12}}|rp\rangle \right) + \left(\langle ps|\frac{1}{r_{12}}|sp\rangle - \langle ps|\frac{1}{r_{12}}|ps\rangle \right) \\ & + \left(\langle qr|\frac{1}{r_{12}}|rq\rangle - \langle qr|\frac{1}{r_{12}}|qr\rangle \right) + \left(\langle qs|\frac{1}{r_{12}}|sq\rangle - \langle qs|\frac{1}{r_{12}}|qs\rangle \right). \end{aligned}$$

The following code sets up the products of these denominator factors.

```
if(l1) x1=dint(ka,kb)
if(l1) x2=dint(kb,ka)
if(.not.l1) x1=dint(kb,ka)
if(.not.l1) x2=dint(ka,kb)
if(l7) x3=dint(i,j)
if(l7) x4=dint(j,i)
if(.not.l7) x3=dint(j,i)
if(.not.l7) x4=dint(i,j)
x5=dint(ka,i)
x6=dint(i,ka)
```

```
x7=dint(kb,i)
x8=dint(i,kb)
x9=dint(ka,j)
x10=dint(j,ka)
x11=dint(kb,j)
x12=dint(j,kb)
y=x5-x6+x7-x8+x9-x10+x11-x12
if(i.ne.j.and.ia.ne.ib) s11=-x1+x2-x3+x4+y
if(i.ne.j.and.ia.eq.ib) s11=-x3+x4+y
if(i.eq.j.and.ia.ne.ib) s11=-x1+x2+y
if(i.eq.j.and.ia.eq.ib) s11=y
s12=-x1-x3+x5-x6+x7+x9+x11-x12
s13=-x1-x3+x5+x7-x8+x9-x10+x11
if(l2) x1=dint(kb,kc)
if(l2) x2=dint(kc,kb)
if(.not.l2) x1=dint(kc,kb)
if(.not.l2) x2=dint(kb,kc)
if(l8) x3=dint(j,k)
if(l8) x4=dint(k,j)
if(.not.l8) x3=dint(k,j)
if(.not.l8) x4=dint(j,k)
x5=dint(kb,j)
x6=dint(j,kb)
x7=dint(kc,j)
x8=dint(j,kc)
x9=dint(kb,k)
x10=dint(k,kb)
x11=dint(kc,k)
x12=dint(k,kc)
y=x5-x6+x7-x8+x9-x10+x11-x12
if(j.ne.k.and.ib.ne.ic) s21=-x1+x2-x3+x4+y
if(j.ne.k.and.ib.eq.ic) s21=-x3+x4+y
if(j.eq.k.and.ib.ne.ic) s21=-x1+x2+y
if(j.eq.k.and.ib.eq.ic) s21=y
s22=-x1-x3+x5-x6+x7+x9+x11-x12
s23=-x1-x3+x5+x7-x8+x9-x10+x11
```

Now the required products of denominator factors corresponding to the Epstein–Nesbet perturbation expansion are formed for each of the possible spin cases. Code to check for vanishing denominators is included so as to avoid overflow. A vanishing denominator causes an error condition via a `goto 902`.

```
iz=2
d=(dijab+s11)*(djkbc+s21)
if(dabs(d).lt.1.0d-10) goto 902
```

```
d1(iz)=1.0d+00/d
d=(dijab+s11)*(djkbc+s22)
if(dabs(d).lt.1.0d-10) goto 902
d2(iz)=1.0d+00/d
d=(dijab+s12)*(djkbc+s22)
if(dabs(d).lt.1.0d-10) goto 902
d3(iz)=1.0d+00/d
d=(dijab+s12)*(djkbc+s21)
if(dabs(d).lt.1.0d-10) goto 902
d4(iz)=1.0d+00/d
d=(dijab+s13)*(djkbc+s23)
if(dabs(d).lt.1.0d-10) goto 902
d5(iz)=1.0d+00/d
```

This completes the formation of the denominators factors.

It remains to construct the numerators in the expressions for the intermediates F^{μ}_{IJKAC}. The logical variables `j1`, `j2`, `j3`, `j4`, `j5` correspond to the five spin cases identified by the index $\mu = 1, 2, 3, 4, 5$. They are initial set to .true. and will be changed to .false. when a particular combination of numerator factors gives rise to a vanishing contribution to the intermediate F^{μ}_{IJKAC}.

```
j1=.true.
j2=.true.
j3=.true.
j4=.true.
j5=.true.
```

Set the logical variables `l1`, `l2`, `l3` and `l4`.

```
l1=(itype.lt.4)
l2=(jtype.lt.4)
l3=((l7.and.ib.gt.ia).or.(.not.l7.and.ia.gt.ib))
l4=((l8.and.ic.gt.ib).or.(.not.l8.and.ib.gt.ic))
```

The code now branches according to the values of logical variables `l1`, `l2`, `l3` and `l4`. Nine distinct cases arise.

```
if(l1) goto 60
if(l2) goto 61
if(l3) goto 70
if(l4) goto 71
```

Case 1: `l1=.false.,l2=.false.,l3=.false.,l4=.false.`

```
s13=v1-v3
s24=v2-v4
t1=s13*s24
t2=s13*v2
```

```
      t3=v1*v2
      t4=v1*s24
      t5=v3*v4
      goto 46
```

Case 2: `l1=.false.,l2=.false.,l3=.false.,l4=.true.`

```
   71 continue
      s13=v1-v3
      s42=v4-v2
      t1=s13*s42
      t2=s13*v4
      t3=v1*v4
      t4=v1*s42
      t5=v3*v2
      goto 46
   70 continue
      if(l4) goto 72
```

Case 3: `l1=.false.,l2=.false.,l3=.true.,l4=.false.`

```
      s31=v3-v1
      s24=v2-v4
      t1=s31*s24
      t2=s31*v2
      t3=v3*v2
      t4=v3*s24
      t5=v1*v4
      goto 46
   72 continue
```

Case 4: `l1=.false.,l2=.false.,l3=.true.,l4=.true.`

```
      s31=v3-v1
      s42=v4-v2
      t1=s31*s42
      t2=s31*v4
      t3=v3*v4
      t4=v3*s42
      t5=v1*v2
      goto 46
   61 continue
      if(l3) goto 75
```

Case 5: `l1=.false.,l2=.true.,l3=.false.`

```
      j1=.false.
      t2=(v1-v3)*v2
```

```
      t3=v1*v2
      j4=.false.
      t5=v3*v2
      goto 46
```

Case 6: `l1=.false.,l2=.true.,l3=.true.`

```
   75 continue
      j1=.false.
      t2=(v3-v1)*v2
      t3=v3*v2
      j4=.false.
      t5=v1*v2
      goto 46
   60 continue
      if(l2) goto 62
      if(l4) goto 77
```

Case 7: `l1=.true.,l2=.false.,l4=.false.`

```
      j1=.false.
      j2=.false.
      t3=v1*v2
      t4=v1*(v2-v4)
      t5=v1*v4
      goto 46
```

Case 8: `l1=.true.,l2=.false.,l4=.true.`

```
   77 continue
      j1=.false.
      j2=.false.
      t3=v1*v4
      t4=v1*(v4-v2)
      t5=v1*v2
      goto 46
```

Case 9: `l1=.true.,l2=.true.`

```
   62 continue
      j1=.false.
      j2=.false.
      t3=v1*v2
      j4=.false.
      t5=t3
   46 continue
```

Now double all components to allow for second spin case obtained by interchanging α and β spins.

```
      t1=t1+t1
      t2=t2+t2
      t3=t3+t3
      t4=t4+t4
      t5=t5+t5
```

The elements of the arrays `f1`, `f2`, `f3`, `f4`, `f5` can now be updated for both the Møller–Plesset expansion and the Epstein–Nesbet series.

```
      do 80 iz=1,nd
      if(j1) f1(ix,iz)=f1(ix,iz)+t1*d1(iz)
      if(j2) f2(ix,iz)=f2(ix,iz)+t2*d2(iz)
      if(j3) f3(ix,iz)=f3(ix,iz)+t3*d3(iz)
      if(j4) f4(ix,iz)=f4(ix,iz)+t4*d4(iz)
      if(j5) f5(ix,iz)=f5(ix,iz)+t5*d5(iz)
   80 continue
   43 continue
```

If $i = k$ or $a = c$ then processing of this block of integrals is completed, otherwise the indices `ia` and `ic` are interchanged and the computation repeated

```
      if(l9.or.l10) goto 40
      isave=ia
      ia=ic
      ic=isave
   41 continue
   40 continue
```

This completes the loop over the virtual orbital index b (`ib`).
Execution continues by calculating the energy components G^{μ}_{IJK}.

```
     if(l9.or.l10) goto 30
      s13=vr1-vr3
      s23=vr2-vr3
      do 31 iz=1,nd
      t1=f1(1,iz)*s13+f1(2,iz)*s23
      t2=f3(1,iz)*s13+f3(2,iz)*s23-(f5(1,iz)+f5(2,iz))
            *vr3
      t3=f2(1,iz)*vr1+f2(2,iz)*vr2
      t4=f4(1,iz)*vr1+f4(2,iz)*vr2
      g1(iz)=g1(iz)+t1+t1
      g2(iz)=g2(iz)+t2+t2
      g3(iz)=g3(iz)+t3+t3
      g4(iz)=g4(iz)+t4+t4
   31 continue
      goto 32
   30 continue
```

```
      s12=vr1-vr2
      do 34 iz=1,nd
      t1=f1(1,iz)*s12
      t2=f3(1,iz)*s12-f5(1,iz)*vr2
      if(l9.and.l10) goto 33
      t3=(f2(1,iz)+f4(1,iz))*vr1
      t4=t3
      t1=t1+t1
      t2=t2+t2
      goto 35
   33 continue
      t3=f2(1,iz)*vr1
      t4=t3
   35 continue
      if(l9.and.l10.and.iz.eq.2) goto 38
      g1(iz)=g1(iz)+t1
      g2(iz)=g2(iz)+t2
   38 continue
      g3(iz)=g3(iz)+t3
      g4(iz)=g4(iz)+t4
      if(.not.(l9.and.l10).or.iz.eq.2) goto 34
      t1=t1+t2
      if(l6) t1=t1+t1
      ediag=ediag+t1
   34 continue
   32 continue
      if(ir.lt.nr) goto 25
      do 39 iz=1,nd
      t1=g1(iz)+g2(iz)
      t2=t1+g3(iz)+g4(iz)
      if(l6) t2=t2+t2
      etotal(iz)=etotal(iz)+t2
      if(.not.l9) goto 39
      if(l6) t1=t1+t1
      etwo(iz)=etwo(iz)+t1
   39 continue
```

The (i,j,k) energy components are stored in the array `gijk( )`.

```
      indx=indx+1
      gijk(1,indx)=g1(1)
      gijk(2,indx)=g1(2)
      indx=indx+1
      gijk(1,indx)=g2(1)
      gijk(2,indx)=g2(2)
```

```
      indx=indx+1
      gijk(1,indx)=g3(1)
      gijk(2,indx)=g3(2)
      indx=indx+1
      gijk(1,indx)=g4(1)
      gijk(2,indx)=g4(2)
    7 continue
      rewind it
      rewind kt
    2 isym0=isym0+nocc
    1 continue
```

This completes the computation of the energy components for each (i,j,k) triple. It remains to rewind the data set `jt` and return control to the calling routine.

```
      rewind jt
      return
```

The following code handles various error conditions.

```
  900 write(io,6000) i,j,k,ir,kr,mi,mj,ni,nj
      go to 999
  901 write(io,6001) i,j,k,ir,kr,mi,mj,ni,nj
      go to 999
  902 write(io,6002) i,j,k,ia,ib,ic
      go to 999
  903 write(io,6003) nr,limit
 6000 format(20h0End of file in ring,5x,9i5)
 6001 format(37h0Expected integrals not found in ring,
                5x,9i5)
 6002 format(28h0Denominator is zero in ring,5x,6i5)
 6003 format(45h0Insufficient store in ring for
                integral list, $2i10)
  999 stop
      end
```

This completes the subroutine for evaluating the third-order “ring” energies.

2.2.6. *Printing of intermediate results:* `gwrt`

The subroutine `gwrt` prints out the intermediate results when required.

```
      subroutine gwrt(io,jt,g,iprnt)
      implicit real*8 (a-h,o-z)
      dimension g(2,1),per(2,55),title(2),ethree(2)
      data title(1)/8h(e0-h0) /,title(2)/8hshifted /
      logical lsym1(550),lsym2(550)
      common/ptind/ ind(60),norb,nnorb,nocc,nnocc,
```

```
      nvirt, $nnvirt
      common/ptsym/ lsym1,lsym2,invlab(60)
      common/ptres/ etwo(2),etotal(2),ediag,eorb(60)
      do 5 id=1,2
      ethree(id)=etotal(id)-etwo(id)
      do 5 ij=1,nnocc
    5 per(id,ij)=0.0d+00
      if(iprnt.ne.0)
     1write (io,6002) title,etotal,etwo,ethree,ediag
      do 2 id=1,2
      isym=0
      index=0
      if(iprnt.ne.0) write (io,6000) title(id)
      do 2 i=1,nocc
      do 2 k=1,i
      do 2 j=1,nocc
      isym=isym+1
      if(lsym1(isym)) goto 2
      if(iprnt.ne.0)
     1write(io,6001) i,j,k,g(id,index+1),i,j,k,
        g(id,index+2),
     $ i,j,k,g(id,index+3),i,j,k,g(id,index+4)
      if(i.ne.k) go to 3
      t1=g(id,index+1)+g(id,index+2)
      if(lsym2(isym)) t1=t1+t1
      if(i.ge.j) ij=(i*i-i)/2+j
      if(i.lt.j) ij=(j*j-j)/2+i
      per(id,ij)=per(id,ij)+t1
    3 index=index+4
    2 continue
      write (jt) etotal,etwo,per
 6000 format(///36h0Components of the ring energy
    with ,a8, $12hdenominators/)
 6001 format(9x,i2,1ha,i2,1ha,i2,1ha,1x,f12.8,5x,i2,
        1ha,i2,
     $       1hb,i2,1ha,1x,f12.8,5x,i2,1ha,i2,1ha,i2,
        1hb,1x,
     $       f12.8,5x,i2,1ha,i2,1hb,i2,1hb,1x,f12.8)
 6002 format(//  1h0,48x,a8,16x,a8/
     $32h0Total third-order ring energy =,
     $8x,f19.14,5x,f19.14/
     $35h0Two-body third-order ring energy =,
     $5x,f19.14,5x,f19.14/
     $37h0Three-body third-order ring energy =,
```

```
$3x,f19.14,5x,f19.14/
$35h0Diagonal third-order ring energy =,
$5x,f19.14//)
 return
 end
```

2.2.7. Input/output subroutine: `rfst`

The input/output subroutine `rfst` performs "fast" unformatted read and write operations, the latter being carried out via the entry `wfst`. `rfst` has arguments `i(n)`, the integer array to be written, `n`, the length of the array, and `kt` the data set to be read from. `wfst` contains the corresponding arguments for a write operation.

```
subroutine rfst(i,n,kt)
integer   i(n)
read (kt) i
return
entry wfst(i,n,kt)
write (kt) i
return
end
```

This completes a literate program for evaluating third-order "ring" energies in the many-body perturbation theory for closed-shell systems within the algebraic approximation.

3. CONCLUDING REMARKS

We have described the *a posteriori* application of literate programming techniques to part of the first published computer program for performing many-body perturbation theory calculations within the algebraic approximation. The program considered in this work was originally published by the author in *Computer Physics Communications* in 1978. It formed part of a set of programs for molecular many-body perturbation theory calculations. The original code was documented by means of "comment" cards within the code together with the companion publication [9].

We have demonstrated in this paper how literate programming methods can do much to facilitate the understanding of computer code by the reader and at the same time bring the coding details into the public domain where they can be critically evaluated and perhaps used as the basis for future studies.

Quantum chemistry is a field of research which depends heavily on computation. The potential of literate programming techniques to accelerate the development of computational quantum chemistry is evident. By placing quantum chemistry computer programs in the public domain in a form which can be easily read and comprehended by the human reader they are placed in the body of scientific knowledge where they can be openly criticized and used constructively.

References

1. J. Ziman, *The Force of Knowledge: The Scientific Dimension of Society*, Cambridge University Press, 1976.
2. J. Ziman, *Public Knowledge: An Essay Concerning the Social Dimension of Science*, Cambridge University Press, 1968.
3. H.M. Quiney and S. Wilson, in *Recent Advances in the Theory of Chemical and Physical Systems*, ed. by J.-P. Julien, J. Maruani, D. Mayou, S. Wilson and G. Delgado-Barrio, Springer, Dordrecht, 2006.
4. D.E. Knuth, Literate Programming, *Comp. J.* **27:2**, 97, 1984.
5. H. Thimbleby, Software – Practice and Experience **33**, 975, 2003.
6. H.M. Quiney and S. Wilson, *Intern. J. Quantum Chem.* **104**, 430, 2005.
7. H.M. Quiney and S. Wilson, *Molec. Phys.* **103**, 389, 2005.
8. H.M. Quiney and S. Wilson, *Literate Programming in Quantum Chemistry: An Introduction*, in preparation.
9. S. Wilson, *Comput. Phys. Commun.* **14**, 91, 1978.
10. D.M. Silver, *Comput. Phys. Commun.* **14**, 71, 1978.
11. D.M. Silver, *Comput. Phys. Commun.* **14**, 81, 1978.
12. S. Wilson, in *Handbook of Molecular Physics and Quantum Chemistry* **2**: *Molecular Electronic Structure*, ed. by S. Wilson, P.F. Bernath and R. McWeeny, Wiley, Chichester, 2003.
13. S. Wilson, *Electron Correlation in Molecules*, International Series of Monographs on Chemistry, Clarendon Press, Oxford, 1984.
14. J. Ziman, *Real Science. What It Is, and What It Means*, Cambridge University Press, 2000.

MULTICOMPONENT MANY-BODY PERTURBATION THEORY FOR THE SIMULTANEOUS DESCRIPTION OF ELECTRONIC AND NUCLEAR MOTION: TOWARDS A PRACTICAL IMPLEMENTATION USING LITERATE PROGRAMMING METHODS

STEPHEN WILSON

Rutherford Appleton Laboratory, Chilton, Oxfordshire OX11 0QX, UK

Abstract The most widely used sequence of approximations in describing the structure and properties of molecules is first to assume the Born-Oppenheimer approximation to separate the electronic and the nuclear motions, then to decouple the electronic problem by assuming an independent particle model, and then to correct the mean field description by taking account of the instantaneous interactions of the electrons. (Nuclear degrees of freedom are often treated classically.) Many-body perturbation theory (MBPT) in its second-order form (designated MP2) remains the most widely used practical technique for describing the effects of electron correlation. MBPT is also invaluable in understanding the relation between different approaches to the electron correlation problem such as configuration interaction and various cluster expansions.

Recent years have seen a growing interest in the simultaneous description of electronic and nuclear motion. The nonadiabatic coupling between the electronic and nuclear motion manifest itself in numerous and rather diverse phenomena. An independent particle model can be formulated in which the averaged interactions between the electrons, between the electrons and the nuclei and between the nuclei are described quantum mechanically. Multicomponent MBPT can then be used to formulate the corresponding correlation problem accounting for electron-electron interactions, electron-nucleus interactions and nucleus–nucleus interactions in either algebraic or diagrammatic terms.

The practical realization of multicomponent MBPT rests on the development of efficient algorithms and the associated computer code. In recent work, we have advocated the use literate programming techniques in the development and publication of computer code for molecular structure calculations. We briefly discuss the application of these methods to the multicomponent many-body perturbation expansion.

S. Lahmar et al. (eds.), Topics in the Theory of Chemical and Physical Systems, 35–61.

1. INTRODUCTION

The vast majority of theoretical molecular structure studies attack the problem in two distinct stages. First the electrons are assumed to move in the field of fixed nuclei, that is, the Born–Oppenheimer approximation [1–3] is made. Solution of the electronic Schrödinger equation yields a potential energy curve or surface which then defines an effective potential in which nuclear motion takes place. The second stage therefore involves the solution of the nuclear Schrödinger equation for the motion of the nuclei in the effective potential generated by the electrons. The study of nuclear motion necessitates the determination of a potential energy hypersurface which in turn requires, in principle, the solution of the electronic Schrödinger equation for all possible nuclear configurations. This problem becomes increasingly intractable as molecular species containing larger numbers of atoms are considered. (For a recent review of the approximate separation of electronic and nuclear motion in the molecular structure problem within the framework of non-relativistic quantum mechanics see the recent work of Sutcliffe [4–8].)

In a recent review, Woolley and Sutcliffe [9] repeat a comments made by Löwdin [10] in 1990

> One of the most urgent problems of modern quantum chemistry is to treat the motions of the atomic nuclei and the electrons on a more or less equivalent basis.

In 1969, Thomas published two papers [11, 12] in which a molecular structure theory was developed without invoking the Born–Oppenheimer approximation. In these publications and two further papers published in 1970 [13, 14], Thomas studied methane, ammonia, water and hydrogen fluoride adding the kinetic energy operators of the protons to the electronic hamiltonian and using Slater-type orbitals centered on the heavier nuclei for the protonic wave functions. Over the years, a number of authors [15–23] have attempted the development of a non-Born–Oppenheimer theory of molecular structure, but problems of accuracy and/or feasibility remain for applications to arbitrary molecular systems.

In 2002, Nakai [24] presented a non-Born–Oppenheimer theory of molecular structure in which molecular orbitals (MO) are used to describe the motion of individual electrons and nuclear orbitals (NO) are introduced each of which describes the motion of single nuclei. Nakai presents an *ab initio* Hartree–Fock theory, which is designated "NO+MO/HF theory", which builds on the earlier work of Tachikawa et al. [25]. In subsequent work published in 2003, Nakai and Sodeyama [26] apply MBPT to the problem of simultaneously describing both the nuclear and electronic components of a molecular system. Their approach will be considered in some detail in this paper as a first step in the development of a literate quantum chemistry program for the simultaneous description of electronic and nuclear motion.

In section 2 we define the total molecular Hamiltonian operator describing both nuclear and electronic motion. The Hartree–Fock theory for nuclei and electrons is presented in section 3 and a many-body perturbation theory which uses this as a reference is developed in section 4. The diagrammatic perturbation theory of nuclei and electrons is reviewed in section 5. In section 6, we turn our attention to the problem of developing a literate quantum chemistry program for the simultaneous description of electronic and nuclear motion. In section 7, we consider the prospects for this area of research and briefly describe future research directions.

2. THE TOTAL MOLECULAR HAMILTONIAN OPERATOR

The total molecular Hamiltonian operator for a system containing N nuclei and n electrons may be written

$$H = T + V \tag{1}$$

where the kinetic energy operator, T, is a sum of a two one-body terms, a nuclear term and an electronic term

$$T = T_n + T_e. \tag{2}$$

The nuclear kinetic energy operator, T_n, is a sum of one-particle operators, that is

$$\begin{aligned} T_n &= -\sum_{P}^{N} \frac{1}{2m_P} \nabla_P^2 \\ &= \sum_{P}^{N} t_P \end{aligned} \tag{3}$$

where the one-particle nuclear kinetic operator is given by

$$t_P = -\frac{1}{2m_P} \nabla_P^2 \tag{4}$$

in which m_P is the mass of the nucleus labelled P. Similarly, the electronic kinetic energy operator, T_e, is a sum of one-electron operators, that is

$$\begin{aligned} T_e &= -\sum_{p}^{n} \frac{1}{2} \nabla_p^2 \\ &= \sum_{p}^{n} t_p \end{aligned} \tag{5}$$

where the one-particle electronic kinetic operator is given by

$$t_p = -\frac{1}{2} \nabla_p^2. \tag{6}$$

The potential energy term, V, is a sum of three two-body terms, the first corresponding to nucleus–nucleus interactions, the second to nucleus–electron interactions and the third to electron-electron interactions.

$$V = V_{nn} + V_{ne} + V_{ee} \tag{7}$$

The nucleus–nucleus interaction term has the form

$$V_{nn} = \sum_{P>Q}^{N} v_{nn}(P, Q)$$

$$(8) \qquad = \sum_{P>Q}^{N} \frac{Z_P Z_Q}{r_{PQ}}$$

where Z_P is the charge associated with nucleus P and r_{PQ} is the distance between nucleus P and nucleus Q. The nucleus–electron interaction term takes the form

$$V_{ne} = \sum_{p}^{n} \sum_{P}^{N} v_{en}(p, P)$$

$$(9) \qquad = \sum_{p}^{n} \sum_{P}^{N} \frac{Z_P}{r_{pP}}$$

where r_{pP} is the distance between nucleus P and electron p. The electron–electron interaction term takes the form

$$V_{ee} = \sum_{p>q}^{n} v_{ee}(p, p)$$

$$(10) \qquad = \sum_{p>q}^{n} \frac{1}{r_{pq}}$$

where r_{pq} is the distance between the electron labelled p and that labelled q.

3. THE HARTREE–FOCK THEORY OF NUCLEI AND ELECTRONS

In order to develop a theory for the motion of both the nuclei and the electrons in a molecule, we write the total Hamiltonian operator, H, as a sum of an unperturbed or zero order Hamiltonian, H_0, and a perturbation, H_1, that is

$$(11) \qquad H = H_0 + \lambda H_1.$$

Here λ is a perturbation parameter which is introduced so as to define the order of different terms in the perturbation series but which is set equal to 1 in order to recover the physical situation.

The unperturbed Hamiltonian operator is based on an independent particle model, that is, a model in which each particle, nucleus or electron, experiences an averaged interaction with the other particles in the system. The unperturbed Hamiltonian operator is a sum of a kinetic energy term and an effective potential energy term

$$(12) \qquad H_0 = T + U$$

The kinetic energy component is the sum of one-particle terms defined in the previous section. The effective potential is a sum of a nuclear and an electron component.

$$U = U_n + U_e \tag{13}$$

It is also a sum of one-particle terms.

The total wave function for a system of nuclei and electrons can be written as a product of a nuclear component

$$\Phi_n = \left\| \varphi_P \varphi_Q \dots \right\| \tag{14}$$

in which φ_P is a single nucleus state function, or NO, and an electronic component

$$\Phi_e = \left\| \varphi_p \varphi_q \dots \right\| \tag{15}$$

in which φ_p is a single electron state function, or electronic orbitals – more usually called a molecular orbital.

The single nucleus state function or nuclear orbital is an eigenfunction of a Hartree–Fock eigenvalue equation for the nuclear motion

$$F_n \varphi_P = \varepsilon_P \varphi_P \tag{16}$$

in which the Fock operator has the form

$$\begin{aligned} F_n &= t_n + \sum_P^N (J_P \mp K_P) + \sum_p^n J_p \\ &= t_n + u_n \end{aligned} \tag{17}$$

where the nuclear Fock potential is

$$u_n = \sum_P^N (J_P \mp K_P) + \sum_p^n J_p \tag{18}$$

J and K denote the Coulomb and exchange operators, respectively. In equation (16), the effective field of the nuclear orbital is due to the motion of the electrons and the remaining nuclei.

The Hartree–Fock equations for the electrons have the form

$$F_e \varphi_p = \varepsilon_p \varphi_p \tag{19}$$

where the Fock operator is given by

$$\begin{aligned} F_e &= t_e + \sum_p^n (J_p - K_p) + \sum_P^N J_P \\ &= t_e + u_e \end{aligned} \tag{20}$$

The effective potential for the electrons is

$$u_e = \sum_p^n (J_p - K_p) + \sum_P^N J_P \tag{21}$$

which includes a mean-field coupling between the electronic and the nuclear motion. In equation (19) the effective field of the electronic (molecular) orbital is due to the motion of the nuclei and the other electrons in the system.

4. THE MANY-BODY PERTURBATION THEORY OF NUCLEI AND ELECTRONS

The unperturbed or zero order Hamiltonian can be rewritten in the form

$$\begin{aligned} H_0 &= T + U \\ &= T_n + T_e + U_n + U_e \\ &= (T_n + U_n) + (T_e + U_e) \\ &= H_{n0} + H_{e0} \end{aligned} \tag{22}$$

where H_{n0} is the unperturbed Hamiltonian describing the motion of the nuclei

$$H_{n0} = T_n + U_n \tag{23}$$

and H_{e0} is the unperturbed Hamiltonian for the motion of the electrons

$$H_{e0} = T_e + U_e. \tag{24}$$

The perturbing operator is the difference between the full Hamiltonian and the zero order Hamiltonian

$$H_1 = H - H_0 \tag{25}$$

so that

$$H_1 = (T + V) - (T + U) \tag{26}$$

and thus

$$\begin{aligned} H_1 &= V - U \\ &= V_{nn} + V_{ne} + V_{ee} - U_n - U_e \end{aligned} \tag{27}$$

Recall that the total molecular Hamiltonian is written

$$H(\lambda) = H_0 + \lambda H_1 \tag{28}$$

where the unperturbed Hamiltonian, $H(0) = H_0$, has eigenvalues E_m and eigenfunctions $|\Phi_m\rangle$

$$H_0 |\Phi_m\rangle = E_m |\Phi_m\rangle \tag{29}$$

The Schrödinger equation for the perturbed system can then be written [27]

$$H(\lambda) |\Psi(\lambda)\rangle = \mathcal{E}(\lambda) |\Psi(\lambda)\rangle \tag{30}$$

where it is assumed that the exact eigenvalue is an analytic function of the perturbation parameter λ and can be expanded in a power series

$$\mathcal{E} = \mathcal{E}(\lambda)$$
$$= \sum_{k=0}^{\infty} E^{(k)} \lambda^k \tag{31}$$

and similarly that the exact eigenfunction is an analytic function of λ and can also be written as a power series

$$|\Psi\rangle = |\Psi(\lambda)\rangle$$
$$= \sum_{k=0}^{\infty} \left|\chi^{(k)}\right\rangle \lambda^k \tag{32}$$

Obviously, the constant term in the power series expansion for $E(\lambda)$ is

$$E^{(0)} = E_0 \tag{33}$$

and the corresponding term in the power series for the exact wave function is

$$\left|\chi^{(0)}\right\rangle = |\Phi_0\rangle . \tag{34}$$

We write

$$\mathcal{E} = E + \Delta E \tag{35}$$

so that the "level shift" is

$$\Delta E = \mathcal{E} - E \tag{36}$$

In order to develop the Rayleigh–Schrödinger perturbation expansion for the energy and the wave function, we define the resolvent

$$R_0 = \frac{Q}{E_0 - H_0} \tag{37}$$

in which Q is the projection operator

$$Q = \sum_{m \neq 0} |\Phi_m\rangle \langle\Phi_m|$$
$$= 1 - |\Phi_0\rangle \langle\Phi_0|$$
$$1 - P \tag{38}$$

and P is its orthogonal complement.

The Rayleigh–Schrödinger perturbation expansion for the energy has the form

$$\Delta E = \sum_{n=1}^{\infty} \langle\Phi_0| H_1 \left[R_0 (H_1 - \Delta E)\right]^{n-1} |\Phi_0\rangle$$
$$= \sum_{n=1}^{\infty} \langle\Phi_0| H_1 \left[R_0 H_1\right]^{n-1} |\Phi_0\rangle + \Omega \tag{39}$$

where Ω represents the "renormalization terms".

The many-body perturbation expansion for the energy takes the form

$$\Delta E = \sum_{n=1}^{\infty} \langle \Phi_0 | H_1 [R_0 H_1]^{n-1} | \Phi_0 \rangle_{\text{linked}} \tag{40}$$

where the subscript "linked" indicates that only terms corresponding to linked diagrams are included in the expansion.

A diagrammatic MBPT describing both nuclei and electrons requires a second quantized formulation. (For a recent review of second quantization see the articles by Pickup [28–30] and by Karwowski [31].) The unperturbed Hamiltonian can be written in second quantized from as follows

$$H_0 = \sum_{P}^{N} \varepsilon_P a_P^{+} a_P + \sum_{p}^{n} \varepsilon_p a_p^{+} a_p \tag{41}$$

where the first term on the right-hand side is associated with the nuclei and the second with the electrons. The ε_P are single particle energies for the nuclei. The ε_p are single particle energies for the electrons. The perturbing operator can be written as a sum of a one-particle and a two-particle part, that is

$$H_1 = H_1^{(1)} + H_1^{(2)} \tag{42}$$

The one-particle component has the form

$$H_1^{(1)} = -\sum_{P,Q}^{N} \langle P | u_n | Q \rangle a_P^{+} a_Q + \sum_{p,q}^{n} \langle p | u_e | q \rangle a_p^{+} a_q \tag{43}$$

where u_n is the Fock operator associated with the motion of the nuclei and u_e is the corresponding operator for the electrons. The two-particle component is

$$\begin{aligned} H_1^{(2)} &= \frac{1}{4} \sum_{P,Q,R,S}^{N} \langle PQ | | RS \rangle a_P^{+} a_Q^{+} a_S a_R \\ &+ \sum_{P,Q}^{N} \sum_{p,q}^{n} \langle Pp | | Qq \rangle a_P^{+} a_p^{+} a_Q a_q \\ &+ \frac{1}{4} \sum_{p,q,r,s}^{n} \langle pq | | rs \rangle a_p^{+} a_q^{+} a_s a_r \end{aligned} \tag{44}$$

where the first term on the right-hand side describes interactions between the nuclei, the second term describes nucleus–electron interactions, and the third term describes interactions between the electrons.

The operators $\left\{ a_P^{+}, a_Q^{+}, \ldots \right\}$ and $\left\{ a_P, a_Q, \right\}$ are, respectively, the creation and annihilation operators for nuclei. These operators satisfy the following relations:

$$\left[a_P^{+}, a_Q \right]_{\pm} \equiv a_P^{+} a_Q \pm a_Q a_P^{+} = \delta_{PQ}$$

$$\left[a_P^+, a_Q^+\right]_\pm \equiv a_P^+ a_Q^+ \pm a_Q^+ a_P^+ = 0$$

$$\left[a_P, a_Q\right]_\pm \equiv a_P a_Q \pm a_Q a_P = 0$$

where the $+$ sign corresponds to nuclei that are fermions and the $-$ sign to those which are bosons. The creation and annihilation operators between different particles, μ and ν, say, satisfy the commutation relations

(45) $$\left[a_\mu^+, a_\nu\right]_- \equiv a_\mu^+ a_\nu - a\nu^+ a_\mu = 0$$

(46) $$\left[a_\mu^+, a_\nu^+\right]_- \equiv a_\mu^+ a_\nu^+ - a\nu^+ a_\mu^+ = 0$$

(47) $$\left[a_\mu, a_\nu\right]_- \equiv a_\mu a_\nu - a_\nu a_\mu = 0$$

The creation and annihilation operators for the electrons are $\left\{a_p^+, a_q^+, ...\right\}$ and $\left\{a_p, a_q, ...\right\}$, respectively. These operators satisfy the anticommutation relations

(48) $$\left[a_p^+, a_q\right]_+ \equiv a_p^+ a_q + a_q a_p^+ = \delta_{pq}$$

(49) $$\left[a_p^+, a_q^+\right]_+ \equiv a_p^+ a_q^+ + a_q^+ a_p^+ = 0$$

(50) $$\left[a_p, a_q\right]_+ \equiv a_p a_q + a_q a_p = 0$$

For ground states and low-lying excited states it is convenient to adopt a particle-hole formalism. We use the indices

(51) $$\{I, J, K, L, ...i, j, k, l, ...\}$$

for occupied single particle state functions and the indices

(52) $$\{A, B, C, D, ..., a, b, c, d, ...\}$$

for unoccupied single particle state functions. The indices

(53) $$\{P, Q, R, S, ...p, q, r, s, ...\}$$

are employed for arbitrary single particle state functions. The normal product of a second quantized operator is written

(54) $$N[...]$$

where the ellipsis denotes an arbitrary product of creation and annihilation operators and involves moving all annihilation operators to the right using the anticommutation and commutation relations given above.

The exact Hamiltonian can be written in normal product form as

(55) $$H^N = H - \langle\Phi_0| H |\Phi_0\rangle$$

The unperturbed Hamiltonian can also be written in normal product form as

$$H_0^N = H_0 - \langle\Phi_0| H_0 |\Phi_0\rangle$$

(56) $$= \sum_P^N \varepsilon_P N\left[a_P^+ a_P\right] + \sum_p^n \varepsilon_p N\left[a_p^+ a_p\right]$$

The perturbing Hamiltonian is written in normal product form using the relation

$$\begin{aligned} H_1^N &= H^N - H_0^N \\ &= H_1^{(1)N} + H_1^{(2)N} \end{aligned} \tag{57}$$

where $H_1^{(1)N}$ and $H_1^{(2)N}$ are the normal product forms of the operators $H_1^{(1)}$ and $H_1^{(2)}$. Explicitly, the one-particle perturbation operator, $H_1^{(1)N}$, can be written as

$$\begin{aligned} H_1^{(1)N} &= \sum_{P,Q}^{N} \left(\left[\sum_{R}^{N} \langle PR | | QR \rangle \right] - \langle P | u_n | Q \rangle \right) N \left[a_P^+ a_Q \right] \\ &+ \sum_{p,q}^{n} \left(\left[\sum_{r}^{n} \langle pr | | qr \rangle \right] - \langle p | u_e | q \rangle \right) N \left[a_p^+ a_q \right] \end{aligned} \tag{58}$$

whilst the two-particle operator, $H_1^{(2)N}$, has the form

$$\begin{aligned} H_1^{(2)N} &= V_{nn}^N + V_{ne}^N + V_{ee}^N \\ &= \sum_{P,Q,R,S}^{N} \langle PQ | | RS \rangle \, N \left[a_P^+ a_Q^+ a_S a_R \right] + \\ &+ \sum_{P,Q}^{N} \sum_{p,q}^{n} \langle Pp | | Qq \rangle \, N \left[a_P^+ a_p^+ a_q a_Q \right] + \\ &+ \sum_{p,q,r,s}^{n} \langle pq | | rs \rangle \, N \left[a_p^+ a_q^+ a_s a_r \right] \end{aligned} \tag{59}$$

Using the normal product unperturbed Hamiltonian, the zero-order Schrödinger equation becomes

$$H_0^N \left| \Phi_m \right\rangle = \Delta E_m^0 \left| \Phi_m \right\rangle \tag{60}$$

whilst the perturbed Schrödinger equation is

$$H^N \left| \Psi \right\rangle = \Delta E \left| \Psi \right\rangle \tag{61}$$

where

$$\Delta E_m^0 = E_m^0 - E_0 \tag{62}$$

and

$$\Delta E = E - E_0 = E^{(1)} + E_{\text{correlation}}$$

5. THE DIAGRAMMATIC PERTURBATION THEORY OF NUCLEI AND ELECTRONS

Diagrammatic methods are well established in handling the electron correlation problem which arising in the description of molecular structure within the Born–Oppenheimer approximation. In fact, for the relativistic electronic structure problem which involves an infinite number of bodies, the use of diagrams becomes almost indispensable; cutting through complicated algebra to expose the essential physics of the various interactions taking place. In lecture notes for the 1980 Coulson Summer School in Theoretical Chemistry, the present author [32] wrote

It should perhaps be stated at this point that the use of diagrams in the many-body perturbation theory is not obligatory. The whole of the theoretical apparatus can be set up in entirely algebraic terms. However, the diagrams are both more physical and easier to handle than the algebraic expressions and it is well worth the effort required to familiarize oneself with the diagrammatic rules and conventions.

This point has been made by many authors. In his lecture notes for the 1972 Oxford Summer School, P.W. Atkins writes [33]

In the early books on quantum theory the pages were covered with integral signs and expression such as

$$\int_{-\infty}^{\infty} dx \psi_n^* (x) H \psi_m (x) . \tag{63}$$

But these soon gave way to the symbol

$$\langle n | H | m \rangle \tag{64}$$

which, as well as being more compact, enables the structure of an equation to be seen more clearly. In recent years, a new change has occurred, and instead of equations containing cumbersome integrals as in (1.1), or Dirac brackets as in (1.2), we now see the same thing written as

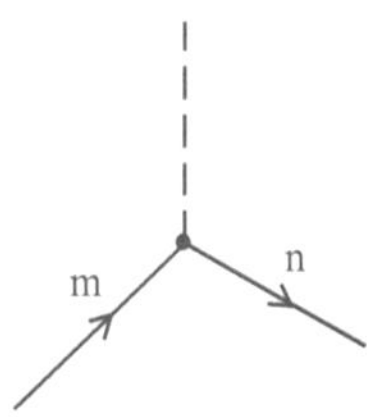

This diagram contains all the information contained in (1.1) and (1.2), but conveys it with remarkable clarity. We see that a system in a state *m is deflected into a state n by the action of the operator H.*

The diagrams are interpreted in terms of the particle-hole formalism. The Fermi level is defined such that all single particle states lying below it are occupied and all above it are unoccupied. In the particle-hole picture, the reference state is taken to be a vacuum state, containing no holes below the Fermi level and no particles above it. Excitation leads to the creation of particle-hole pairs, with particles above the Fermi level and holes below it.

The diagrammatic method can be extend to system containing both nuclei and electrons by defining nuclear and electronic vertices. A nuclear vertex is represented by an open dot ∘ whereas an electronic vertex is represented a filled dot •. When it is necessary to describe the nuclear vertex associated with a particular nuclear species corresponding to a specific element or mass number then the details are written close to the relevant open dot. The basic components of the diagrams which are used to

○ – – – –X	one-nucleus operator
• – – – –X	one-electron operator
○ – – – – ○	nucleus-nucleus interaction operator
○ – – – – •	nucleus-electron interaction operator
• – – – – •	electron-electron interaction operator

Figure 1. Basic components of the diagrams for a system of nuclei and electrons.

describe processes in the particle-hole formalism are summarized in Figure 1. All changes in the state of a many-body system are caused by an interaction which is described by an operator. This operator may be a one-particle operator or a two-particle operator. A one-particle operator is represented by a horizontal dashed interaction line terminated by a cross. For the one-nucleus operator the other end of the horizontal dashed line is terminated by an open dot ○, that is

○ – – – –x

whereas for the one-electron operator a filled dot • is used, that is

• – – – –x.

Three types of two-particle interactions can occur. Each is represented by a horizontal dashed interaction line. For the nucleus-nucleus interaction this line is terminated by open dots, ○, at both end, that is

○ – – – –○

The nucleus-electron interaction is represented by a horizontal dashed line terminated by an open dot, ○, at one end and a filled dot, •, at the other

○ – – – –•

The electron-electron interaction is represented by a horizontal dashed line terminated by filled dots, •, at both ends, that is

• – – – –•

In the convention which we shall follow here the two-electron interaction includes permutation of the two electrons. The two-nucleus interaction does not include permutation of the nuclei. The nucleus creation operators, a_P^+, a_Q^+, ..., are represented by arrows leaving the nucleus vertices. Similarly, the electron creation operators, a_p^+, a_q^+, ..., are represented by arrows leaving the electron vertices. The nucleus annihilation operators, a_P, a_Q, ..., are represented by arrows directed into the nucleus vertices. The electron annihilation operators, a_p, a_q, ..., are similarly represented by arrows directed into the electron vertices. Upward arrows represent "particle" lines whereas "downward" directed arrows "hole" lines. It should be noted that "particle" and "hole" lines may not connect a vertex corresponding to a nucleus to one associated with an electron and *vice versa*. Equally, "particle" and "hole" lines may not connect vertices corresponding to different elements or difference mass numbers.

5.1. Types of interaction

The types of interactions which can arise in the diagrammatic perturbation theory of nuclei and electrons can be classified according to the number and type of particles involved. First we subdivide the interactions into one-particle and two-particle types.

5.1.1. One-particle interactions

Obviously, the one-particle interactions can be subdivided into those involving a nucleus and those involving an electron. We consider each type in turn.

One-nucleus interactions The one-nucleus interactions that can arise in energy diagrams are classified in Figure 2 according to the level of excitation involved. There is a total of four diagrams of this type. Two of these diagrams do not involve any change in the level of excitation. They involve the interaction of the one-nucleus operator with either a hole line or a particle line associate with a nuclear orbital. One diagram in Figure 2 involves a single de-excitation; that is, the destruction of a nuclear particle-hole pair. The remaining diagram in Figure 2 involves a single excitation; that is, the creation of a nuclear particle-hole pair.

One-electron interactions The one-electron interactions that can arise in energy diagrams are classified in Figure 3 according to the level of excitation involved. There are four diagrams of this type. Two of these diagrams do not involve any change in the level of excitation. They involve the interaction of the one-electron operator with either a hole line or a particle line associate with an electronic orbital. One diagram in Figure 3 involves a single de-excitation; that is, the destruction of a electronic particle-hole pair. The remaining diagram in Figure 3 involves a single excitation; that is, the creation of an electronic particle-hole pair.

5.1.2. Two-particle interactions

The two-particle interactions can be subdivided into those between nuclei, those between nuclei and electrons and those between electrons.

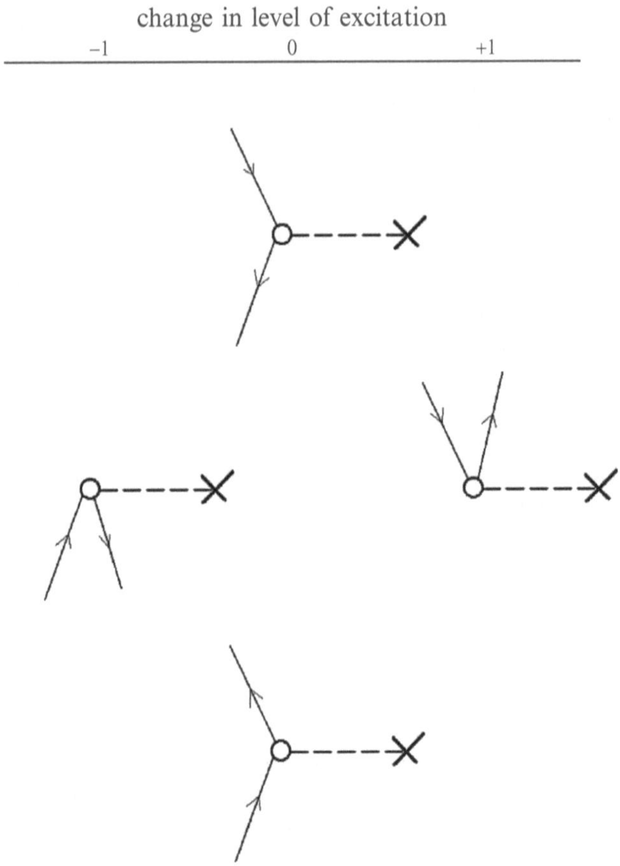

Figure 2. Classification of one-nucleus interactions that can arise in energy diagrams according to the level of excitation involved.

Nucleus-nucleus interactions Nucleus–nucleus interactions that can arise in energy diagrams are classified in Figure 4 according to the level of excitation involved. If the nuclei are identical then the matrix elements include a permutation operator which interchanges the coordinates of the two nuclei. There is a total of fourteen types of interaction between nuclei.

Four of these involve a "bubble" or self-energy. They can be classified in the same way as the one-nucleus interaction; so one bubble diagram involves a single excitation, two bubble diagrams involve no change in the level of excitation and the remaining bubble diagram involves a single de-excitation.

Of the remaining ten nucleus–nucleus interaction diagrams, one involves a double de-excitation; that is, the destruction of two nuclear particle-hole pairs, two diagrams involve a single de-excitation; that is, the destruction of one nuclear particle-hole pair, four diagrams involve no change in the level of excitation, two diagrams involve

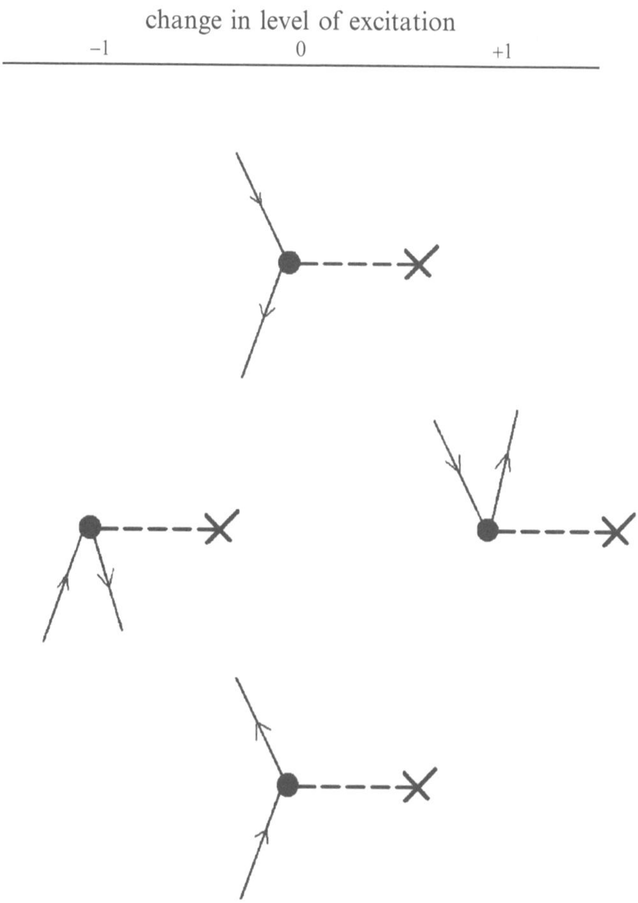

Figure 3. Classification of one-electron interactions that can arise in energy diagrams according to the level of excitation involved.

a single excitation; that is, the creation of a nuclear particle-hole pair, and, finally, one diagram involve a double excitation; that is, the creation of two nuclear particle-hole pairs.

Nucleus-electron interactions Nucleus–electron interactions that can arise in energy diagrams are classified in Figure 5 according to the level of excitation involved. There is a total of twenty-four types of interaction between nuclei.

Eight of these involve a "bubble" or self-energy. They can be classified in the same way as the one-particle interaction; so two bubble diagram involves a single excitation, four bubble diagrams involve no change in the level of excitation and the remaining two bubble diagrams involves a single de-excitation.

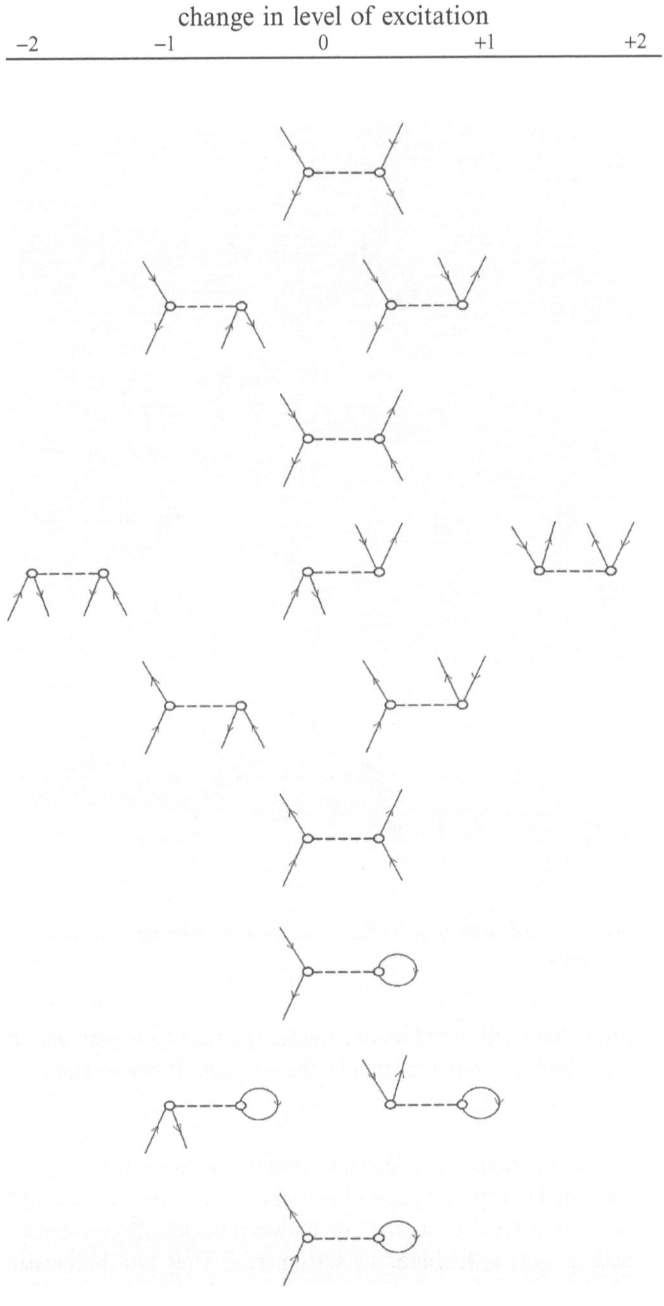

Figure 4. Classification of two-nucleus interactions that can arise in energy diagrams according to the level of excitation involved.

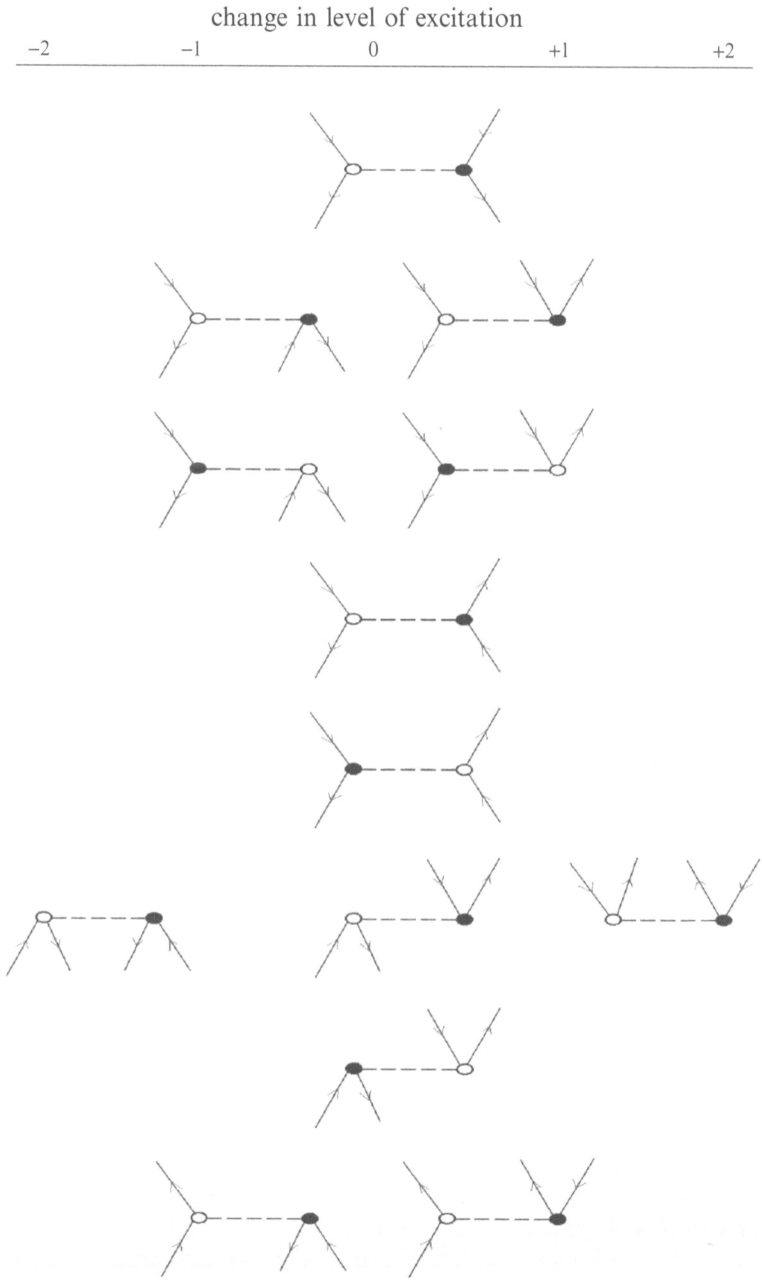

Figure 5a. Classification of nucleus–electron interactions that can arise in energy diagrams according to the level of excitation involved.

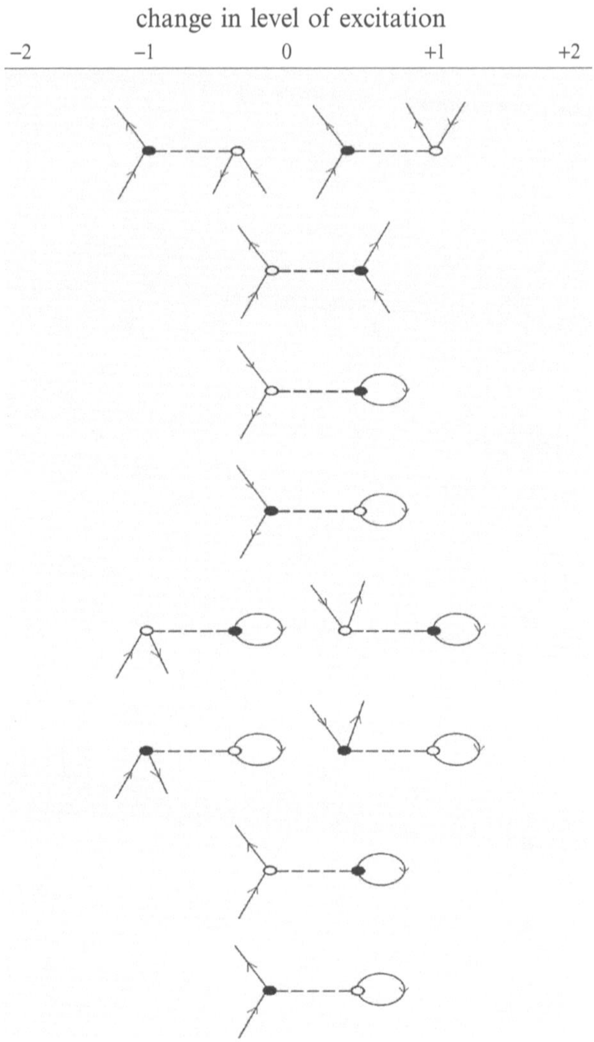

Figure 5b. Classification of nucleus–electron interactions that can arise in energy diagrams according to the level of excitation involved.

Of the remaining sixteen nucleus–nucleus interaction diagrams, one involves a double de-excitation; that is, the destruction of one nuclear particle-hole pair and one electron particle-hole pair, four diagrams involve a single de-excitation; that is, the destruction of either a nuclear particle-hole pair or an electron particle-hole pair, six diagrams involve no change in the level of excitation, four diagrams involve a single excitation; that is, the creation of a nuclear particle-hole pair or electron particle-hole pair, and, finally, one diagram involve a double excitation; that is, the creation of a nuclear particle-hole pair and an electron particle-hole pair.

Electron-electron interactions Electron–electron interactions that can arise in energy diagrams are classified in Figure 6 according to the level of excitation involved. Since the electrons are identical the matrix elements include a permutation operator which interchanges the coordinates of the two electrons. There is a total of fourteen types of interaction between electrons.

Four of these involve a "bubble" or self-energy. They can be classified in the same way as the one-electron interaction; so one bubble diagram involves a single excitation, two bubble diagrams involve no change in the level of excitation and the remaining bubble diagram involves a single de-excitation.

Of the remaining ten electron–electron interaction diagrams, one involves a double de-excitation; that is, the destruction of two electronic particle-hole pairs, two diagrams involve a single de-excitation; that is, the destruction of one electronic particle-hole pair, four diagrams involve no change in the level of excitation, two diagrams involve a single excitation; that is, the creation of an electronic particle-hole pair, and, finally, one diagram involve a double excitation; that is, the creation of two electronic particle-hole pairs.

5.2. First-order diagrammatic perturbation theory of nuclei and electrons

There are five first-order energy terms in the diagrammatic perturbation theory expansion for the motion of nuclei and electrons. The diagrams are shown in Figure 7. Two of the diagrams in Figure 7 describe interactions with the mean field potential. The remaining three first-order diagrams describe the interaction of the nuclei, of the electrons, and of the nuclei with the electrons.

5.3. Second-order diagrammatic perturbation theory of nuclei and electrons

The second-order energy diagrams can be usefully subdivided into those involving only one-particle perturbations, those involving both one- and two-particle perturbations and those involving two-particle perturbations. We consider each of these classes of diagrams in turn.

5.3.1. Components involving a one-particle perturbation

There are only two second-order diagrams involving the one-particle mean field potential. They are displayed in Figure 8. The left-hand diagram arises from interaction of the electrons with the mean field. The right-hand diagrams arises from the interaction of the nuclei with the mean field.

5.3.2. Components involving one- and two-particle perturbations

The second-order energy diagrams involving both one- and two-particle perturbations are collected in Figure 9. There is a total of eight such diagrams – four containing a mean field interaction for electrons whilst the remaining four contain a mean field interaction for nuclei. Of the four diagrams containing a mean field interaction

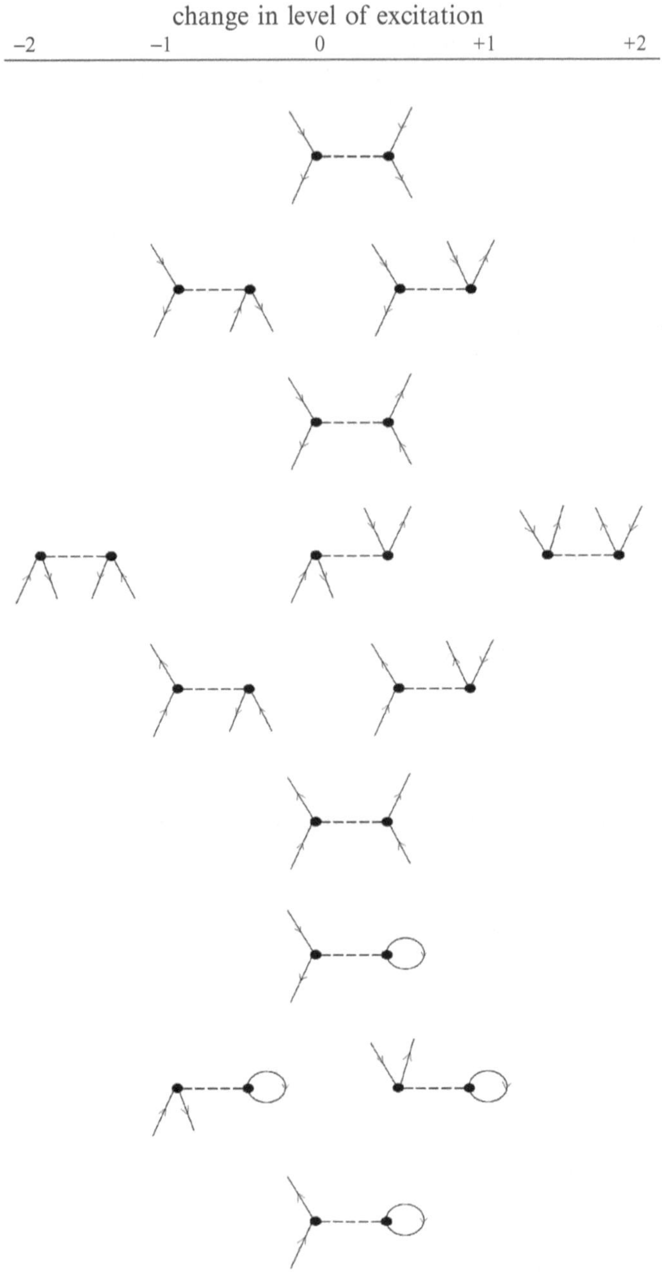

Figure 6. Classification of two-electron interactions that can arise in energy diagrams according to the level of excitation involved.

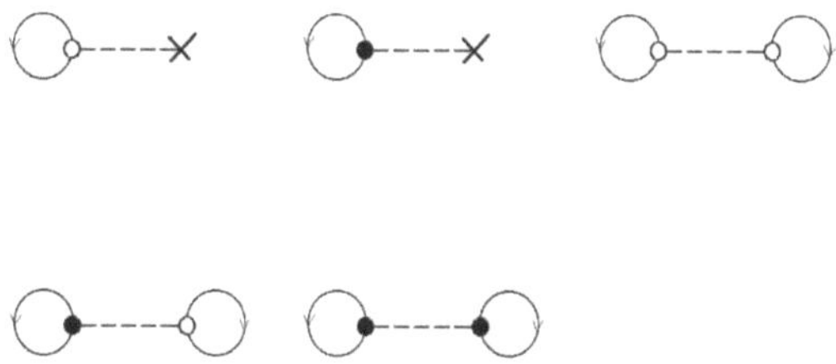

Figure 7. First-order diagrams in the perturbation theory of nuclei and electrons.

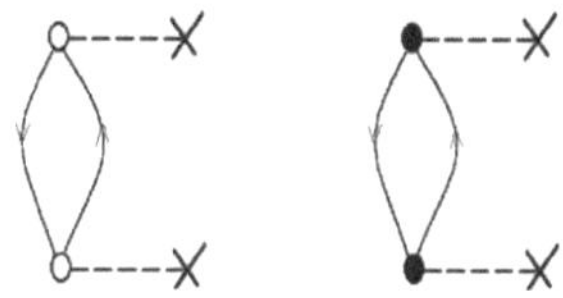

Figure 8. Second-order diagrams involving a one-particle interaction in the perturbation theory of nuclei and electrons.

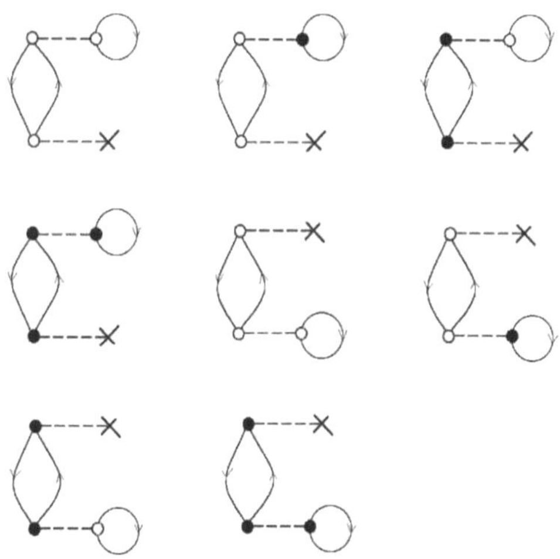

Figure 9. Second-order diagrams involving a one- and two-particle interactions in the perturbation theory of nuclei and electrons.

for electrons, one contains a "bubble" interaction involving an electron, one contains a "bubble" interaction involving a nucleus, and the remaining two diagrams are obtained by "time reversal". The four diagrams containing a mean field interaction for nuclei can be classified in a similar fashion.

5.3.3. Components involving two-particle perturbations

The second-order energy diagrams involving two-particle perturbations are collected in Figure 10. There is a total of eleven second-order diagrams of this type; eight

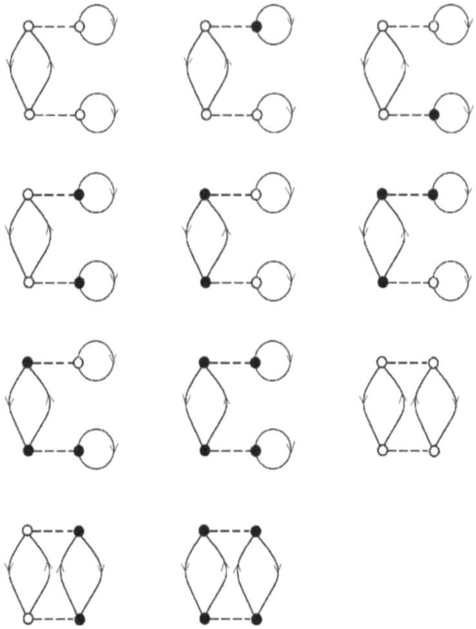

Figure 10. Second-order diagrams involving two-particle interactions in the perturbation theory of nuclei and electrons.

containing "bubble" interactions and three without such interactions. Of the eight diagrams containing "bubble" interactions, four contain particle and hole lines associated with electrons extending between the two interactions whereas the remaining four diagrams contain similar lines describing nuclei. The four diagrams in each of these subsets are distinguished by the nature of the "bubble interaction": electrons in both "bubbles", nuclei in the upper "bubble" and electrons in the lower one, and vice versa, and nuclei in both "bubbles".

5.4. Third-order diagrammatic perturbation theory of nuclei and electrons

Some of examples of third-order energy diagrams in the perturbation theory of nuclear and electronic motion are displayed in Figure 11. The energy diagrams in the top row describe interactions between electrons only. The diagrams in the middle row describe interactions between electrons and nuclei. The energy diagrams in the bottom row describe only interactions between nuclei.

5.5. Fourth-order diagrammatic perturbation theory of nuclei and electrons

Some of the examples of the fourth-order energy diagrams in the perturbation theory of nuclear and electronic motion are displayed in Figure 12. The three diagrams in the top row are associated with excited states which are only doubly excited with respect

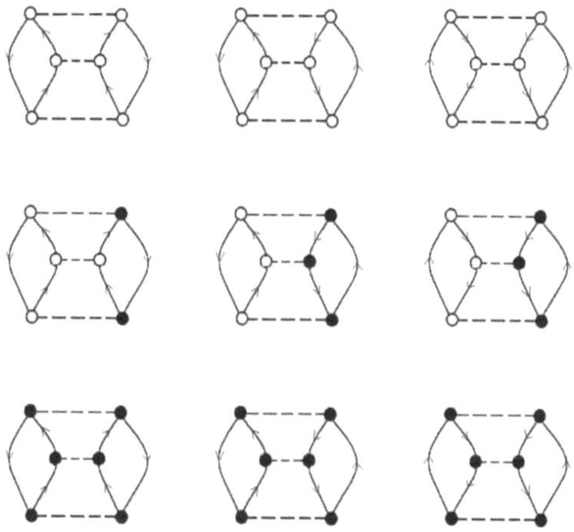

Figure 11. Some examples of third-order energy diagrams in the perturbation theory of nuclei and electrons.

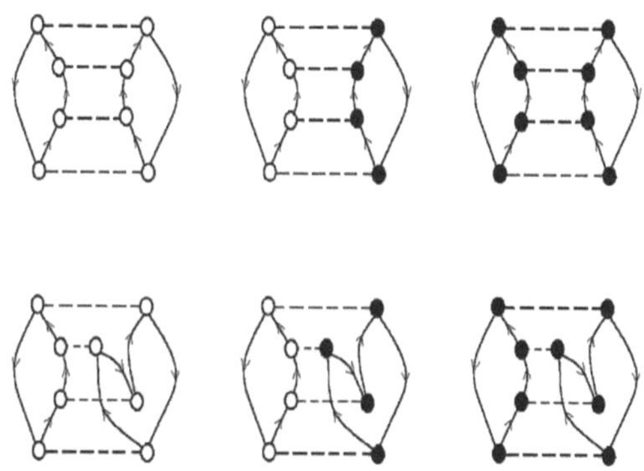

Figure 12. Some examples of fourth-order energy diagrams in the perturbation theory of nuclei and electrons.

to the reference function. The first of these diagrams involves only electrons. The second-energy diagram involves interactions between electrons and nuclei. The third diagram on the top row describes interactions between nuclei. The three diagrams in the bottom row of Figure 12 involve a triply excited intermediate state. Again, the first diagram describes the interaction only between electrons. The second diagram is associated with electron-nucleus correlation. The third diagram in the lower row describes nucleus-nucleus correlation effects.

6. TOWARDS A LITERATE PROGRAM

In a companion paper [34], we give an example of the *a posteriori* application of literate programming techniques to a quantum chemistry computer program - a program for MBPT electronic structure calculations. Such calculations have been a mainstay of modern quantum chemistry for thirty years now. Reviews of both developments of many-body Møller–Plesset theory and it applications have been provided in contributions to the Royal Society of Chemistry's Specialist Periodical Reports [35–39].

It is not our purpose to give a detailed description of literate programming and its application in quantum chemistry. We restrict our attention here to those details which are essential to our present purpose – the development of a literate quantum chemistry program for the simultaneous description of electronic and nuclear motion in molecules. For further details of literate programming methods we refer the reader to the original publication of Knuth [40]. For further details of applications of literate programming in quantum chemistry we refer the reader to our previous work [41–44] as well as our paper in the present volume [34].

A literate program consists of tightly coupled computer code and associated documentation. Code and documentation are contained within the same file, called a `.web` file, from which code and documentation can be extracted using the `tangle` and `weave` commands, respectively. If LaTeX is used for documentation and FORTRAN for code then the structure of a literate program `.web` file is as shown in Figure 13. The `.web` file consists of alternate fragments of LaTeX and FORTRAN separated by the symbol "@". The LaTeX source code may generate Tables and Figures as well as text and equations. The FORTRAN may contain "comments", but these should not replace the documentation contained in the LaTeX fragments. The present work is written in LaTeX2e and is being used as the basis for a literate program.

7. PROSPECTS

We began this paper by recalling Löwdin's statement [10], made in 1990, that there is an urgent need to describe the motions of electrons and nuclei in a more or less equivalent manner. We have demonstrated in this paper that significant progress towards this end has been achieved over recent years. The work of Nakai and Sodeyama [26], in particular, has established a firm foundation for the approach in which electrons and nuclei are treated on a more or less equivalent footing and further progress can be expected. In very recent work, Nakai *et al* [45] have presented a translation-free and rotation-free Hamiltonian for use in nuclear orbital plus molecular orbital theory. In a comment on this recent paper by Nakai *et al.* Sutcliffe [46] has suggested that their "chosen rotational term is not unique and is not valid over all regions of space". We are exploring an approach [47] based on the work of Hubač *et al.* [48–50] in which the total molecular Hamiltonian is subjected to two canonical transformations – a normal coordinate transformation and a momentum transformation. Progress will be reported in due course.

We emphasize again that the ability to perform practical calculations which go beyond the Born–Oppenheimer approximation is important not only for accurate studies of small molecular species but also in studies of larger molecules, including biomolecules, for which the number of nuclear configurations which have to be considered in the usual Born–Oppenheimer-based approach can become very large indeed. As an example of recent studies which underline the need for studies beyond

@

LaTeX source code describing the C code which follows below

...

@

C source code corresponding to the descriptive text above

...

@

LaTeX source code describing the C code which follows below

...

@

C source code corresponding to the descriptive text above

...

@

LaTeX source code describing the C code which follows below

...

@

C source code corresponding to the descriptive text above

...

@

LaTeX source code describing the C code which follows below

...

...

Figure 13. Structure of a web file, after Knuth (*The Computer Journal,* 1984, **27:2**, 97). The web file consists of alternating fragments of LaTeX source code, which may generate Tables and Figures as well as text and equations, and C source code corresponding to the LaTeX code fragments.

the Born–Oppenheimer approximation in small molecules, we point to recent work on the CH_5^+ ion [51, 52]. As for large molecules, the importance of intra- and intermolecular proton scrambling in amino acids and proteins is widely recognized.

References

1. M. Born and J.R. Oppenheimer, *Ann. der Phys.* **84**, 457, 1927.
2. S.M. Blinder, *English translation of* M. Born and J.R. Oppenheimer, 1927, *Ann. der Phys.* 84, 457 (with emendations by B.T. Sutcliffe and W. Geppert) in *Handbook of Molecular Physics and Quantum Chemistry*, vol. 1: Fundamentals, ed. S. Wilson, P.F. Bernath, and R. McWeeny, Wiley, Chichester, 2003.
3. M. Born and K. Huang, *Dynamical Theory of Crystal Lattices*, Oxford University Press, Oxford, 1955.
4. B.T. Sutcliffe, Coordinate systems and transformations, in *Handbook of Molecular Physics and Quantum Chemistry*, vol. 1: Fundamentals, ed. S. Wilson, P.F. Bernath and R. McWeeny, chapter 31, Wiley, Chichester, 2003.
5. B.T. Sutcliffe, Molecular Hamiltonians, in *Handbook of Molecular Physics and Quantum Chemistry*, vol. 1: Fundamentals, ed. S. Wilson, P.F. Bernath and R. McWeeny, chapter 32, Wiley, Chichester, 2003.
6. B.T. Sutcliffe, Potential energy curves and surfaces, in *Handbook of Molecular Physics and Quantum Chemistry*, vol. 1: Fundamentals, ed. S. Wilson, P.F. Bernath and R. McWeeny, chapter 34, Wiley, Chichester, 2003.
7. B.T. Sutcliffe, Molecular structure and bonding, in *Handbook of Molecular Physics and Quantum Chemistry*, vol. 1: Fundamentals, ed. S. Wilson, P.F. Bernath and R. McWeeny, chapter 35, Wiley, Chichester, 2003.
8. B.T. Sutcliffe, Breakdown of the Born–Oppenheimer approximation, in *Handbook of Molecular Physics and Quantum Chemistry*, vol. 1: Fundamentals, ed. by S. Wilson, P.F. Bernath and R. McWeeny, chapter 36, Wiley, Chichester, 2003.
9. R.G. Woolley and B.T. Sutcliffe, in *Fundamental World of Quantum Chemistry*. A tribute to the memory of P.-O. Löwdin ed. E.J. Brandas and E.S. Kryachko, Kluwer Academic, Dordrecht, 2003.
10. P.-O. Löwdin, *J. Molec. Struct. Theochem* **230**, 13, 1991.
11. I.L. Thomas, *Phys. Rev.* **185**, 90, 1969.
12. I.L. Thomas, *Chem. Phys. Lett.* **1**, 705, 1969.
13. I.L. Thomas, *Phys. Rev.* A **2**, 1200, 1970.
14. I.L. Thomas, *Phys. Rev.* A **3**, 1200, 1970.
15. W. Kolos and L. Wolniewicz, *Rev. Mod. Phys.* 35, 473, 1963.
16. D.M. Bishop, *Molec. Phys.* **28**, 1397, 1974.
17. D.M. Bishop and L.M. Cheung, *Phys. Rev.* A **16**, 640, 1977.
18. B.A. Pettite, *Chem. Phys. Lett.* **130**, 399, 1986.
19. H.J. Monkhorst, *Phys. Rev.* A **36**, 1544, 1987.
20. H. Nagao, K. Kodama, Y. Shigeta, H. Kawabe, K. Nishikawa, M. Makano and K. Yamaguchi, *Int. J. Quantum Chem.* **60**, 45, 1996.
21. Y. Shigeta, Y. Ozaki, K. Kodama, H. Nagao, H. Kawabe and K. Nishikawa, *Int. J. Quantum Chem.* **69**, 629, 1998.
22. Y. Shigeta, H. Takahashi, S. Yamanaka, M. Mitani, H. Nagao, K. Yamaguchi, *Int. J. Quantum Chem.* **70**, 659, 1998.
23. Y. Shigeta, H. Nagao, K. Nishikawa and K. Yamaguchi, *J. Chem. Phys.* **111**, 6171, 1999.
24. H. Nakai, *Int. J. Quantum Chem.* **86**, 511, 2002.
25. M. Tachikawa, K. Mori, H. Nakai and K. Iguchi, *Chem. Phys. Lett.* **290**, 437, 1998.
26. H. Nakai and K. Sodeyama, *J. Chem. Phys.* **118**, 1119, 2003.
27. S. Wilson, Perturbation theory, in *Handbook of Molecular Physics and Quantum Chemistry*, vol. 2: Molecular electronic structure, ed. S. Wilson, P.F. Bernath and R. McWeeny, chapter 8, Wiley, Chichester, 2003.

28. B.T. Pickup, Classical field theory and second quantization, in *Handbook of Molecular Physics and Quantum Chemistry*, vol. 1: Fundamentals, ed. S. Wilson, P.F. Bernath and R. McWeeny, chapter 26, Wiley, Chichester, 2003.
29. B.T. Pickup, The occupation number representation and the many-body problem, in *Handbook of Molecular Physics and Quantum Chemistry*, vol. 1: Fundamentals, ed. S. Wilson, P.F. Bernath and R. McWeeny, chapter 27, Wiley, Chichester, 2003.
30. B.T. Pickup, Second quantization and Lie algebra, in *Handbook of Molecular Physics and Quantum Chemistry*, vol. 1: Fundamentals, ed. S. Wilson, P.F. Bernath and R. McWeeny, chapter 28, Wiley, Chichester, 2003.
31. J.A Karwowski, Spectral density distribution moments, in *Handbook of Molecular Physics and Quantum Chemistry*, vol. 1: Fundamentals, ed. S. Wilson, P.F. Bernath and R. McWeeny, chapter 29, Wiley, Chichester, 2003.
32. S. Wilson, Lecture Notes, 1980, unpublished.
33. P.W. Atkins, Lecture Notes, 1972, unpublished.
34. S. Wilson, Literate many-body perturbation theory programming: Third-order "ring" diagrams, *this volume*.
35. S. Wilson, in *Specialist Periodical Reports: Theoretical Chemistry*, vol. 4; Senior Reporters C. Thomson and R.N. Dixon, Royal Society of Chemistry, London, 1981.
36. S. Wilson, in *Specialist Periodical Reports: Chemical Modelling- Applications and Theory*, vol. 1, Senior Reporter A. Hinchliffe, Royal Society of Chemistry, London, 2000.
37. S. Wilson, in *Specialist Periodical Reports: Chemical Modelling- Applications and Theory*, vol. 2, Senior Reporter A. Hinchliffe, Royal Society of Chemistry, London, 2002.
38. S. Wilson, in *Specialist Periodical Reports: Chemical Modelling- Applications and Theory*, vol. 3, Senior Reporter A. Hinchliffe, Royal Society of Chemistry, London, 2004.
39. S. Wilson, in *Specialist Periodical Reports: Chemical Modelling- Applications and Theory*, vol. 4, Senior Reporter A. Hinchliffe, Royal Society of Chemistry, London, 2006.
40. D.E. Knuth, *Literate Programming, The Computer Journal* **27:2**, 97, 1984.
41. H.M. Quiney and S. Wilson, Intern. *J. Quantum Chem.* **104**, 430, 2005.
42. H.M. Quiney and S. Wilson, *Molec. Phys* **103**, 389, 2005.
43. H.M. Quiney and S. Wilson, in *Recent Advances in the Theory of Chemical and Physical Systems*, ed. J.-P. Julien, J. Maruani, D. Mayou, S. Wilson, and G. Delgado–Barrio, Springer, Dordrecht, 2006.
44. H.M. Quiney and S. Wilson, *Literate Programming In Quantum Chemistry: An Introduction*, in preparation.
45. H. Nakai, M. Hoshino, K. Miyamato and S. Hyodo, *J. Chem. Phys.* **122**, 164101, 2005.
46. B.T. Sutcliffe, *J. Chem. Phys.* **123**, 237101, 2005.
47. I. Hubač and S. Wilson, *work in progress*
48. I. Hubač and M. Svrček, In *Methods in Computational Chemistry* **4**, Molecular Vibrations, ed. S. Wilson, Plenum Press, New York, 1992.
49. I. Hubač and M. Svrček, In *Methods in Computational Molecular Physics*, ed. S. Wilson and G.H.F. Diercksen, NATO ASI Series B: Physics **293** Plenum Press, New York 1992.
50. I. Hubač, P. Babinec, M. Polášek, J. Urban, P. Mach, J. Mášik, and J. Leszczyński, in *Quantum Systems in Chemistry and Physics: Basic Problems and Model Systems*, Progress in Theoretical Chemistry and Physics 2, ed. A. Hernandez-Laguna, J. Maruani, R. McWeeny, and S. Wilson, Kluwer Academic, Dordrecht, 2000, p. 383.
51. K.C. Thompson, D.L. Crittenden and M.J.T. Jordan, *J. Am. Chem. Soc.* **127**, 4954, 2005.
52. O. Asvany, P. Kumar P, B. Redlich, I. Hegemann, S. Schlemmer, and D. Marx, *Science* **309**, 1219, 2005.

RENORMALIZED COUPLED-CLUSTER METHODS: THEORETICAL FOUNDATIONS AND APPLICATION TO THE POTENTIAL FUNCTION OF WATER

PIOTR PIECUCH[1,2], MARTA WŁOCH[1], AND ANTÓNIO J.C. VARANDAS[2]

[1] *Department of Chemistry, Michigan State University, East Lansing, Michigan 48824, USA*
[2] *Departamento de Química, Universidade de Coimbra, P-3049 Coimbra Codex, Portugal*

Abstract Conventional single-reference methods fail when bond breaking and other situations characterized by larger non-dynamical correlation effects are examined. In consequence, the adequate treatment of molecular potential energy surfaces involving significant bond rearrangements has been the domain of expert multi-reference methods. The question arises if one can develop practical single-reference procedures that could be applied to at least some of the most frequent multi-reference situations, such as single and double bond dissociations. This question is addressed in the present paper by examining the performance of the conventional and renormalized coupled-cluster (CC) methods in calculations of the potential enery surface of the water molecule. A comparison with the results of the highly accurate internally contracted multi-reference configuration interaction calculations including the quasi-degenerate Davidson correction (MRCI(Q)) and the spectroscopically accurate potential energy surface of water resulting from the use of the energy switching (ES) approach indicates that the relatively inexpensive completely renormalized (CR) CC methods with singles (S), doubles (D), and a non-iterative treatment of triples (T) or triples and quadruples (TQ), such as CR-CCSD(T), CR-CCSD(TQ), and the recently developed rigorously size extensive extension of the CR-CCSD(T), termed CR-CC(2,3), provide considerable improvements in the results of conventional CCSD(T) and CCSD(TQ) calculations at larger internuclear separations. It is shown that the CR-CC(2,3) results *a posteriori* corrected for the effect of quadruply excited clusters (the CR-CC(2,3)+Q approach) can compete with the highly accurate MRCI(Q) data. The excellent agreement between the CR-CC(2,3)+Q and MRCI(Q) results suggests ways of improving the global potential energy surface of water resulting from the use of the ES approach in the regions of intermediate bond stretches and intermediate and higher energies connecting the region of the global minimum with the asymptotic regions. In addition to the examination of the performance of the CR-CCSD(T), CR-CCSD(TQ), CR-CC(2,3), and CR-CC(2,3)+Q approaches, we provide a thorough review of the method of moments of CC equations

S. Lahmar et al. (eds.), Topics in the Theory of Chemical and Physical Systems, 63–121.

(MMCC), as applied to ground electronic states, including the most recent biorthogonal formulation of MMCC theory employing the left eigenstates of the similarity-transformed Hamiltonian, and other mathematical and physical concepts that lie behind all renormalized CC approximations. In particular, we discuss the similarities and differences between the older CR-CCSD(T) and CR-CCSD(TQ) approximations and the recently formulated size extensive renormalized CC methods, such as CR-CC(2,3), and open questions that emerge in the process of designing higher-order schemes based on the biorthogonal MMCC formalism, such as CR-CC(2,4), which describe the combined effect of triples (already present in CR-CC(2,3) calculations) and quadruples in a proper manner.

1. INTRODUCTION

The key to a successful description of molecular potential energy surfaces involving bond breaking is an accurate and balanced treatment of dynamical and non-dynamical correlation effects. Conventional single-reference coupled-cluster (CC) methods [1–5] (cf. Refs. [6–16] for selected reviews), such as CCSD[T] [17, 18] and CCSD(T) [19], in which non-iterative corrections due to triply excited clusters are added to the CCSD (CC singles and doubles) [20–22] energy, and their extensions to quadruply excited clusters through the CCSD(TQ$_\text{f}$) [23] and CCSD(TQ),b [24] approaches, provide an accurate description of dynamical correlation effects, which dominate electron correlations in the closed-shell regions of potential energy surfaces, but they fail, often dramatically, when the bond breaking, biradicals, and other situations characterized by larger non-dynamical correlation effects are investigated (see, e.g. Refs. [8, 11–15, 24–58] for representative examples). Traditionally, the adequate treatment of ground- and excited-state potential energy surfaces along bond breaking coordinates and other cases of electronic quasi-degeneracies has been the domain of expert multi-reference methods, but multi-reference approaches have their limitations as well. For example, the low-order multi-reference perturbation theory (MRPT) methods, such as the popular second-order CASPT2 approach [59–62], may encounter serious difficulties with balancing dynamical and non-dynamical correlations in studies of chemical reaction pathways and relative energetics of systems characterized by a varying degree of biradical character [53, 63], while the more robust multi-reference configuration interaction (MRCI) approaches, including the highly successful internally contracted MRCI approach with quasi-degenerate Davidson corrections (the MRCI(Q) method [64, 65]), are often prohibitively expensive. In the case of MRPT, one may be able to improve the results by using larger active spaces or by switching to genuine multi-state MRPT theories (see, e.g. Refs. [66, 67] and references therein), but there are cases where choosing an active space for MRPT calculations becomes a major challenge [53] and where the use of multi-state MRPT methods may result in the emergence of additional problems, such as intruders [67–75]. Going beyond the second order in MRPT calculations or performing MRCI(Q) calculations with large active spaces to improve accuracies is far from being routine and is usually prohibitively expensive, which shows that there are problems in multi-reference theories that are not easily solvable, in spite of the fact that multi-reference methods are specifically designed to address important issues of bond breaking and electronic quasi-degeneracies. One should also remember about other shortcomings of MRCI, such as lack of size extensivity (the fraction of the correlation energy recovered rather substantially decreases for larger molecules; CASPT2 and some other MRPT methods are not size extensive either [76]), its great computational cost, and the requirement that the user of MRCI must specify several parameters, such as active orbitals or reference determinants and numerical thresholds for neglecting unimportant electron configurations. At present, a universally applicable and accepted methodology for choosing these parameters is absent [77,78]. In fact, there is no general, commonly adopted, and systematic procedure of improving the results of MRCI calculations (other than making the reference space larger), since all of the existing implementations of MRCI are based on a single MRCISD (MRCI singles and doubles) approximation, in which additional simplifications are made, with the Davidson or other similar *a posteriori* corrections added to full or approximate MRCISD energies to account for the missing correlations and lack of size extensivity.

In principle, one could address many of the above limitations of the MRPT and MRCI methods by turning to multi-reference CC approaches (cf. Refs. [8, 79, 80] for selected reviews and references therein for more information). Indeed, significant progress has been made toward the development of practical MRCC schemes, particularly after the introduction of the elegant and very promising concept of general model space MRCC approaches based on the Jeziorski-Monkhorst ansatz [81] by Li and Paldus [82–87] (which might be combined with the non-iterative triples or triples and quadruples corrections discussed in Refs. [80, 88–90] and an approximate treatment of core-virtual excitations described in Refs. [80, 91]) and the state-selective active-space methods by Adamowicz, Piecuch, and their collaborators [26,28,29,35,47,92–111], which have already been adopted and successfully applied by several research groups [112–117], in addition to groups of Adamowicz and Piecuch, and which can provide very accurate results for ground- and excited-state potential energy surfaces of closed- and open-shell systems. However, in spite of tremendous progress in MRCC methodology and in spite of the existence of highly efficient MRPT and MRCI algorithms in popular electronic structure packages, all multi-reference methods require experienced expert users and one often faces difficulties in obtaining consistent results, since in all multi-reference calculations each molecular system requires an individual practical decision on the active space. In many situations, this is not a major issue, but there are, probably equally many, challenging systems, such as the dicopper compounds that have recently been examined by Cramer *et al.* [53] and Rode and Werner [63] in the context of the ongoing studies of oxygen activation by copper metalloenzymes, where active spaces that one should use in multi-reference calculations to obtain reasonable results are far too large to be manageable, at least at this time. In this respect, practical single-reference procedures that could be applied to at least some of the most frequent multi-reference situations, such as single and double bond dissociations, biradicals, and excited states dominated by two-electron transitions, and that could provide a balanced description of dynamical and non-dynamical correlation effects with a more or less black-box effort would be an important step toward widespread progress. In this work, we discuss our recent attempts to develop and test such procedures, focusing on the ground-state renormalized CC methods [11–14,24,33–38,41,43–48,51–53] which have recently been incorporated (cf., e.g. Refs. [45, 118]) in the GAMESS package [119]. We refer the reader to the previous volumes of *Progress in Theoretical Chemistry and Physics* [13, 15], earlier reviews [11, 12, 14, 33], and original Refs. [24, 34, 43–46, 120–123] for the detailed information about the renormalized CC methods for ground as well as excited states and formal aspects of the method of moments of CC equations (MMCC) [11, 24, 34, 43–46, 89, 120, 121], on which all renormalized CC methods are based. In this paper, we focus on systematically testing a variety of conventional and renormalized CC methods, including the recently developed size extensive CR-CC(2,3) approach [45, 46, 48] and its augmented CR-CC(2,3)+Q version, in which the CR-CC(2,3) energies are approximately corrected for the dominant quadruples effects. We also provide an updated overview of the ground-state MMCC theory, including the recently developed biorthogonal MMCC formalism [45, 46], which leads to the CR-CC(2,3) approach and which could not be reviewed in the earlier volumes of *Progress in Theoretical Chemistry and Physics*. This includes the discussion of interesting formal aspects that emerge in the process of designing higher-order schemes based on the biorthogonal MMCC formalism, such as CR-CC(2,4), which would describe the combined effect of triples (already present in CR-CC(2,3) calculations) and quadruples in a proper manner.

All renormalized CC methods can be regarded as a new generation of non-iterative single-reference CC approaches that are designed to improve the results of the CCSD(T) and CCSD(TQ) calculations in the bond breaking/biradical regions of molecular potential energy surfaces, while preserving the ease-of-use and the relatively low computer cost of the CCSD(T) and CCSD(TQ) approximations. It has been demonstrated, by us and others, that the basic renormalized CC approach, termed CR-CCSD(T) (completely renormalized CCSD(T) method) [11–14, 24, 33, 34], in which a simple non-iterative correction due to triply excited clusters is added to the CCSD energy, provides very good results for single bond breaking [11–14, 33–35, 37, 38, 41, 44, 47, 118, 124], and reaction pathways involving biradicals and similar cases of electronic quasi-degeneracies [50–53, 67, 125, 126], eliminating the failures of the conventional CCSD(T) and CCSD(TQ) methods in those multi-reference situations. The CR-CCSD(TQ) extension [11–14, 24, 33, 34] of CR-CCSD(T), in which a correction due to triply and quadruply excited clusters is added to the CCSD energy, provides further improvements in the results for single bond breaking [12–14, 33, 35, 37], while helping to obtain reasonable accuracies in cases of multiple bond stretching or breaking [12–14, 24, 32–34, 40, 43–45, 48] (cf., also, Ref. [58] for the analogous findings for the approximate versions of CR-CCSD(TQ)). However, with an exception of one study of a global potential energy surface of the BeFH system [41], none of the previous calculations have examined the performance of the CR-CCSD(T) and CR-CCSD(TQ) approaches in large-basis-set calculations for different potential energy surface cuts of a triatomic molecule. Moreover, none of the earlier studies employing larger basis sets have compared the results of the CR-CCSD(T) calculations with those obtained with the recently formulated rigorously size extensive modification of CR-CCSD(T), termed CR-CC(2,3) [45, 46, 48]. The CR-CCSD(T) and CR-CCSD(TQ) methods provide great improvements in the poor description of bond breaking by the CCSD(T) and CCSD(TQ) methods, but they do it at the expense of slightly violating the size extensivity of CC theory (at the level of 0.5–1% of the correlation energy or changes of the correlation energy when reaction pathways are examined [12]). Moreover, the results of the CR-CCSD(T) and CR-CCSD(TQ) calculations for closed-shell molecules near the equilibrium geometries, where renormalization is not necessary, can be somewhat less accurate than those obtained in the corresponding CCSD(T) and CCSD(TQ) calculations if the number of electrons is larger. The issue of restoring strict size extensivity in CR-CC calculations can be dealt with in two different ways. The first method is to employ the idea of locally renormalized CC (LR-CC) approaches, which rely on the numerator-denominator connected form of the MMCC energy expansion [44]. The resulting methods, such as LR-CCSD(T) and LR-CCSD(TQ) [44, 127], provide size extensive results when localized orbitals are employed [44]. It is not yet clear how the localization of orbitals affects the accuracy of the LR-CCSD(T) and LR-CCSD(TQ) results, but the idea of local renormalization is worth further exploration. The second, in our view more robust, approach to size extensivity of CR-CC methods is to employ the CR-CC(m_A,m_B) approximations, such as CR-CC(2,3), which are based on the recently derived biorthogonal form of the MMCC theory [45, 46]. The CR-CC(2,3) approach, in which, in analogy to CR-CCSD(T), a non-iterative correction due to triply excited clusters is added to CCSD energy, provides a size extensive description without the need to localize orbitals [45, 46]. According to the initial benchmark calculations reported in Refs. [45, 46, 48], the CR-CC(2,3) approach provides the results which are competitive or at least as good as those obtained with CCSD(T) for closed-shell molecules near the equilibrium geometries, while improving the already reasonable results of the CR-CCSD(T) calculations in the biradical/bond breaking regions. The CR-CC(2,3) approach is also more accurate than the LR-CCSD(T) [44, 127] and CCSD(2)$_T$ [42] approaches, which both aim at eliminating the failures of CCSD(T) in the biradical/bond breaking regions. As shown in Refs. [45, 46, 48], the CR-CC(2,3) method seems to provide the results of the full CCSDT (CC singles, doubles, and triples) [128, 129] quality when bond breaking is examined. It is, therefore, interesting to investigate how the CR-CC(2,3) approach performs when a few different potential energy surface cuts of a triatomic molecule are examined, particularly that the aforementioned CCSD(2)$_T$ method can be regarded as the simplest approximation to the full CR-CC(2,3) scheme (the CCSD(2)$_T$ approach is equivalent to the CR-CC(2,3), A approach [45, 46, 48]). As explained in Refs. [45, 46], the CR-CC(2,3) approach can be extended to quadruple excitations through the CR-CC(2,4) theory, but the CR-CC(2,4) method has not been implemented yet and we may have to investigate issues such as the coupling of triples and quadruples in the definition of the CR-CC(2,4) energy corrections which was heuristically ignored in the original papers on the biorthogonal MMCC formalism and CR-CC(m_A,m_B) methods. On the other hand, a highly accurate description of potential energy surfaces along bond breaking coordinates that might help various spectroscopic and dynamical studies may require the inclusion of quadruples in addition to triples that are already well described by the CR-CC(2,3) theory. Thus, in this paper we examine the effect of quadruples on the CR-CC(2,3) results by adding the *a posteriori* corrections due to quadruply excited clusters, extracted from the CR-CCSD(TQ) calculations, to the CR-CC(2,3) energies. The resulting approximation is referred to as the CR-CC(2,3)+Q scheme.

The CR-CCSD(T), CR-CCSD(TQ), CR-CC(2,3), and CR-CC(2,3)+Q methods are carefully tested by using three important cuts of the global potential energy surface of the water molecule. Those cuts are: (i) the dissociation of a single O–H bond which correlates with the $\mathrm{H}(1s\ ^2S) + \mathrm{OH}(X\ ^2\Pi)$ asymptote, (ii) the simultaneous dissociation of both O–H bonds of the water molecule correlating with the $2\mathrm{H}(1s\ ^2S) + \mathrm{O}(2p^4\ ^3P)$ channel, and (iii) the C_{2v} dissociation pathway of the water molecule into $\mathrm{H}_2(X\ ^1\Sigma_g^+)$ and $\mathrm{O}(2p^4\ ^1D)$. The CR-CCSD(T), CR-CCSD(TQ), CR-CC(2,3), and CR-CC(2,3)+Q results and the corresponding CCSD, CCSD(T), and CCSD(TQ) results, all obtained with the basis sets of the aug-cc-pCVXZ (X = D, T, Q) quality [130, 131], are compared with the results of the large-scale MRCI(Q) calculations, also carried out in this work, and the spectroscopically accurate global potential energy surface of water resulting from the use of the energy switching (ES) approach [132].

We chose the water molecule as our benchmark system for a number of reasons. Clearly, water is among the most important molecules and a prototype system for a variety of spectroscopic and reaction dynamics studies. There are many applications involving water molecule in which the knowledge of reliable potential energy surface is required. Selected examples of such applications include the spectrum of the water vapor, which is important for the understanding of the absorption and retention of sunlight in Earth's atmosphere and physics of other planets and stars [133–142], and combustion studies involving hot steam. For example, the $\mathrm{O}(2p^4\ ^1D) + \mathrm{H}_2(X\ ^1\Sigma_g^+) \rightarrow \mathrm{OH}(X\ ^2\Pi) + \mathrm{H}(1s\ ^2S)$ reaction, which takes place on the ground-state potential energy surface of water, is known to play a significant role in combustion and atmospheric chemistry [143, 144]. Two of the dissociation pathways examined in this work, namely, the dissociation of a single O–H bond and the C_{2v} dissociation path of the water molecule into $\mathrm{H}_2(X\ ^1\Sigma_g^+)$ and $\mathrm{O}(2p^4\ ^1D)$ are directly related to this important reaction. The water molecule has received considerable attention in recent years due to several attempts to produce the spectroscopically and dynamically accurate global potential energy surface using *ab initio* and other theoretical means [132, 139, 145–149]. One such attempt has resulted in the ES potential function used in this work [132]. By comparing various CR-CC and MRCI(Q) data with the energies provided by the ES potential function, we may suggest ways of improving the ES and similar potentials in higher-energy regions where no precise or well understood spectroscopic or *ab initio* data are available. Thus, in addition to testing CR-CC methods, we may contribute to the ongoing effort to produce the high accuracy water surface which could be used in a variety of spectroscopic and dynamical applications.

The paper is organized as follows. In section 2, we provide the relevant background information about the MMCC formalism and overview the CR-CC methods employed in this study. In section 3, we examine the performance of various CC and CR-CC methods in calculations for the three cuts of the water potential energy surface described above and compare the results with those obtained with MRCI(Q) and the ES potential function. Finally, in section 4, we provide the concluding remarks.

2. METHOD OF MOMENTS OF COUPLED-CLUSTER EQUATIONS AND RENORMALIZED COUPLED-CLUSTER APPROACHES

The main idea of the MMCC formalism and of the resulting CR–CC methods [11–15, 24, 33, 34, 39, 40, 43–46, 48, 120–123] is that of the non-iterative energy corrections which, when added to the energies obtained in the standard CC or equation-of-motion CC [150–153] calculations, recover the exact, full CI, energies of the electronic states of interest (for the genuine multi-reference extension of the MMCC formalism, which is not discussed here, see Refs. [80, 89, 90]). Thus, all MMCC methods, including the ground-state CR-CCSD(T) [11–14, 24, 33, 34], CR-CCSD(TQ) [11–14, 24, 33, 34], CR-CC(2,3) [45, 46, 48], and CR-CC(2,3)+Q approaches discussed and tested in this work, in which relatively simple non-iterative corrections due to triples or triples and quadruples are added to CCSD energies, preserve the conceptual and computational simplicity of the traditional non-iterative CC methods, such as CCSD(T) or CCSD(TQ), while offering us a new way of controlling the quality of CC results by directly focusing on the quantity of interest, which is the difference between the full CI and CC energies. By dealing with the remanent errors that occur in the standard CC (e.g. CCSD) calculations, which we estimate by using the explicit relationships between the CC and full CI energies defining the MMCC theory, we can obtain significant improvements in the results in situations such as bond breaking, where the usual arguments based on many-body perturbation theory (MBPT) that are used to design the CCSD(T), CCSD(TQ), and similar methods fail due to the divergent behavior of the MBPT series in cases of electronic quasi-degeneracies.

In the ground-state MMCC theory, considered in this work, we focus on the non-iterative energy correction

$$(1) \qquad \delta_0^{(A)} \equiv E_0 - E_0^{(A)},$$

which, when added to the energy $E_0^{(A)}$, obtained in the standard single-reference CC calculations, referred to as method A and defined by the truncated cluster operator

(2) $$T^{(A)} = \sum_{n=1}^{m_A} T_n,$$

where

(3) $$T_n = \sum_{\substack{i_1 < \cdots < i_n \\ a_1 < \cdots < a_n}} t_{a_1 \ldots a_n}^{i_1 \ldots i_n} \, a^{a_1} \cdots a^{a_n} a_{i_n} \cdots a_{i_1}$$

is the n-body component of $T^{(A)}$, recovers the corresponding exact, i.e. full CI energy E_0. In the CCSD case, which interest us here most, $m_A = 2$; in general, $m_A \leq N$, where N is the number of correlated electrons in a system of interest. Here and elsewhere in this chapter, we use the usual notation where $i_1, i_2, \ldots$ or $i, j, \ldots$ are the spin-orbitals occupied in the reference determinant $|\Phi\rangle$ and $a_1, a_2, \ldots$ or $a, b, \ldots$ are the unoccupied spin-orbitals. The a^p (a_p) operators are the creation (annihilation) operators associated with the spin-orbitals $|p\rangle$.

2.1. Preliminaries: basic elements of coupled-cluster theory

We assume that the reader is familiar with basic concepts of CC theory, such as the exponential wave function ansatz defining all single-reference CC methods. However, for the consistency of this presentation, let us recall that the ground-state wave function associated with the CC method A has the form

(4) $$|\Psi_0^{(A)}\rangle = e^{T^{(A)}} |\Phi\rangle,$$

where the cluster operator $T^{(A)}$, defined by Eq. (2), is obtained by solving the usual system of CC equations,

(5) $$\langle \Phi_{i_1 \ldots i_n}^{a_1 \ldots a_n} | \bar{H}^{(A)} | \Phi \rangle = 0,$$

where $n = 1, \ldots, m_A$,

(6) $$\bar{H}^{(A)} = e^{-T^{(A)}} H e^{T^{(A)}} = (H e^{T^{(A)}})_C$$

is the similarity-transformed Hamiltonian of the CC theory corresponding to approximation A, subscript C designates the connected part of the corresponding operator expression, and $|\Phi_{i_1 \ldots i_n}^{a_1 \ldots a_n}\rangle \equiv a^{a_1} \cdots a^{a_n} a_{i_n} \cdots a_{i_1} |\Phi\rangle$ are the n-tuply excited determinants relative to $|\Phi\rangle$. Once the system of non-linear polynomial equations, Eq. (5), is solved for the cluster components T_n or cluster amplitudes $t_{a_1 \ldots a_n}^{i_1 \ldots i_n}$, $n = 1, \ldots, m_A$, that define them, we calculate the CC energy $E_0^{(A)}$ as follows:

(7) $$E_0^{(A)} = \langle \Phi | \bar{H}^{(A)} | \Phi \rangle.$$

In the case of many-electron systems, which are described by the Hamiltonians H containing up to two-body interactions, the CC energy $E_0^{(A)}$ is determined using only the T_1 and T_2 clusters,

$$(8)\qquad E_0^{(A)} = \langle\Phi|[H(1+T_1+T_2+\tfrac{1}{2}T_1^2)]_C|\Phi\rangle,$$

independent of the truncation scheme used to define $T^{(A)}$ (assuming that $m_A \geq 2$).

The CR-CC(2,3) method [45, 46, 48] tested in this work, which is based on the biorthogonal formulation of MMCC theory [45, 46], requires that in addition to the usual ket or right CC state $|\Psi_0^{(A)}\rangle$, Eq. (4), one considers the corresponding bra or left (or dual) CC state of method A [15, 16, 152, 153],

$$(9)\qquad \langle\tilde{\Psi}_0^{(A)}| = \langle\Phi|L^{(A)}e^{-T^{(A)}},$$

which satisfies the normalization condition,

$$(10)\qquad \langle\tilde{\Psi}_0^{(A)}|\Psi_0^{(A)}\rangle = \langle\Phi|L^{(A)}|\Phi\rangle = 1.$$

The $L^{(A)}$ operator entering Eq. (9) and satisfying the normalization condition given by Eq. (10) is a desexcitation operator defined as follows:

$$(11)\qquad L^{(A)} = \mathbf{1} + \Lambda^{(A)},$$

where **1** is a unit operator and $\Lambda^{(A)}$ is the "lambda" operator of the analytic gradient CC theory [154],

$$(12)\qquad \Lambda^{(A)} = \sum_{n=1}^{m_A} \Lambda_n.$$

The many-body components of $\Lambda^{(A)}$,

$$(13)\qquad \Lambda_n = \sum_{\substack{i_1<\cdots<i_n\\ a_1<\cdots<a_n}} \lambda_{i_1\ldots i_n}^{a_1\ldots a_n}\, a^{i_1}\cdots a^{i_n} a_{a_n}\cdots a_{a_1},$$

and the corresponding amplitudes $\lambda_{i_1\ldots i_n}^{a_1\ldots a_n}$ are obtained by solving the linear system of equations [15, 16, 153],

$$(14)\qquad \langle\Phi|\bar{H}^{(A)}|\Phi_{i_1\ldots i_n}^{a_1\ldots a_n}\rangle + \langle\Phi|\Lambda^{(A)}\,\bar{H}^{(A)}|\Phi_{i_1\ldots i_n}^{a_1\ldots a_n}\rangle = E_0^{(A)}\lambda_{i_1\ldots i_n}^{a_1\ldots a_n},$$

in which $n = 1, \ldots, m_A$ and $E_0^{(A)}$ is the energy of CC method A, Eq. (7).

2.2. Generalized moments of CC equations, MMCC functional, and MMCC expansions for the exact ground-state energy

The main purpose of all ground-state MMCC calculations, including the CR-CC methods considered in this work, is to estimate correction $\delta_0^{(A)}$, Eq. (1), such that the resulting energy, defined as

$$(15)\qquad E_0^{(\mathrm{MMCC})} = E_0^{(A)} + \delta_0^{(A)},$$

is as close as possible to the corresponding full CI energy E_0. In order to do this and end up, at the same time, with practical computational schemes, we have to come up with explicit many-body expansions of $\delta_0^{(A)}$ in terms of the quantities that one can easily extract from the usual CC (e.g. CCSD) calculations.

In order to understand what quantities provided by the CC theory may be needed to recover the full CI ground-state energy from the results of approximate CC calculations, such as CCSD, we should recall that the system of single-reference CC equations, Eq. (5), is formally obtained by inserting the exact CC wave function $|\Psi_0\rangle = e^T|\Phi\rangle$ into the electronic Schrödinger equation,

$$H|\Psi_0\rangle = E_0|\Psi_0\rangle, \tag{16}$$

premultiplying both sides of Eq. (16) on the left by e^{-T} to obtain the connected cluster form of the Schrödinger equation [3, 4, 8, 10, 11, 79],

$$\bar{H}|\Phi\rangle = E_0|\Phi\rangle, \tag{17}$$

with

$$\bar{H} = e^{-T} H e^T = (He^T)_C, \tag{18}$$

and projecting Eq. (17), in which $T = T^{(A)}$, onto the excited determinants $|\Phi_{i_1\ldots i_n}^{a_1\ldots a_n}\rangle$ corresponding to the excitations included in $T^{(A)}$. This general prescription how to derive the equations for all standard CC methods, which was introduced by Čížek [3,4], implies that projections of the connected cluster form of the Schrödinger equation, Eq. (17), on the excited determinants $|\Phi_{i_1\ldots i_n}^{a_1\ldots a_n}\rangle$, i.e.

$$\mathfrak{M}_{a_1\ldots a_n}^{i_1\ldots i_n}(m_A) = \langle\Phi_{i_1\ldots i_n}^{a_1\ldots a_n}|\bar{H}^{(A)}|\Phi\rangle, \tag{19}$$

represent the most fundamental quantities for the CC theory. These projections define the generalized moments of CC equations [11–15, 24, 33, 34, 45, 46, 120, 121] (for a discussion of the relationship between the method of moments of Krylov [155] used in various areas of mathematical physics and the single-reference CC theory, see Ref. [156]). In the language of the method of moments of Krylov [155], we might say that the standard CC equations, Eq. (5), are obtained by requiring that all moments $\mathfrak{M}_{a_1\ldots a_n}^{i_1\ldots i_n}(m_A)$ with $n = 1, \ldots, m_A$ vanish. However, the use of the generalized moments of CC equations, Eq. (19), does not have to end there. Clearly, once the cluster operator $T^{(A)}$, Eq. (2), is determined by zeroing moments $\mathfrak{M}_{a_1\ldots a_n}^{i_1\ldots i_n}(m_A)$ with $n = 1, \ldots, m_A$, we can calculate the remaining moments $\mathfrak{M}_{a_1\ldots a_n}^{i_1\ldots i_n}(m_A)$ with $n > m_A$. Since moments $\mathfrak{M}_{a_1\ldots a_n}^{i_1\ldots i_n}(m_A)$ with $n = 1, \ldots, m_A$ are needed to determine the CC energy $E_0^{(A)}$, it is natural to expect that the remaining moments $\mathfrak{M}_{a_1\ldots a_n}^{i_1\ldots i_n}(m_A)$ with $n > m_A$, which correspond to projections of the connected cluster form of the Schrödinger equation on the excited determinants $|\Phi_{i_1\ldots i_n}^{a_1\ldots a_n}\rangle$ with $n > m_A$ that are normally disregarded in the CC calculations defining approximation A, can be used to determine the difference $\delta_0^{(A)}$ between the exact, full CI, energy E_0 and $E_0^{(A)}$. In particular, if we want to recover the full CI energy E_0 by adding the correction $\delta_0^{(A)}$ to the CCSD energy (the $m_A = 2$ case), we must calculate the generalized moments of the CCSD equations, i.e. the projections of these equations on triply, quadruply, pentuply, and hextuply excited determinants or

$$\mathfrak{M}_{a_1\ldots a_n}^{i_1\ldots i_n}(2) = \langle\Phi_{i_1\ldots i_n}^{a_1\ldots a_n}|\bar{H}^{(\mathrm{CCSD})}|\Phi\rangle, \quad n = 3-6, \tag{20}$$

where

$$\bar{H}^{(\text{CCSD})} = e^{-(T_1+T_2)} H e^{T_1+T_2} = (H e^{T_1+T_2})_C \quad (21)$$

is the similarity-transformed Hamiltonian of the CCSD approach (T_1 and T_2 are the singly and doubly excited cluster components obtained in the CCSD calculations). The projections of the CCSD equations on higher-than-hextuply excited determinants do not have to be calculated, since for Hamiltonians containing up to two-body interactions used in electronic structure calculations the generalized moments $\mathfrak{M}_{a_1\ldots a_n}^{i_1\ldots i_n}(2)$ with $n > 6$ vanish. This is, in fact, a general feature of the generalized moments of CC equations: usually, moments $\mathfrak{M}_{a_1\ldots a_n}^{i_1\ldots i_n}(m_A)$ are non-zero for a relatively small range of n values. Thus, we will use symbol N_A to define the highest value of n for which the corresponding n-body moments $\mathfrak{M}_{a_1\ldots a_n}^{i_1\ldots i_n}(m_A)$ are non-zero. For example, if the CC method A is a standard CCSD approach, in which m_A is set at 2, N_A equals 6 (in this case, moments $\mathfrak{M}_{a_1\ldots a_n}^{i_1\ldots i_n}(m_A)$ also vanish for $n = 1$ and 2, since, by definition, the ground-state CCSD equations that are used to determine the corresponding T_1 and T_2 clusters are obtained by zeroing moments $\mathfrak{M}_{a_1\ldots a_n}^{i_1\ldots i_n}(2)$ with $n = 1, 2$; cf. Eq. (5)).

The exact formula for the correction $\delta_0^{(A)}$ which, when added to the CC energy $E_0^{(A)}$ of method A recovers the full CI energy E_0, in terms of the generalized moments $\mathfrak{M}_{a_1\ldots a_n}^{i_1\ldots i_n}(m_A)$ with $n = m_A + 1, \ldots, N_A$, can be given a few alternative, but equivalent forms [11–15, 24, 33, 34, 43–46, 120, 121]. In this paper, we focus on the original MMCC expansion introduced by Piecuch and Kowalski in Refs. [11, 24, 34] and the ground-state biorthogonal MMCC theory introduced by Piecuch and Włoch [45] and Piecuch *et al.* [46] (see, also, Refs. [15, 48]). All MMCC expansions for ground electronic states are derived by considering the asymmetric energy expression

$$E[\Psi] = \langle \Psi | H e^{T^{(A)}} | \Phi \rangle / \langle \Psi | e^{T^{(A)}} | \Phi \rangle, \quad (22)$$

or its slightly modified form obtained by subtracting the CC energy $E_0^{(A)}$ from $E[\Psi]$, i.e.

$$\Lambda[\Psi] = \langle \Psi | (H - E_0^{(A)}) e^{T^{(A)}} | \Phi \rangle / \langle \Psi | e^{T^{(A)}} | \Phi \rangle. \quad (23)$$

The latter expression defines the MMCC functional. These expressions were introduced in the original MMCC work [34], extended to excited and multi-reference states in Refs. [89, 120] (cf. Ref. [12] for a pedagogical overview), and later exploited in Refs. [157–160] to examine the MMCC-based energy-corrected CC methods, which belong to a family of the CI-corrected MMCC methods [11, 12, 14, 15, 33, 39, 120, 121]. Clearly, $E[\Psi]$, Eq. (22), gives the exact, full CI, ground-state energy E_0, independent of the truncation level m_A defining $T^{(A)}$, when $|\Psi\rangle$ entering Eq. (22) is a full CI ground-state wave function $|\Psi_0\rangle$. Similarly, the MMCC functional $\Lambda[\Psi]$, Eq. (23), gives the difference between the exact energy E_0 and the CC energy $E_0^{(A)}$, which defines the exact value of $\delta_0^{(A)}$, when $|\Psi\rangle$ entering Eq. (23) is replaced by $|\Psi_0\rangle$,

$$\Lambda[\Psi_0] \equiv \langle \Psi_0 | (H - E_0^{(A)}) e^{T^{(A)}} | \Phi \rangle / \langle \Psi_0 | e^{T^{(A)}} | \Phi \rangle = E_0 - E_0^{(A)} \equiv \delta_0^{(A)}. \quad (24)$$

The energy expansions that define all MMCC theories are obtained by expressing $\Lambda[\Psi_0]$, Eq. (24), in terms of moments $\mathfrak{M}_{a_1 \ldots a_n}^{i_1 \ldots i_n}(m_A)$ with $n = m_A + 1, \ldots, N_A$. This can be done in various ways [11–15, 33, 34, 43–46, 120, 121]. From now on, we focus on two particular forms of $\delta_0^{(A)}$, which are relevant to the CR-CC approaches examined in this work.

The CR-CCSD(T) and CR-CCSD(TQ) methods are obtained by considering the truncated form of the following MMCC expansion for $\delta_0^{(A)}$ [11–14, 24, 33, 34]:

$$\delta_0^{(A)} = \sum_{n=m_A+1}^{N} \sum_{k=m_A+1}^{\min(n,N_A)} \langle \Psi_0 | C_{n-k}(m_A)\, M_k(m_A) | \Phi \rangle / \langle \Psi_0 | e^{T^{(A)}} | \Phi \rangle. \tag{25}$$

Here,

$$C_{n-k}(m_A) = (e^{T^{(A)}})_{n-k} \tag{26}$$

are the $(n-k)$-body components of the CC wave operator $e^{T^{(A)}}$, defining method A, $|\Psi_0\rangle$ is the exact ground-state wave function, and

$$M_k(m_A) = \sum_{\substack{i_1 < \cdots < i_k \\ a_1 < \cdots < a_k}} \mathfrak{M}_{a_1 \ldots a_k}^{i_1 \ldots i_k}(m_A)\, a^{a_1} \cdots a^{a_k} a_{i_k} \cdots a_{i_1}, \tag{27}$$

with the generalized moments $\mathfrak{M}_{a_1 \ldots a_k}^{i_1 \ldots i_k}(m_A)$ defined by Eq. (19). The $C_{n-k}(m_A)$ quantities are trivial to generate. The zero-body term, $C_0(m_A)$, equals 1; the one-body term, $C_1(m_A)$, equals T_1; the two-body term, $C_2(m_A)$, equals $T_2 + \frac{1}{2}T_1^2$ if $m_A \geq 2$; the three-body term $C_3(m_A)$ equals $T_1 T_2 + \frac{1}{6}T_1^3$ if $m_A = 2$ and $T_3 + T_1 T_2 + \frac{1}{6}T_1^3$ if $m_A \geq 3$, etc. As shown in the next subsection, the $\mathfrak{M}_{a_1 \ldots a_k}^{i_1 \ldots i_k}(m_A)$ moments entering the CR-CCSD(T) and CR-CCSD(TQ) corrections to the CCSD energy (the $m_A = 2$ case) can be given the relatively simple form as well. In general, the formula for the correction $\delta_0^{(\mathrm{CCSD})}$, which must be added to the CCSD energy $E_0^{(\mathrm{CCSD})}$ to recover the exact energy E_0, is

$$\delta_0^{(\mathrm{CCSD})} = \sum_{n=3}^{N} \sum_{k=3}^{\min(n,6)} \langle \Psi_0 | C_{n-k}(2)\, M_k(2) | \Phi \rangle / \langle \Psi_0 | e^{T_1+T_2} | \Phi \rangle, \tag{28}$$

where

$$M_k(2) = \sum_{\substack{i_1 < \cdots < i_k \\ a_1 < \cdots < a_k}} \mathfrak{M}_{a_1 \ldots a_k}^{i_1 \ldots i_k}(2)\, a^{a_1} \cdots a^{a_k} a_{i_k} \cdots a_{i_1}, \tag{29}$$

with the CCSD moments $\mathfrak{M}_{a_1 \ldots a_k}^{i_1 \ldots i_k}(2)$ defined by Eq. (20).

As one can see, the above formula for the correction $\delta_0^{(A)}$, Eq. (25), or its CCSD analog, Eq. (28), is a complete many-body expansion including up to $n = N$-body terms, where N is the number of all correlated electrons (fermions) in a system. Although moments $\mathfrak{M}_{a_1 \ldots a_k}^{i_1 \ldots i_k}(m_A)$ with $m_A > N_A$ vanish and, hence, do not contribute to Eq. (25) (cf. $\sum_{k=m_A+1}^{\min(n,N_A)}$ in Eq. (25) or $\sum_{k=3}^{\min(n,6)}$ in Eq. (28)), no particular form of the exact state $|\Psi_0\rangle$, on which we project in Eqs. (25) and (28), is assumed, so that one has to use the entire full CI expansion of $|\Psi_0\rangle$ if we want to determine the exact values of $\delta_0^{(A)}$ and $\delta_0^{(\mathrm{CCSD})}$ using Eqs. (25) and (28). This may have some advantages in practice, since we can use a variety of different forms of $|\Psi_0\rangle$ in approximate calculations based on Eqs. (25) and (28), which may or may not be related to CC theory, resulting in the renormalized CC methods [11–15,24,33,34,36,122,123], such as CR-CCSD(T) and CR-CCSD(TQ) [11–14, 24, 33, 34], as well as the externally corrected MMCC schemes [11, 12, 14, 15, 33, 39, 48, 49, 120, 121], including the aforementioned energy-corrected CC approaches [157–160] (for information about a broad class of the externally corrected CC methods, pioneered by Paldus and collaborators, in which CC and non-CC concepts are combined together to improve single-reference CC results in the presence of quasi-degeneracies, see Refs. [8,18,161–168]). One may wonder, however, if it is possible to rewrite Eq. (25) or its CCSD analog, Eq. (28), in a form of a compact many-body expansion which terminates at the $n = N_A$-body terms, compatible with the highest many-body rank of the corresponding moments $\mathfrak{M}_{a_1\dots a_k}^{i_1\dots i_k}(m_A)$, independent of the number of electrons in a system. The biorthogonal formulation of the MMCC theory, introduced in Refs. [45, 46], on which the CR-CC(2,3) and other CR-CC(m_A,m_B) approximations considered in Refs. [45, 46, 48] are based, leads to such a compact expansion.

In the biorthogonal MMCC formalism, we calculate the correction $\delta_0^{(A)}$, Eq. (1), as follows [45,46]:

$$\delta_0^{(A)} = \sum_{n=m_A+1}^{N_A} \langle\Phi|\mathscr{L}_n M_n(m_A)|\Phi\rangle = \sum_{n=m_A+1}^{N_A} \sum_{\substack{i_1<\dots<i_n\\ a_1<\dots<a_n}} \ell_{i_1\dots i_n}^{a_1\dots a_n}\, \mathfrak{M}_{a_1\dots a_n}^{i_1\dots i_n}(m_A). \tag{30}$$

The $M_n(m_A)$ operators, which enter the first form of Eq. (30), are defined by Eq. (27) and, again, $\mathfrak{M}_{a_1\dots a_n}^{i_1\dots i_n}(m_A)$ are the generalized moments of CC equations of method A defined by Eq. (19). The $\mathscr{L}_n$ operators in Eq. (30) are the n-body components of the desexcitation operator $\mathscr{L}$, which parameterizes the exact, full CI bra state $\langle\Psi_0|$ as follows:

$$\langle\Psi_0| = \langle\Phi|\mathscr{L}\, e^{-T^{(A)}}. \tag{31}$$

We have

$$\mathscr{L} = \sum_{n=0}^{N} \mathscr{L}_n, \tag{32}$$

where

$$\mathscr{L}_n = \sum_{\substack{i_1<\dots<i_n\\ a_1<\dots<a_n}} \ell_{i_1\dots i_n}^{a_1\dots a_n}\, a^{i_1}\cdots a^{i_n} a_{a_n}\cdots a_{a_1} \tag{33}$$

and $\ell^{a_1 \ldots a_n}_{i_1 \ldots i_n}$ are the amplitudes defining $\mathscr{L}_n$. In order to guarantee that Eq. (30) represents the exact difference between the full CI and CC energies, Eq. (1), we must require that the full CI bra state $\langle \Psi_0|$, Eq. (31), is normalized as

$$(34) \qquad \langle \Psi_0 | \Psi_0^{(A)} \rangle \equiv \langle \Psi_0 | e^{T^{(A)}} | \Phi \rangle = 1,$$

where $|\Psi_0^{(A)}\rangle$ is the CC ket state of method A, Eq. (4). Clearly, the operator $\mathscr{L}$ that produces the exact, full CI bra state $\langle \Psi_0|$ according to Eq. (31), subject to the normalization condition given by Eq. (34), always exists.

As one can see, the non-iterative correction $\delta_0^{(A)}$, Eq. (30), is defined in terms of the n-body components of $\mathscr{L}$ with $n > m_A$. These components originate from the following decomposition of the operator $\mathscr{L}$:

$$(35) \qquad \mathscr{L} = \mathscr{L}^{(A)} + \delta\mathscr{L}^{(A)},$$

where

$$(36) \qquad \mathscr{L}^{(A)} = \sum_{n=0}^{m_A} \mathscr{L}_n$$

and

$$(37) \qquad \delta\mathscr{L}^{(A)} = \sum_{n=m_A+1}^{N} \mathscr{L}_n.$$

The normalization condition defined by Eq. (34) can be rewritten as

$$(38) \qquad \langle \Phi | \mathscr{L}^{(A)} | \Phi \rangle = 1,$$

so that the zero-body component $\mathscr{L}_0 = \mathbf{1}$, where $\mathbf{1}$ is a unit operator. The fact that one does not have to consider the $\mathscr{L}_n$ components with $n > N_A$ in Eq. (30) is a consequence of the property that, by definition of symbol N_A, moments $\mathfrak{M}^{i_1 \ldots i_n}_{a_1 \ldots a_n}(m_A)$ with $n > N_A$ vanish. In particular, the formula for the correction $\delta_0^{(\mathrm{CCSD})}$, which must be added to the CCSD energy $E_0^{(\mathrm{CCSD})}$ to recover the exact energy E_0, assumes now the following very compact form which terminates at the hextuply excited terms:

$$(39) \qquad \delta_0^{(\mathrm{CCSD})} = \sum_{n=3}^{6} \langle \Phi | \mathscr{L}_n \, M_n(2) | \Phi \rangle = \sum_{n=3}^{6} \sum_{\substack{i_1 < \cdots < i_n \\ a_1 < \cdots < a_n}} \ell^{a_1 \ldots a_n}_{i_1 \ldots i_n} \, \mathfrak{M}^{i_1 \ldots i_n}_{a_1 \ldots a_n}(2),$$

with $M_n(2)$ defined by Eq. (29) and $\mathfrak{M}^{i_1 \ldots i_n}_{a_1 \ldots a_n}(2)$ representing the CCSD moments (cf. Eq. (20)). In analogy to the general expression, Eq. (30), the $\mathscr{L}_n$ operators in Eq. (39) are the n-body components of the desexcitation operator $\mathscr{L}$, which parameterizes the exact, full CI bra state $\langle \Psi_0|$ as follows:

$$(40) \qquad \langle \Psi_0 | = \langle \Phi | \mathscr{L} \, e^{-T_1 - T_2}$$

and $\ell^{a_1 \ldots a_n}_{i_1 \ldots i_n}$ are the amplitudes defining $\mathscr{L}_n$. As in the general case, we must require that the zero-body component of $\mathscr{L}$ entering Eq. (40), $\mathscr{L}_0$, is a unit operator.

The compact nature of the biorthogonal MMCC expansions, Eqs. (30) and (39), as described above, which even in the exact case terminate at the n-body terms with the n values that are usually much smaller than the number of correlated electrons N, constitutes one of the most essential differences between the original MMCC theory, represented by Eqs. (25) and (28), which are full CI-like expansions including up to N-body terms, and the biorthogonal MMCC theory. Let us, therefore, end this subsection by pointing out other differences between Eqs. (25) and (30). The older MMCC expression for $\delta_0^{(A)}$, Eq. (25), uses the projections on the exact, full CI, state $|\Psi_0\rangle$, which we approximate in actual applications by truncating it at some excitation level [11–14, 24, 33, 34, 39, 48, 49, 120–123], whereas the more recent biorthogonal MMCC formula, Eq. (30), is expressed in terms of the n-body components of the desexcitation operator $\mathscr{L}$ with $m_A < n \leq N_A$, which we can estimate by relating them, for example, to the many-body components of the left eigenstate of the similarity-transformed Hamiltonian, Eq. (11), of the CC theory of interest [45,46]. Operator $\mathscr{L}$ parameterizes the exact bra state $\langle\Psi_0|$ via Eq. (31), subject to the normalization condition defined by Eq. (38), which enables us to eliminate the overlap denominator term $\langle\Psi_0|e^{T^{(A)}}|\Phi\rangle$ from Eq. (25). It is the presence of this overlap denominator term in the original MMCC theory defined by Eq. (25), which renormalizes the correction $\delta_0^{(A)}$ to the CC energy $E_0^{(A)}$ due to higher-than-m_A-body excitations (e.g. triples in the CCSD, $m_A = 2$, case), that often leads to the desired improvements in the CCSD(T) and other standard CC results for potential energy surfaces along bond breaking coordinates [11, 24, 34] and biradicals [51, 52]. The calculation of the overlap denominator terms $\langle\Psi_0|e^{T^{(A)}}|\Phi\rangle$ for the approximate wave functions $|\Psi_0\rangle$ defining various MMCC approximations, such as CR-CCSD(T) or CR-CCSD(TQ), which are nothing else but the truncated forms of $|\Psi_0\rangle$ at some excitation level n ($n = 3$ in the CR-CCSD(T) case and $n = 4$ in the CR-CCSD(TQ) case; see section 2.3), constitutes a small fraction of the total computer effort related to the determination of approximate corrections $\delta_0^{(A)}$, but usually these denominators introduce small size extensivity errors, estimated at $\sim$0.5$-$1% of the correlation energy (changes in the correlation energy when the reaction pathways are examined) in the ground-state calculations [12]. The incorporation of the overlap denominator term $\langle\Psi_0|e^{T^{(A)}}|\Phi\rangle$ into the numerator of Eq. (25) through the use of the suitably chosen parameterization of the exact bra state $\langle\Psi_0|$, Eq. (31), and the conveniently chosen normalization of $\langle\Psi_0|$, Eq. (34), as is done in the biorthogonal MMCC formalism [45, 46], has an advantage that at least the basic and most practical ground-state approximations resulting from the biorthogonal MMCC expansions, Eqs. (30) and (39), such as the CR-CC(2,3) method proposed in Refs. [45, 46] and discussed in section 2.3, are rigorously size extensive (see Ref. [46] for a numerical example). Without going into every detail, we might simply state that the CC bra state $\langle\tilde{\Psi}_0^{(A)}|$, Eq. (9), and the corresponding ket CC state $|\Psi_0^{(A)}\rangle$, Eq. (4), satisfy the biorthonormality relation given by Eq. (10), so that in the exact, full CI, limit the $\langle\tilde{\Psi}_0^{(A)}|$ bra state becomes the renormalized form of the ket state $|\Psi_0^{(A)}\rangle$. This is reflected in the normalization of the exact bra state $\langle\tilde{\Psi}_0|$, Eq. (34), used in the biorthogonal MMCC theory, which should be contrasted with the fact that, in general, $\langle\Psi_0^{(A)}|\Psi_0^{(A)}\rangle \neq 1$ for $|\Psi_0^{(A)}\rangle$ defined by Eq. (4). The more precise mathematical argument that shows how the overlap denominator terms $\langle\Psi_0|e^{T^{(A)}}|\Phi\rangle$ can be absorbed by the numerator of Eq. (25), resulting in the biorthogonal MMCC expansion, Eq. (30), in which the many-body components of the CC wave operator $e^{T^{(A)}}$, i.e. the $C_{n-k}(m_A)$ terms entering Eq. (25), are simultaneously eliminated through a suitable resummation procedure, will be discussed in the future work [169]. It is also worth mentioning that the fact that the $C_{n-k}(m_A)$ terms entering Eq. (25), which appear at a given $M_k(m_A)$ term in Eq. (25) and are no longer present in Eq. (30) since they are properly summed up to all orders, has a positive effect on the accuracy of the CR-CC(2,3) calculations, when we compare them with the corresponding CR-CCSD(T) calculations. Indeed, the $\mathscr{L}_3 M_3(2)$ term defining the CR-CC(2,3) triples correction (see section 2.3) is essentially equivalent to the sum of all $C_{n-3}(m_A)M_3(2)$ $(n = 3, 4, \ldots)$ contributions in Eq. (28), including $M_3(2)$, $T_1 M_3(2)$, $(T_2 + \frac{1}{2}T_1^2)M_3(2)$, etc. The CR-CCSD(T) approximation uses only the bare $M_3(2)$ component. Thus, in addition to restoring size extensivity, the CR-CC(2,3) approach provides improvements in the CR-CCSD(T) results [45, 46, 48]. This interesting feature will be discussed in detail in future work [169].

2.3. Renormalized coupled-cluster approaches: CR-CCSD(T), CR-CCSD(TQ), CR-CC(2,3), and CR-CC(2,3)+Q

There are two issues that have to be addressed before one can use Eqs. (25) or (28) in practical calculations. First of all, the exact MMCC corrections $\delta_0^{(A)}$ and $\delta_0^{(\mathrm{CCSD})}$, Eqs. (25) and (28), respectively, have the form of long many-body expansions involving all n-tuply excited configurations with $n = m_A + 1, \ldots, N$, where N is the number of correlated electrons in a system. Thus, in order to propose the computationally inexpensive MMCC methods, we have to truncate the many-body expansions for $\delta_0^{(A)}$ or $\delta_0^{(\mathrm{CCSD})}$ at some, preferably low, excitation level m_B. This leads to the so-called MMCC(m_A, m_B) schemes [11–15,24,33,34,39,48,120,121]. The CR-CCSD(T) and CR-CCSD(TQ) methods [11–14,24,33,34], reviewed and tested in this work, are the MMCC(m_A, m_B) schemes with $m_A = 2$ and $m_B = 3$ (the CR-CCSD(T) case) or 4 (the CR-CCSD(TQ) case). Second of all, the wave function $|\Psi_0\rangle$ that enters the exact Eqs. (25) or (28) is a full CI ground state, which we usually do not know (if we knew the exact $|\Psi_0\rangle$ state, we would not have to perform any calculations!). Thus, in order to propose the computationally tractable approaches based on the MMCC theory defined by Eqs. (25) and (28), we must approximate $|\Psi_0\rangle$ in some way as well. The CR-CCSD(T) and CR-CCSD(TQ) methods employ the low-order MBPT-like expressions to define $|\Psi_0\rangle$ [11–14,24,33,34].

Similar remarks apply to practical computational schemes, such as CR-CC(2,3), based on the biorthogonal MMCC theory, as defined by Eqs. (30) and (39). In this case, to avoid the calculation of the entire set of moments of CC equations for a given

CC approximation A, which in the case of correcting the CCSD energy by Eq. (39) includes up to the hextuply excited moments $\mathfrak{M}^{ijklmn}_{abcdef}(2)$, it may be prudent to truncate Eqs. (30) or (39) at some lower-order moments, such as $\mathfrak{M}^{ijk}_{abc}(2)$ or $\mathfrak{M}^{ijkl}_{abcd}(2)$. Truncation of Eq. (30) at terms containing moments $\mathfrak{M}^{i_1 \ldots i_n}_{a_1 \ldots a_n}(m_A)$ with $n = m_B$ leads to the biorthogonal analogs of the MMCC(m_A, m_B) schemes, designated as the MMCC$(m_A, m_B)_{\mathscr{L}}$ approximations [45, 46, 48]. The CR-CC(2,3) approach of Refs. [45, 46] is an example of the MMCC$(m_A, m_B)_{\mathscr{L}}$ truncation scheme with $m_A = 2$ and $m_B = 3$. The second issue that needs to be addressed in all biorthogonal MMCC calculations is that we do not know the exact $\mathscr{L}$ operators that define the full CI wave functions $|\Psi_0\rangle$ through Eqs. (31) or (40) and that enter Eqs. (30) or (39) for $\delta_0^{(A)}$ or $\delta_0^{(\text{CCSD})}$, respectively. Thus, we must use some approximate forms of the relevant many-body components of $\mathscr{L}$ that enter the MMCC$(m_A, m_B)_{\mathscr{L}}$ energy expressions. In the CR-CC(2,3) and other CR-CC(m_A, m_B) approaches [45, 46], we use the left eigenstates of the similarity-transformed Hamiltonian $\bar{H}^{(A)}$, i.e. the $\langle\Phi|L^{(A)}$ states generated by the desexcitation operator $L^{(A)}$, Eq. (11), obtained by solving the left CC problem, Eq. (14), to construct an approximate form of $\delta_0^{(A)}$ or $\delta_0^{(\text{CCSD})}$.

Let us begin with the MMCC(m_A, m_B) schemes and the resulting CR-CCSD(T) and CR-CCSD(TQ) approximations. In all MMCC(m_A, m_B) methods, we enforce the CI-like truncation of the complete many-body expansion for $\delta_0^{(A)}$, Eq. (25), by limiting ourselves to wave functions $|\Psi_0\rangle$ in that do not contain higher-than-m_B-tuply excited components relative to $|\Phi\rangle$. The resulting MMCC(m_A, m_B) energies, $E_0(m_A, m_B)$, are given by the following expression [11–15, 24, 33, 34, 39, 48, 120, 121]:

$$E_0(m_A, m_B) = E_0^{(A)} + \delta_0(m_A, m_B), \tag{41}$$

where $E_0^{(A)}$ is the energy obtained with the CC method A and

$$\delta_0(m_A, m_B) = \sum_{n=m_A+1}^{m_B} \sum_{k=m_A+1}^{n} \langle\Psi_0|C_{n-k}(m_A) M_k(m_A)|\Phi\rangle / \langle\Psi_0|e^{T^{(A)}}|\Phi\rangle \tag{42}$$

is the relevant MMCC correction. The non-zero corrections $\delta_0(m_A, m_B)$ are obtained only when $m_B > m_A$. When $m_B = N$ and when $|\Psi_0\rangle$ is the exact ground state, the MMCC(m_A, m_B) energy $E_0(m_A, m_B)$ becomes equivalent to the exact, full CI, energy.

In this article, we focus on the MMCC(m_A, m_B) schemes with $m_A = 2$, which can be used to correct the CCSD energy. Two truncation schemes of this type are particularly useful, namely, MMCC(2,3), which results in the CR-CCSD(T) method, and MMCC(2,4), which leads to the CR-CCSD(TQ),x (x = a, b) approximations. In the MMCC(2,3) and MMCC(2,4) approaches, we add the relevant energy corrections, $\delta_0(2, 3)$ and $\delta_0(2, 4)$, respectively, to the CCSD energy $E_0^{(\text{CCSD})}$ to obtain the following total energies [11–15, 24, 33, 34, 39, 48, 120, 121]:

$$E_0(2, 3) = E_0^{(\text{CCSD})} + \langle\Psi_0|M_3(2)|\Phi\rangle / \langle\Psi_0|e^{T_1+T_2}|\Phi\rangle, \tag{43}$$

$$E_0(2,4) = E_0^{(\mathrm{CCSD})} + \langle\Psi_0|\{M_3(2) + [M_4(2) + T_1M_3(2)]\}|\Phi\rangle / \langle\Psi_0|e^{T_1+T_2}|\Phi\rangle \,, \tag{44}$$

where $M_3(2)$ and $M_4(2)$ are defined by Eq. (29). The MMCC(2,3) approach requires that we determine moments $\mathfrak{M}_{abc}^{ijk}(2)$, which correspond to projections of the CCSD equations on triply excited determinants $|\Phi_{ijk}^{abc}\rangle$ (Eq. (20) with $n = 3$). The MMCC(2,4) approach requires that we determine moments $\mathfrak{M}_{abc}^{ijk}(2)$ and $\mathfrak{M}_{abcd}^{ijkl}(2)$. The latter moments represent the projections of the CCSD equations on quadruply excited determinants $|\Phi_{ijkl}^{abcd}\rangle$ (Eq. (20) with $n = 4$). Assuming that the Hamiltonian does not contain higher-than-two-body interactions, we can write the following, relatively simple, expressions for moments $\mathfrak{M}_{abc}^{ijk}(2)$ and $\mathfrak{M}_{abcd}^{ijkl}(2)$, in terms of the T_1 and T_2 clusters (cf. Eq. (20)):

$$\mathfrak{M}_{abc}^{ijk}(2) = \langle\Phi_{ijk}^{abc}|[H_N(T_2+T_1T_2+\tfrac{1}{2}T_2^2+\tfrac{1}{2}T_1^2T_2+\tfrac{1}{2}T_1T_2^2+\tfrac{1}{6}T_1^3T_2)]_C|\Phi\rangle, \tag{45}$$

$$\mathfrak{M}_{abcd}^{ijkl}(2) = \langle\Phi_{ijkl}^{abcd}|[H_N(\tfrac{1}{2}T_2^2 + \tfrac{1}{2}T_1T_2^2 + \tfrac{1}{6}T_2^3 + \tfrac{1}{4}T_1^2T_2^2)]_C|\Phi\rangle, \tag{46}$$

where $H_N = H - \langle\Phi|H|\Phi\rangle$ is the Hamiltonian in the normal-ordered form.

Equations (43) and (44) provide a formal basis for designing the CR-CCSD(T) and CR-CCSD(TQ),x (x = a, b) approximations. The CR-CCSD(T) method is an example of the MMCC(2,3) approach, in which the wave function $|\Psi_0\rangle$ entering Eq. (43) is replaced by the expression, which is reminiscent of the MBPT(2)[SDT] wave function (recall that the second-order MBPT wave function is the lowest-order wave function that provides information about triply excited clusters). This expression is [11–14, 24, 33, 34]

$$|\Psi_0^{(\mathrm{CR\text{-}CCSD(T)})}\rangle = (1 + T_1 + T_2 + T_3^{[2]} + Z_3)|\Phi\rangle, \tag{47}$$

where T_1 and T_2 are the singly and doubly excited clusters obtained in the CCSD calculations, the

$$T_3^{[2]}|\Phi\rangle = R_0^{(3)}(V_NT_2)_C|\Phi\rangle \tag{48}$$

term is an approximation of the connected triples (T_3) contribution, which is correct through second order, and

$$Z_3|\Phi\rangle = R_0^{(3)}V_NT_1|\Phi\rangle \tag{49}$$

is the disconnected triples correction, which is responsible for the difference between the [T] and (T) triples corrections defining the standard CCSD[T] and CCSD(T) methods. We use the notation, in which $R_0^{(3)}$ designates the three-body component of the MBPT reduced resolvent and V_N is the two-body part of H_N. After replacing the wave function $|\Psi_0\rangle$ in the MMCC(2,3) energy formula, Eq. (43), by $|\Psi_0^{(\mathrm{CR\text{-}CCSD(T)})}\rangle$, Eq. (47), we obtain the following compact and computationally convenient expression for the CR-CCSD(T) energy [12–14]:

$$E_0^{(\mathrm{CR\text{-}CCSD(T)})} = E_0^{(\mathrm{CCSD})} + N^{(\mathrm{CR(T)})}/D^{(\mathrm{T})}, \tag{50}$$

where

$$N^{(\text{CR(T)})} = \langle \Psi_0^{(\text{CR-CCSD(T)})} | M_3(2) | \Phi \rangle = \langle \Phi | (T_3^{[2]} + Z_3)^\dagger M_3(2) | \Phi \rangle \tag{51}$$

and

$$\begin{aligned} D^{(\text{T})} &= \langle \Psi_0^{(\text{CR-CCSD(T)})} | e^{T_1+T_2} | \Phi \rangle \\ &= 1 + \langle \Phi | T_1^\dagger T_1 | \Phi \rangle + \langle \Phi | T_2^\dagger \left(T_2 + \tfrac{1}{2} T_1^2 \right) | \Phi \rangle \\ &\quad + \langle \Phi | (T_3^{[2]} + Z_3)^\dagger (T_1 T_2 + \tfrac{1}{6} T_1^3) | \Phi \rangle. \end{aligned} \tag{52}$$

The above equations allow us to see that the CR-CCSD(T) approach reduces to the standard CCSD(T) method, when the overlap denominator $D^{(\text{T})}$ in Eq. (50) is replaced by 1 and moments $\mathfrak{M}_{abc}^{ijk}(2)$ that enter Eq. (51) are replaced by the lead term $\langle \Phi_{ijk}^{abc} | (V_N T_2)_C | \Phi \rangle$ (cf. Eq. (45)). Indeed, by performing the above simplifications in Eq. (50), we obtain

$$E_0^{(\text{CCSD(T)})} = E_0^{(\text{CCSD})} + \langle \Phi | (T_3^{[2]} + Z_3)^\dagger (V_N T_2)_C | \Phi \rangle, \tag{53}$$

which is the well-known formula for the CCSD(T) energy [19]. As explained in Refs. [11, 34], the approximation of the $D^{(\text{T})}$ denominator by 1 can be viewed as a justified step from the point of view of MBPT, since $D^{(\text{T})}$ equals 1 plus terms which are at least of the second order in the perturbation V_N (see Eq. (52); if we ignore 1, the lowest-order term in Eq. (52) is $\langle \Phi | T_2^\dagger T_2 | \Phi \rangle$; since T_2 contains the first-order contributions in V_N, $\langle \Phi | T_2^\dagger T_2 | \Phi \rangle$ is at least of the second order). One can, in fact, expect that for the closed-shell molecules at or near their equilibrium geometries, where the MBPT series usually converges and T_1 and T_2 amplitudes are small, the $D^{(\text{T})}$ denominator, Eq. (52), defining the CR-CCSD(T) approach, is very close to 1. This is what we observe in actual calculations if the number of electrons in a system is not too large [12]. The situation changes when chemical bonds are stretched or broken and when the biradical systems are examined. In the region of stretched nuclear geometries and in the region of a significant increase of biradical character, where the MBPT series is usually strongly divergent, the $D^{(\text{T})}$ denominator, Eq. (52), can be significantly greater than 1 (cf., e.g. Refs. [11, 24, 34, 51]). For example, when a single bond is broken, the leading T_2 cluster amplitude is often close to -1. Thus, assuming that other amplitudes are small, the $D^{(\text{T})}$ denominator becomes close to $1 + \langle \Phi | T_2^\dagger T_2 | \Phi \rangle \approx 1 + (-1) \times (-1) = 2$, damping the excessively negative triples correction of CCSD(T) by a factor of ≈ 2 (in reality, this factor will be somewhat greater than 2, since other T_2 and T_1 amplitudes are not zero [34]). This increase in the value of $D^{(\text{T})}$, as one proceeds from the closed-shell region of the potential energy surface to the bond breaking or biradical region, is the main reason for the excellent performance of the CR-CCSD(T) method in improving the poor CCSD(T) results in cases of large non-dynamical correlation effects. The overlap of the $|\Psi_0^{\text{CR-CCSD(T)}}\rangle$ and CCSD wave functions, defining the denominator $D^{(\text{T})}$, plays a role of a natural damping factor, which renormalizes the excessively large and, thus, completely unphysical values of the non-iterative triples corrections of

CCSD(T) at larger internuclear separations and in the biradical regions of potential energy surfaces. Because of the use of intermediate normalization, no such denominators are present in the conventional CCSD(T) theory, resulting in the pathological description of potential energy surfaces in the bond breaking and biradical regions (cf. Refs. [8, 11–15, 24–58]; see, also, the next section).

The idea of renormalizing the CCSD(T) method via the MMCC formalism, as described above, can be easily extended to the CCSD(TQ) case. The resulting CR-CCSD(TQ) approaches are examples of the MMCC(2,4) scheme, defined by Eq. (44), in which we improve the results of the CCSD calculations by adding the non-iterative corrections $\delta_0(2,4)$, defined in terms of moments $\mathfrak{M}_{abc}^{ijk}(2)$ and $\mathfrak{M}_{abcd}^{ijkl}(2)$, to the CCSD energies. Two variants of the CR-CCSD(TQ) method, labeled by the extra letters "a" and "b", are particularly useful. The CR-CCSD(TQ),a and CR-CCSD(TQ),b energies are calculated in the following manner [11–14, 24, 33, 34]:

$$E_0^{(\text{CR-CCSD(TQ)},x)} = E_0^{(\text{CCSD})} + \langle \Psi_0^{(\text{CR-CCSD(TQ)},x)} | \{ M_3(2) + [T_1 M_3(2) + M_4(2)] \} | \Phi \rangle / \langle \Psi_0^{(\text{CR-CCSD(TQ)},x)} | e^{T_1 + T_2} | \Phi \rangle, \tag{54}$$

where $x = \text{a}$ or b and

$$|\Psi_0^{(\text{CR-CCSD(TQ)},\text{a})}\rangle = |\Psi_0^{(\text{CR-CCSD(T)})}\rangle + \tfrac{1}{2} T_2 T_2^{(1)} |\Phi\rangle \tag{55}$$

and

$$|\Psi_0^{(\text{CR-CCSD(TQ)},\text{b})}\rangle = |\Psi_0^{(\text{CR-CCSD(T)})}\rangle + \tfrac{1}{2} T_2^2 |\Phi\rangle \tag{56}$$

are the wave functions $|\Psi_0\rangle$ used to define the relevant MMCC(2,4) energy corrections, with $T_2^{(1)}$ representing the first-order MBPT estimate of the T_2 cluster and $|\Psi_0^{(\text{CR-CCSD(T)})}\rangle$ given by Eq. (47). In analogy to $|\Psi_0^{(\text{CR-CCSD(T)})}\rangle$, the wave functions $|\Psi_0^{(\text{CR-CCSD(TQ)},\text{a})}\rangle$ and $|\Psi_0^{(\text{CR-CCSD(TQ)},\text{b})}\rangle$, Eqs. (55) and (56), respectively, are the MBPT(2)-like expressions, obtained by augmenting $|\Psi_0^{(\text{CR-CCSD(T)})}\rangle$ that defines the CR-CCSD(T) approach by the lowest-order contributions due to quadruple excitations. As in the case of the CR-CCSD(T) method, we can rewrite the above expressions for the CR-CCSD(TQ),a and CR-CCSD(TQ),b energies in the computationally convenient, compact form, namely [12–14],

$$E_0^{(\text{CR-CCSD(TQ)},x)} = E_0^{(\text{CCSD})} + N^{(\text{CR(TQ)},x)} / D^{(\text{TQ}),x} \quad (x = \text{a, b}), \tag{57}$$

where

$$N^{(\text{CR(TQ)},\text{a})} = N^{(\text{CR(T)})} + \tfrac{1}{2} \langle \Phi | T_2^\dagger (T_2^{(1)})^\dagger [T_1 M_3(2) + M_4(2)] | \Phi \rangle, \tag{58}$$

$$N^{(\text{CR(TQ)},\text{b})} = N^{(\text{CR(T)})} + \tfrac{1}{2} \langle \Phi | (T_2^\dagger)^2 [T_1 M_3(2) + M_4(2)] | \Phi \rangle, \tag{59}$$

$$D^{(\text{TQ}),\text{a}} = D^{(\text{T})} + \tfrac{1}{2} \langle \Phi | T_2^\dagger (T_2^{(1)})^\dagger (\tfrac{1}{2} T_2^2 + \tfrac{1}{2} T_1^2 T_2 + \tfrac{1}{24} T_1^4) | \Phi \rangle, \tag{60}$$

$$D^{(\text{TQ}),\text{b}} = D^{(\text{T})} + \tfrac{1}{2} \langle \Phi | (T_2^\dagger)^2 (\tfrac{1}{2} T_2^2 + \tfrac{1}{2} T_1^2 T_2 + \tfrac{1}{24} T_1^4) | \Phi \rangle, \tag{61}$$

with $N^{(\mathrm{CR(T)})}$ and $D^{(\mathrm{T})}$ defined by Eqs. (51) and (52), respectively. In analogy to the CR-CCSD(T) approach, the CR-CCSD(TQ),a and CR-CCSD(TQ),b methods reduce to the standard approximations of the CCSD(TQ) type, such as the CCSD($\mathrm{TQ_f}$) method of Ref. [23] or the closely related CCSD(TQ),b approach [24, 36], if we simplify or slightly modify the expressions for the numerators $N^{(\mathrm{CR(TQ),a})}$ and $N^{(\mathrm{CR(TQ),b})}$, and replace the denominators $D^{(\mathrm{TQ}),\mathrm{a}}$ and $D^{(\mathrm{TQ}),\mathrm{b}}$ by 1. For example, we obtain the CCSD(TQ),b energy expression [24,36],

$$\begin{aligned} E_0^{(\mathrm{CCSD(TQ),b})} &= E_0^{(\mathrm{CCSD})} + \langle \Phi | (T_3^{[2]} + Z_3)^\dagger (V_N T_2)_C | \Phi \rangle \\ &\quad + \tfrac{1}{2} \langle \Phi | (T_2^\dagger)^2 [V_N (\tfrac{1}{2} T_2^2 + T_3^{[2]})]_C | \Phi \rangle, \end{aligned} \tag{62}$$

by replacing moments $\mathfrak{M}_{abc}^{ijk}(2)$ and $\mathfrak{M}_{abcd}^{ijkl}(2)$, which enter the CR-CCSD(TQ),b energy, Eq. (57), by $\langle \Phi_{ijk}^{abc} | (V_N T_2)_C | \Phi \rangle$ and $\langle \Phi_{ijkl}^{abcd} | [V_N (\tfrac{1}{2} T_2^2 + T_3^{[2]})]_C | \Phi \rangle$, respectively, where $T_3^{[2]}$ is defined by Eq. (48), while approximating $D^{(\mathrm{TQ}),\mathrm{b}}$ by 1 and neglecting the $T_1 M_3(2)$ term in Eq. (59). Again, the key difference between the standard methods of the CCSD(TQ) type and the CR-CCSD(TQ),x approximations is the presence of the $D^{(\mathrm{TQ}),x}$ overlap denominators in Eq. (57), which increase their values much above 1 in the bond breaking and biradical regions and damp, in this way, the excessively large corrections due to triples and quadruples that lead to pathological shapes of potential energy surfaces in situations involving large non-dynamical correlation effects (cf., e.g. Refs. [12–14, 24, 32–37, 40, 43, 44, 58]).

As already mentioned, the CR-CCSD(T) and CR-CCSD(TQ) methods provide great improvements in the poor description of bond breaking and biradicals by the CCSD(T) and CCSD(TQ) methods, but they do it at the expense of slightly violating the rigorous size extensivity of CC theory. Moreover, the overlap denominators $\langle \Psi_0 | e^{T_1 + T_2} | \Phi \rangle$, which enter the CR-CCSD(T) and CR-CCSD(TQ) expressions, may somewhat overdamp the corrections due to triples or triples and quadruples in the equilibrium region, when compared to the standard CCSD(T) and CCSD(TQ) approaches, when the number of correlated electrons in a system becomes large [12]. Although none of these features of the CR-CCSD(T) and CR-CCSD(TQ) methods create serious problems in molecular applications, even when the number of correlated electrons in a system is on the order of 50–80 (cf., e.g. Refs. [52, 53] for the successful, large scale applications of the CR-CCSD(T) and CR-CCSD(TQ) approaches to systems with up to 80 correlated electrons), which is a consequence of the facts that bulk of the correlation energy, represented by the CCSD part of the CR-CCSD(T) and CR-CCSD(TQ) energies, is calculated in a size extensive manner and the triples or the triples and quadruples corrections of the CR-CCSD(T) and CR-CCSD(TQ) methods are constructed with the T_1 and T_2 clusters originating from the strictly size extensive CCSD calculations, it is useful to contemplate alternative formulations of renormalized CC methods, which are rigorously size extensive and which do not reduce the accuracy of the standard CCSD(T) and CCSD(TQ) calculations near the equilibrium geometries of larger systems, where renormalization is not necessary. The biorthogonal MMCC formalism, as defined by Eqs. (30) and (39), and the aforementioned MMCC$(m_A, m_B)_{\mathscr{L}}$ truncation schemes, which result in the family of the CR-CC(m_A, m_B) methods including CR-CC(2,3), provide such alternatives.

In the MMCC$(m_A, m_B)_{\mathscr{L}}$ approximations and the ensuing CR-CC(m_A, m_B) approaches, we calculate the total ground-state energy as [45, 46, 48]

$$E_0(m_A, m_B)_{\mathscr{L}} = E_0^{(A)} + \delta_0(m_A, m_B)_{\mathscr{L}}, \tag{63}$$

where

$$\begin{aligned} \delta_0(m_A, m_B)_{\mathscr{L}} &= \sum_{n=m_A+1}^{m_B} \langle \Phi | \mathscr{L}_n \, M_n(m_A) | \Phi \rangle \\ &= \sum_{n=m_A+1}^{m_B} \sum_{\substack{i_1 < \cdots < i_n \\ a_1 < \cdots < a_n}} \ell_{i_1 \ldots i_n}^{a_1 \ldots a_n} \, \mathfrak{M}_{a_1 \ldots a_n}^{i_1 \ldots i_n}(m_A) \end{aligned} \tag{64}$$

is the relevant correction to the CC energy $E_0^{(A)}$ due to higher-than-m_A-tuply excited clusters. The excitation level m_B satisfies the condition $m_A < m_B \leq N_A$, since N_A is the highest value of n for which moments $\mathfrak{M}_{a_1 \ldots a_n}^{i_1 \ldots i_n}(m_A)$ are nonzero and m_B must be greater than m_A to obtain a non-zero value of $\delta_0(m_A, m_B)_{\mathscr{L}}$. As explained earlier, in the CCSD, $m_A = 2$, case, $N_A = 6$, which means that the highest MMCC(2,m_B) scheme is MMCC(2,6).

An example of the MMCC$(m_A, m_B)_{\mathscr{L}}$ method is the MMCC$(2,3)_{\mathscr{L}}$ approximation, in which, in analogy to the MMCC(2,3) truncation scheme defined by Eq. (43), we correct the results of the CCSD calculations by adding the triples correction

$$\delta_0(2, 3)_{\mathscr{L}} = \sum_{\substack{i<j<k \\ a<b<c}} \ell_{ijk}^{abc} \, \mathfrak{M}_{abc}^{ijk}(2) \tag{65}$$

to the CCSD energy $E_0^{(\mathrm{CCSD})}$. As in the case of MMCC(2,3), the only moments that are needed in the MMCC$(2,3)_{\mathscr{L}}$ calculations are the CCSD moments $\mathfrak{M}_{abc}^{ijk}(2)$ corresponding to triple excitations, which one can calculate using Eq. (45). In the more complete MMCC$(2,4)_{\mathscr{L}}$ scheme, which might be regarded as a biorthogonal analog of the MMCC(2,4) approach defined by Eq. (44), we add the following correction due to the combined effect of triples and quadruples,

$$\delta_0(2, 4)_{\mathscr{L}} = \sum_{\substack{i<j<k \\ a<b<c}} \ell_{ijk}^{abc} \, \mathfrak{M}_{abc}^{ijk}(2) + \sum_{\substack{i<j<k<l \\ a<b<c<d}} \ell_{ijkl}^{abcd} \, \mathfrak{M}_{abcd}^{ijkl}(2), \tag{66}$$

to the CCSD energy.

In analogy to the CR-CCSD(T) and CR-CCSD(TQ) methods, in order to propose practical computational methods, based on the MMCC$(m_A, m_B)_{\mathscr{L}}$ truncation schemes such as MMCC$(2,3)_{\mathscr{L}}$, we have to come up with reasonably accurate approximations for the amplitudes $\ell_{i_1 \ldots i_n}^{a_1 \ldots a_n}$ with $n = m_A + 1, \ldots, m_B$, which enter Eq. (64). These amplitudes define the many-body components of the desexcitation operator $\mathscr{L}$ which generates the similarity-transformed form of the exact, full CI, bra state $\langle \Psi_0 |$ via Eq. (31). The following observation is useful in this context. The operator $\mathscr{L}$ or, more precisely, the A part of $\mathscr{L}$, i.e. $\mathscr{L}^{(A)}$, defined by Eq. (36), is similar to the desexcitation operator $L^{(A)}$, Eq. (11), which defines the left CC state of method A, $\langle \tilde{\Psi}_0^{(A)} |$, Eq. (9). For example, both operators satisfy the same normalization condition, given by Eq. (10) for $L^{(A)}$ and Eq. (38) for $\mathscr{L}^{(A)}$. Clearly, they do not satisfy identical systems of equations, but there are similarities between systems of equations defining $\mathscr{L}$ and $L^{(A)}$. Indeed, the exact $\ell_{i_1 \ldots i_n}^{a_1 \ldots a_n}$ amplitudes defining the operator $\mathscr{L}$ and, through Eq. (31), the full CI bra state $\langle \Psi_0 |$, satisfy the left eigenvalue equation involving $\bar{H}^{(A)}$,

$$\langle \Phi | \mathscr{L} \, \bar{H}^{(A)} = E_0 \, \langle \Phi | \mathscr{L}, \tag{67}$$

which is equivalent to the adjoint form of the Schrödinger equation, $\langle \Psi_0 | H = E_0 \, \langle \Psi_0 |$, in the entire N-electron Hilbert space if $\langle \Phi | \mathscr{L} \, e^{-T^{(A)}}$ is the exact bra state $\langle \Psi_0 |$. The $\lambda_{i_1 \ldots i_n}^{a_1 \ldots a_n}$ amplitudes, which define the $\Lambda^{(A)}$ component of $L^{(A)}$, are obtained by solving the left ground-state eigenvalue problem involving $\bar{H}^{(A)}$, Eq. (14), which is similar to Eq. (67), although we must immediately point out that the left eigenvalue problem for $\Lambda^{(A)}$ or $L^{(A)}$ is an eigenvalue problem in the subspace of the N-electron Hilbert space spanned by the reference determinant $|\Phi\rangle$ and the excited determinants $|\Phi_{i_1 \ldots i_n}^{a_1 \ldots a_n}\rangle$ with $n = 1, \ldots, m_A$. Thus, only when $m_A = N$, the $L^{(A)}$ and $\mathscr{L}^{(A)}$ operators become identical and satisfy the same equations. There is, however, a similarity between the A part of the exact operator $\mathscr{L}$, i.e. $\mathscr{L}^{(A)}$, Eq. (36), and the desexcitation operator $L^{(A)}$, Eq. (11), defining the left CC state of method A, which provides us with a basis for obtaining the approximate values of the n-body components of $\mathscr{L}$ with $n > m_A$ that define the remainder of $\mathscr{L}^{(A)}$, i.e. $\delta\mathscr{L}^{(A)}$ (cf. Eq. (37)), and, ultimately, the energy correction $\delta_0^{(A)}$, Eq. (30). This observation is exploited in formulating the CR-CC(2,3) and other CR-CC(m_A, m_B) approaches [45, 46].

We begin with the CR-CC(2,3) method. The CR-CC(2,3) method is a special case of the MMCC(2,3)$_{\mathscr{L}}$ truncation scheme, in which the ℓ_{ijk}^{abc} amplitudes entering Eq. (65) are approximated by [45, 46, 48]

$$\begin{aligned}
\tilde{\ell}_{ijk}^{abc}(\text{CCSD}) &= \langle \Phi | L^{(\text{CCSD})} \, \bar{H}^{(\text{CCSD})} | \Phi_{ijk}^{abc} \rangle / D_{abc}^{ijk}(\text{CCSD}) \\
&= \langle \Phi | \Lambda^{(\text{CCSD})} \, \bar{H}^{(\text{CCSD})} | \Phi_{ijk}^{abc} \rangle / D_{abc}^{ijk}(\text{CCSD}) \\
&= \langle \Phi | [(\Lambda_1 \bar{H}_2^{(\text{CCSD})})_{DC} + (\Lambda_2 \bar{H}_1^{(\text{CCSD})})_{DC} \\
&\quad + (\Lambda_2 \bar{H}_2^{(\text{CCSD})})_{C}] | \Phi_{ijk}^{abc} \rangle / D_{abc}^{ijk}(\text{CCSD}),
\end{aligned} \tag{68}$$

where Λ_1 and Λ_2 are the one- and two-body components of the “lambda” operator $\Lambda^{(\text{CCSD})}$ of the analytic gradient CCSD theory, obtained by solving Eq. (14) in which $m_A = 2$, $L^{(\text{CCSD})} = \mathbf{1} + \Lambda^{(\text{CCSD})}$, and

$$
\begin{aligned}
D_{abc}^{ijk}(\text{CCSD}) &= E_0^{(\text{CCSD})} - \langle \Phi_{ijk}^{abc} | \bar{H}^{(\text{CCSD})} | \Phi_{ijk}^{abc} \rangle \\
&= - \langle \Phi_{ijk}^{abc} | \bar{H}_1^{(\text{CCSD})} | \Phi_{ijk}^{abc} \rangle \\
&\quad - \langle \Phi_{ijk}^{abc} | \bar{H}_2^{(\text{CCSD})} | \Phi_{ijk}^{abc} \rangle \\
&\quad - \langle \Phi_{ijk}^{abc} | \bar{H}_3^{(\text{CCSD})} | \Phi_{ijk}^{abc} \rangle, \qquad (69)
\end{aligned}
$$

where $\bar{H}_1^{(\text{CCSD})}$, $\bar{H}_2^{(\text{CCSD})}$, and $\bar{H}_3^{(\text{CCSD})}$ represent the one, two-, and three-body components of the similarity-transformed Hamiltonian $\bar{H}^{(\text{CCSD})}$ of the CCSD theory, Eq. (21), and subscript DC refers to the disconnected part of the corresponding operator product. The final formula for the ground-state CR-CC(2,3) energy is [45,46,48] (cf. Eqs. (63) and (64) for $m_A = 2$ and $m_B = 3$)

$$
E_0^{(\text{CR-CC(2,3)})} = E_0^{(\text{CCSD})} + \delta_0^{(\text{CR(T)}_{\mathscr{L}})}(\text{CCSD}), \qquad (70)
$$

where the triples correction $\delta_0^{(\text{CR(T)}_{\mathscr{L}})}(\text{CCSD})$ is calculated as (cf. Eq. (65))

$$
\delta_0^{(\text{CR(T)}_{\mathscr{L}})}(\text{CCSD}) = \sum_{\substack{i<j<k \\ a<b<c}} \tilde{\ell}_{ijk}^{abc}(\text{CCSD}) \, \mathfrak{M}_{abc}^{ijk}(2), \qquad (71)
$$

with moments $\mathfrak{M}_{abc}^{ijk}(2)$ and amplitudes $\tilde{\ell}_{ijk}^{abc}(\text{CCSD})$ given by Eqs. (45) and (68), respectively.

We can extend the above expressions to higher-order CR-CC(m_A, m_B) approaches [45, 46, 169], such as CR-CC(2,4), but since we have not implemented the CR-CC(m_A, m_B) methods other than CR-CC(2,3), we only make a few remarks about how to design such methods in general and what are the remaining open issues in this area that will have to be addressed in the future. This will also help us to explain where the above equations for $\tilde{\ell}_{ijk}^{abc}(\text{CCSD})$, which enter the CR-CC(2,3) energy formula, come from. The basic idea behind all CR-CC(m_A, m_B) approaches is to determine the approximate values of the relevant amplitudes $\ell_{i_1 \ldots i_n}^{a_1 \ldots a_n}$, $n = m_A + 1, \ldots, m_B$ (ℓ_{ijk}^{abc} in the CR-CC(2,3) case), which enter the MMCC$(m_A, m_B)_{\mathscr{L}}$ correction $\delta_0(m_A, m_B)_{\mathscr{L}}$, Eq. (64), by considering the exact form of the similarity-transformed bra Schrödinger equation, Eq. (67), which we right-project on the excited determinants $|\Phi_{i_1 \ldots i_n}^{a_1 \ldots a_n}\rangle$ corresponding to the amplitudes $\ell_{i_1 \ldots i_n}^{a_1 \ldots a_n}$ with $n = m_A + 1, \ldots, m_B$ that we want to determine. Next, based on the aforementioned similarity between the A part of the exact operator $\mathscr{L}$, i.e. $\mathscr{L}^{(A)}$, Eq. (36), and the desexcitation operator $L^{(A)}$, Eq. (11), defining the left CC state of method A, we approximate the exact operator $\mathscr{L}$ in the resulting equation as follows:

$$
\mathscr{L} \approx L^{(A)} + \sum_{n=m_A+1}^{m_B} \tilde{\mathscr{L}}_n^{(A)}, \qquad (72)
$$

where $L^{(A)}$, defined by Eq. (11), is obtained by solving the linear system of equations, Eq. (14), and

$$\tilde{\mathscr{L}}_n^{(A)} = \sum_{\substack{i_1 < \cdots < i_n \\ a_1 < \cdots < a_n}} \tilde{\ell}_{i_1 \ldots i_n}^{a_1 \ldots a_n}(A)\, a^{i_1} \cdots a^{i_n} a_{a_n} \cdots a_{a_1}, \tag{73}$$

where $n = m_A + 1, \ldots, m_B$, are the approximate n-body components of $\mathscr{L}$ that we want to determine to ultimately calculate the energy correction $\delta_0(m_A, m_B)_{\mathscr{L}}$, Eq. (64). This leads to the following system of equations for the approximate $\tilde{\ell}_{i_1 \ldots i_n}^{a_1 \ldots a_n}(A)$ amplitudes that one might eventually use in the CR-CC(m_A, m_B) calculations [45,46]:

$$\langle \Phi | L^{(A)} \bar{H}^{(A)} | \Phi_{i_1 \ldots i_n}^{a_1 \ldots a_n} \rangle + \sum_{m=m_A+1}^{m_B} \langle \Phi | \tilde{\mathscr{L}}_m^{(A)} \bar{H}^{(A)} | \Phi_{i_1 \ldots i_n}^{a_1 \ldots a_n} \rangle = E_0\, \tilde{\ell}_{i_1 \ldots i_n}^{a_1 \ldots a_n}(A), \tag{74}$$

where $n = m_A + 1, \ldots, m_B$. In order to solve this system, while keeping the costs of calculations at the low level, one has to introduce additional approximations. One possibility is to rewrite Eq. (74) in the following manner [169]:

$$\begin{aligned} \langle \Phi | L^{(A)} \bar{H}^{(A)} | \Phi_{i_1 \ldots i_n}^{a_1 \ldots a_n} \rangle + & \sum_{m=m_A+1\ (m \neq n)}^{m_B} \langle \Phi | \tilde{\mathscr{L}}_m^{(A)} \bar{H}^{(A)} | \Phi_{i_1 \ldots i_n}^{a_1 \ldots a_n} \rangle \\ & + \langle \Phi | \tilde{\mathscr{L}}_n^{(A)} \bar{H}^{(A)} | \Phi_{i_1 \ldots i_n}^{a_1 \ldots a_n} \rangle = E_0\, \tilde{\ell}_{i_1 \ldots i_n}^{a_1 \ldots a_n}(A), \end{aligned} \tag{75}$$

where we separated the $m = n$ term in the summation over m, which corresponds to the many-body rank of the amplitude $\tilde{\ell}_{i_1 \ldots i_n}^{a_1 \ldots a_n}(A)$ on the right-hand side of Eq. (74), from the rest of the summation over m. By replacing the expensive $\langle \Phi | \tilde{\mathscr{L}}_n^{(A)} \bar{H}^{(A)} | \Phi_{i_1 \ldots i_n}^{a_1 \ldots a_n} \rangle$ term on the left-hand side of Eq. (75) by its diagonal part, $\tilde{\ell}_{i_1 \ldots i_n}^{a_1 \ldots a_n}(A) \langle \Phi_{i_1 \ldots i_n}^{a_1 \ldots a_n} | \bar{H}^{(A)} | \Phi_{i_1 \ldots i_n}^{a_1 \ldots a_n} \rangle$, and by approximating the exact energy E_0 in the resulting equation by the CC energy of method A, i.e. $E_0^{(A)}$, we obtain the following simplified form of Eq. (75), which may be useful in actual CR-CC(m_A, m_B) calculations [169]:

$$\begin{aligned} \langle \Phi | L^{(A)} \bar{H}^{(A)} | \Phi_{i_1 \ldots i_n}^{a_1 \ldots a_n} \rangle + & \sum_{m=m_A+1\ (m \neq n)}^{m_B} \langle \Phi | \tilde{\mathscr{L}}_m^{(A)} \bar{H}^{(A)} | \Phi_{i_1 \ldots i_n}^{a_1 \ldots a_n} \rangle \\ & = D_{a_1 \ldots a_n}^{i_1 \ldots i_n}(A)\, \tilde{\ell}_{i_1 \ldots i_n}^{a_1 \ldots a_n}(A), \end{aligned} \tag{76}$$

where

$$D_{a_1 \ldots a_n}^{i_1 \ldots i_n}(A) = E_0^{(A)} - \langle \Phi_{i_1 \ldots i_n}^{a_1 \ldots a_n} | \bar{H}^{(A)} | \Phi_{i_1 \ldots i_n}^{a_1 \ldots a_n} \rangle. \tag{77}$$

The amplitudes $\tilde{\ell}_{i_1 \ldots i_n}^{a_1 \ldots a_n}(A)$, $n = m_A + 1, \ldots, m_B$, obtained by solving Eq. (77), can be used to define the CR-CC(m_A, m_B) energies by replacing the exact $\ell_{i_1 \ldots i_n}^{a_1 \ldots a_n}$ amplitudes in the MMCC(m_A, m_B)$_{\mathscr{L}}$ correction $\delta_0(m_A, m_B)_{\mathscr{L}}$, Eq. (64), by their approximate $\tilde{\ell}_{i_1 \ldots i_n}^{a_1 \ldots a_n}(A)$ counterparts.

The above description provides the general framework for designing the hierarchy of CR-CC(m_A, m_B) approaches, but there are various details which will require future study. For example, the above system of equations for $\tilde{\ell}_{i_1 \ldots i_n}^{a_1 \ldots a_n}(A)$, Eq. (76), has an advantage in that it eliminates the main problem of having to deal with the most expensive blocks of the similarity-transformed Hamiltonian along the diagonal (e.g. the triples–triples, quadruples–quadruples, etc. blocks of $\bar{H}^{(\mathrm{CCSD})}$), which we approximated in Eq. (76) by the diagonal matrix elements $\langle\Phi_{i_1\ldots i_n}^{a_1\ldots a_n}|\bar{H}^{(A)}|\Phi_{i_1\ldots i_n}^{a_1\ldots a_n}\rangle$ entering $D_{a_1\ldots a_n}^{i_1\ldots i_n}(A)$, but one has to propose now an efficient computational procedure how to handle the off-diagonal terms $\langle\Phi|\tilde{\mathscr{L}}_m^{(A)}\,\bar{H}^{(A)}|\Phi_{i_1\ldots i_n}^{a_1\ldots a_n}\rangle$ with $m \neq n$ in Eq. (76), which couple different excitations from the range of m values $m = m_A + 1, \ldots, m_B$ (for example, triples and quadruples in the CR-CC(2,4) scheme). In the simplest approach, suggested in Refs. [45, 46], one can ignore the $\langle\Phi|\tilde{\mathscr{L}}_m^{(A)}\,\bar{H}^{(A)}|\Phi_{i_1\ldots i_n}^{a_1\ldots a_n}\rangle$ coupling terms with $m \neq n$ in Eq. (76) altogether, but our recent numerical experiments indicate that this may not be the best idea if the objective is to obtain the highest possible accuracy with methods such as CR-CC(2,4), where coupling terms involving triples and quadruples appear [169]. This is, in fact, understandable, since triples calculated alone, as is done in the CR-CC(2,3) approach where quadruples are simply neglected (meaning, decoupled from triples and assumed to be very small), and triples that are part of the CR-CC(2,4) correction $\delta_0(2,4)$, which describes the combined effect of triples and quadruples, are, generally, different objects (the difference between them grows as the importance of quadruples increases). We have not yet solved the problem of the best way of handling the $\langle\Phi|\tilde{\mathscr{L}}_m^{(A)}\,\bar{H}^{(A)}|\Phi_{i_1\ldots i_n}^{a_1\ldots a_n}\rangle$ coupling terms with $m \neq n$ in the CR-CC(m_A, m_B) approaches with $m_B \geq m_A + 2$, including the CR-CC(2,4) method, so in the following we focus on the CR-CC(2,3) approach and simpler, *a posteriori*, ways of correcting the CR-CC(2,3) results for quadruple excitations using the information provided by the CR-CCSD(TQ) approach.

The good news is that in the CR-CC(2,3) approach and all CR-CC(m_A, m_B) methods with $m_B = m_A + 1$ the situation is rather straightforward, since the $\sum_{m=m_A+1\ (m\neq n)}^{m_B}\langle\Phi|\tilde{\mathscr{L}}_m^{(A)}\,\bar{H}^{(A)}|\Phi_{i_1\ldots i_n}^{a_1\ldots a_n}\rangle$ coupling term in Eq. (76) is zero in this case. Thus, in the CR-CC($m_A, m_A + 1$) methods, including CR-CC(2,3), we can determine the relevant $\tilde{\ell}_{i_1\ldots i_n}^{a_1\ldots a_n}(A)$ amplitudes using the equation [45, 46],

$$\tilde{\ell}_{i_1\ldots i_n}^{a_1\ldots a_n}(A) = \langle\Phi|L^{(A)}\,\bar{H}^{(A)}|\Phi_{i_1\ldots i_n}^{a_1\ldots a_n}\rangle / D_{a_1\ldots a_n}^{i_1\ldots i_n}(A), \tag{78}$$

where $L^{(A)}$ is defined by Eq. (11) and $D_{a_1\ldots a_n}^{i_1\ldots i_n}(A)$ is given by Eq. (77). One can immediately recognize that Eq. (68) for the desexcitation amplitudes $\tilde{\ell}_{ijk}^{abc}(\mathrm{CCSD})$ entering the CR-CC(2,3) triples energy correction $\delta_0^{(\mathrm{CR(T)}_{\mathscr{L}})}(\mathrm{CCSD})$, Eq. (71), is a special case of Eq. (78), written for $m_A = 2$. This shows how the basic equations defining the recently proposed CR-CC(2,3) approach emerge from the more general considerations.

Before discussing different variants of the CR-CC(2,3) method and the CR-CC(2,3)+Q approach, in which the CR-CC(2,3) energy is corrected for the effect of quadruples, we should mention that the above expression for the amplitudes $\tilde{\ell}_{i_1\ldots i_n}^{a_1\ldots a_n}(A)$, Eq. (78), which is used in the CR-CC($m_A, m_A + 1$) calculations, has to be modified if one of the indices $i_1, \ldots, i_n$ or $a_1, \ldots, a_n$ corresponds to the orbital which is degenerate with some other orbitals. In that case, we should replace Eq. (78) by a more elaborate expression in which, instead of simply using the diagonal matrix elements $\langle\Phi_{i_1\ldots i_n}^{a_1\ldots a_n}|\bar{H}^{(A)}|\Phi_{i_1\ldots i_n}^{a_1\ldots a_n}\rangle$ that enter $D_{a_1\ldots a_n}^{i_1\ldots i_n}(A)$, one solves a small system of linear equations, similar to Eq. (76), where all amplitudes $\tilde{\ell}_{i_1\ldots i_n}^{a_1\ldots a_n}(A)$ involving indices of degenerate spin-orbitals are coupled together through the off-diagonal matrix elements of $\bar{H}^{(A)}$ involving the n-tuply excited determinants $|\Phi_{i_1\ldots i_n}^{a_1\ldots a_n}\rangle$ that carry the indices of degenerate spin-orbitals. This remark applies, in particular, to the formula for the amplitudes $\tilde{\ell}_{ijk}^{abc}$(CCSD), Eq. (68), defining the CR-CC(2,3) energy correction $\delta_0^{(\mathrm{CR(T)}\mathscr{L})}$(CCSD), which should be modified in cases of orbital degeneracies. Without taking care of this issue, the CR-CC(2,3) energy correction $\delta_0^{(\mathrm{CR(T)}\mathscr{L})}$(CCSD) and similar corrections defining the CR-CC($m_A, m_A + 1$) energies are no longer strictly invariant with respect to the rotations among degenerate orbitals. Although all of our numerous tests indicate that changes in the values of $\delta_0^{(\mathrm{CR(T)}\mathscr{L})}$(CCSD) due to the rotations among degenerate orbitals do not exceed 0.1 millihartree when we uncritically use Eq. (68) for all amplitudes $\tilde{\ell}_{ijk}^{abc}$(CCSD), it is better (and safer) to calculate amplitudes $\tilde{\ell}_{ijk}^{abc}$(CCSD) involving degenerate orbitals in a different way, by considering their couplings through the appropriate off-diagonal matrix elements of the similarity-transformed Hamiltonian, as described above. Clearly, if the molecule has at most an Abelian symmetry, so that there are no orbital degeneracies, one can apply Eq. (68) and its generalization, Eq. (78), to all amplitudes $\tilde{\ell}_{ijk}^{abc}$(CCSD) and $\tilde{\ell}_{i_1\ldots i_n}^{a_1\ldots a_n}(A)$, without risking any formal problems.

Equations (68)–(71) describe the most complete variant of the CR-CC(2,3) approach which, in analogy to some of our earlier publications, such as Refs. [14, 122, 123], can also be designated by an additional letter D (e.g. CR-CC(2,3),D). Other variants can be suggested by considering approximate forms of the denominator D_{abc}^{ijk}(CCSD), Eq. (69) [48]. For example, variant C is obtained by ignoring the last, three-body, term in Eq. (69) and the B variant of CR-CC(2,3) is obtained by ignoring the last two terms, leaving the one-body contribution $-\langle\Phi_{ijk}^{abc}|\bar{H}_1^{(\mathrm{CCSD})}|\Phi_{ijk}^{abc}\rangle$ in D_{abc}^{ijk}(CCSD) only. Finally, variant A of the CR-CC(2,3) approach is obtained by replacing the denominator D_{abc}^{ijk}(CCSD), Eq. (69), by the usual MBPT denominator for triple excitations, $(\epsilon_i + \epsilon_j + \epsilon_k - \epsilon_a - \epsilon_b - \epsilon_c)$, where ϵ_p's are the spin-orbital energies. In this paper, we focus on the most complete variant D, for which we do not use any additional letter, and the simplified variant A, which we continue to call CR-CC(2,3),A. Similar variants could be introduced for other CR-CC(m_A, m_B) approaches.

The CR-CC(2,3) and other CR-CC(m_A, m_B) methods have several interesting features. In particular, they reduce to the previously formulated non-iterative CC methods if we make additional approximations [45, 46]. For example, the CR-CC(2,3) approach reduces to the CCSD(T) method if we replace the denominator D_{abc}^{ijk}(CCSD), Eq. (69), in Eq. (68) by the spin-orbital energy difference $(\epsilon_i + \epsilon_j + \epsilon_k - \epsilon_a - \epsilon_b - \epsilon_c)$ (as is done in the CR-CC(2,3),A approximation), neglect the $(\Lambda_2\bar{H}_1^{(\mathrm{CCSD})})_{DC}$ term in Eq. (68), which is at least the fourth-order term in MBPT if the Hartree-Fock reference is employed, replace the $(\Lambda_1 \bar{H}_2^{(\text{CCSD})})_{DC}$ and $(\Lambda_2 \bar{H}_2^{(\text{CCSD})})_C$ terms in the resulting expression, which appear in the third and second orders of MBPT, respectively, by $(T_1^\dagger V_N)_{DC}$ and $(T_2^\dagger V_N)_C$, where T_1 and T_2 are obtained in the CCSD calculations, and approximate moment $\mathfrak{M}_{abc}^{ijk}(2)$ by the lead term $\langle \Phi_{ijk}^{abc} | (V_N T_2)_C | \Phi \rangle$. Thus, the CR-CC(2,3) approach provides a rigorous justification for the CCSD(T) method (see, e.g. Ref. [170] for a related discussion). The simplified CR-CC(2,3),A variant of CR-CC(2,3), defined above, is equivalent to the CCSD(2)$_T$ method of Ref. [42]. The higher-order CR-CC(2,4) approach, in which we add a correction due to triples and quadruples to the CCSD energy, as in Eq. (66), reduces to the CCSD(2) or CC(2)PT(2) method of Refs. [42, 171] in a similar manner, if we use Eq. (78) to calculate the relevant $\tilde{\ell}_{ijk}^{abc}(\text{CCSD})$ and $\tilde{\ell}_{ijkl}^{abcd}(\text{CCSD})$ amplitudes. The analogous relationships exist between the CR-CC(2,4) method and the CCSD(2) approach of Refs. [54–57]. Essentially, the CCSD(2)$_T$ and CCSD(2) methods of Refs. [42, 54–57, 171] rely on the Møller–Plesset-type forms of the corresponding denominators $D_{abc}^{ijk}(\text{CCSD})$ and $D_{abcd}^{ijkl}(\text{CCSD})$, respectively, whereas the CR-CC(2,3) and other CR-CC(m_A, m_B) schemes discussed here rely on the Epstein–Nesbet-type denominators, such as $D_{abc}^{ijk}(\text{CCSD})$, although we could obviously consider alternative forms of these denominators as well, as implied by the above considerations. Just like CCSD(2)$_T$ and CCSD(2), the CR-CC(2,3) and CR-CC(2,4) methods are rigorously size extensive. This has been illustrated numerically in Ref. [46].

As already pointed out, we have not implemented the CR-CC(2,4) approach yet and there remain open issues, such as the role of terms that couple triples and quadruples in a system of equations for the relevant $\tilde{\ell}_{ijk}^{abc}(\text{CCSD})$ and $\tilde{\ell}_{ijkl}^{abcd}(\text{CCSD})$ amplitudes, as defined by Eq. (76), where $m_A = 2$ and $m_B = 4$ (the $\langle \Phi | \tilde{\mathscr{L}}_4^{(\text{CCSD})} \bar{H}^{(\text{CCSD})} | \Phi_{ijk}^{abc} \rangle$ and $\langle \Phi | \tilde{\mathscr{L}}_3^{(\text{CCSD})} \bar{H}^{(\text{CCSD})} | \Phi_{ijkl}^{abcd} \rangle$ terms in Eq. (76)). Thus, for the time being, we consider two heuristic models of accounting for connected quadruple excitations, termed CR-CC(2,3)+Q(a) and CR-CC(2,3)+Q(b), in which the CR-CC(2,3) energy (the complete CR-CC(2,3),D approximation) is corrected for the effect of quadruples in the following manner:

$$(79)\quad E_0^{(\text{CR-CC(2,3)+Q(a)})} = E_0^{(\text{CR-CC(2,3)})} + \left[E_0^{(\text{CR-CCSD(TQ),a})} - E_0^{(\text{CR-CCSD(T)})} \right],$$

$$(80)\quad E_0^{(\text{CR-CC(2,3)+Q(b)})} = E_0^{(\text{CR-CC(2,3)})} + \left[E_0^{(\text{CR-CCSD(TQ),b})} - E_0^{(\text{CR-CCSD(T)})} \right],$$

where $E_0^{(\text{CR-CCSD(TQ),a})}$ and $E_0^{(\text{CR-CCSD(TQ),b})}$ are the CR-CCSD(TQ),a and CR-CCSD(TQ),b energies, defined by Eq. (57), and $E_0^{(\text{CR-CCSD(T)})}$ is the CR-CCSD (T) energy, Eq. (50). In other words, we use the difference between the CR-CCSD(TQ),a or CR-CCSD(TQ),b and CR-CCSD(T) energies to estimate the effect of quadruples. As shown in the earlier part of this subsection, the CR-CCSD(TQ),a or CR-CCSD(TQ),b methods are the natural extensions of CR-CCSD(T) to quadruple excitations, so we expect the CR-CC(2,3)+Q(a) and CR-CC(2,3)+Q(b) methods to describe the combined effect of triples and quadruples in a reasonable manner. Since the CR-CCSD(T) and CR-CCSD(TQ) methods are quite robust in the bond breaking region, the estimate of the effects due to quadruples provided by the difference between the CR-CCSD(TQ),a or CR-CCSD(TQ),b and CR-CCSD(T) energies should be quite accurate, even when the internuclear separations become large. We expect the CR-CC(2,3)+Q(b) method to be somewhat more robust than the CR-CC(2,3)+Q(a) approach, since the CR-CCSD(TQ),a scheme uses the simplest possible estimate of T_2 clusters, originating from the first order of MBPT, in the corresponding energy expression, whereas the CR-CCSD(TQ),b method relies on the more accurate values of these clusters obtained in the CCSD calculations (cf., e.g. Eqs. (55) and (56)).

Before ending this section and discussing the results for the water molecule, let us make a few remarks about the computer costs characterizing the CR-CCSD(T), CR-CCSD(TQ), CR-CC(2,3), and CR-CC(2,3)+Q calculations. As already explained, the CR-CCSD(T) and CR-CC(2,3) approaches are closely related to the standard CCSD(T) theory. Similarly, the CR-CCSD(TQ) and CR-CC(2,3)+Q methods represent the renormalized extensions of the standard CCSD(TQ) approximations, such as CCSD(TQ$_{\mathrm{f}}$) [23] and CCSD(TQ),b [24, 36]. These rather simple relationships between the conventional CCSD(T) and CCSD(TQ) methods and their CR-CC counterparts imply that computer costs of the CR-CC calculations are essentially identical to the costs of the corresponding standard CC calculations. Thus, in analogy to the CCSD(T) method, the CR-CCSD(T) and CR-CC(2,3) approaches are $n_o^3 n_u^4$ procedures in the non-iterative steps involving triples and $n_o^2 n_u^4$ procedures in the iterative CCSD steps (n_o and n_u are, respectively, the numbers of occupied and unoccupied orbitals used in the correlated calculations). To be more precise, the cost of calculating the triples corrections of CR-CCSD(T) and CR-CC(2,3), if we ignore the CCSD steps, is exactly twice the cost of calculating the triples correction of CCSD(T). The CR-CC(2,3) method is slightly more expensive than CR-CCSD(T), since in addition to the iterative $n_o^2 n_u^4$ steps of CCSD, one has to use similar steps to obtain the Λ_1 and Λ_2 components of the CCSD "lambda" operator, but the overall costs of the CR-CC(2,3) and CR-CCSD(T) calculations are similar, particularly that the calculation of the Λ_1 and Λ_2 components constitutes a rather small fraction of the time spent on the CCSD iterations for T_1 and T_2, since Λ_1 and Λ_2 are obtained by solving a linear system of equations, Eq. (14), using the previously determined matrix elements of the similarity-transformed Hamiltonian of the CCSD approach. The fact that one is, so to speak, "forced" to calculate the CCSD "lambda" amplitudes in the CR-CC(2,3) calculations has an advantage since, in addition to highly accurate CR-CC(2,3) energies, one gets an immediate access to the first-order reduced density matrices, calculated at the CCSD level of theory, i.e. [6, 15, 16, 152, 153],

$$(81) \qquad \gamma_q^p = \langle \Phi | (1 + \Lambda_1 + \Lambda_2) \, \overline{a^p a_q} | \Phi \rangle,$$

where $\overline{a^p a_q} = e^{-T_1 - T_2} \, a^p a_q \, e^{T_1 + T_2} = (a^p a_q \, e^{T_1 + T_2})_C$, and all kinds of one-electron properties resulting from these matrices, which are determined using the T_1 and T_2 clusters obtained in CCSD calculations and Λ_1 and Λ_2 desexcitation operators that define the left CCSD ground state. In analogy to the non-iterative triples corrections, the cost of the CR-CCSD(TQ) calculations is similar to the cost of the CCSD(TQ$_f$) or CCSD(TQ),b calculations (the CCSD(TQ$_f$) and CCSD(TQ),b methods are $n_o^3 n_u^4$ procedures in the triples part and $n_o^2 n_u^5$ procedures in steps involving the T_4 cluster contributions). Again, if we put aside the iterative CCSD steps, the CR-CCSD(TQ),a and CR-CCSD(TQ),b approaches are only twice as expensive as the conventional CCSD(TQ$_f$) and CCSD(TQ),b methods in the steps involving the non-iterative corrections to the CCSD energy. Thus, one can almost always perform the CR-CCSD(T), CR-CCSD(TQ), CR-CC(2,3), and CR-CC(2,3)+Q calculations if the corresponding CCSD(T) and CCSD(TQ) calculations are affordable. This is an important remark since, just like CCSD(T) and CCSD(TQ), the CR-CCSD(T), CR-CCSD(TQ), CR-CC(2,3), and CR-CC(2,3)+Q procedures are more robust in the bond breaking situations, while being the relatively easy-to-use single-reference procedures which can be used by experts as well as non-experts. As shown in the next section, the CR-CC methods, particularly CR-CC(2,3) and CR-CC(2,3)+Q, can compete with the CCSD(T) and CCSD(TQ) methods in the equilibrium region, while being often as effective as the MRCI(Q) approach when chemical bonds are stretched.

3. NUMERICAL RESULTS: PROBING THE POTENTIAL ENERGY SURFACE OF THE WATER MOLECULE WITH THE STANDARD AND RENORMALIZED COUPLED-CLUSTER METHODS

In order to illustrate the performance of the renormalized CC methods discussed in section 2 and explore the potential benefits of exploiting the recently proposed size extensive CR-CC(2,3) approach, which will eventually be extended to quadruples via the CR-CC(2,4) approximation, we performed the CR-CCSD(T), CR-CCSD(TQ),a, CR-CCSD(TQ),b, CR-CC(2,3),A, CR-CC(2,3) ($\equiv$CR-CC(2, 3), D), CR-CC(2,3)+Q(a), and CR-CC(2,3)+Q(b) calculations for the three cuts of the global potential energy surface of the water molecule. Those cuts are: (i) the dissociation of a single O–H bond, which correlates with the H($1s\ ^2S$) + OH($X\ ^2\Pi$) asymptote, (ii) the simultaneous dissociation of both O–H bonds, which correlates with the 2H($1s\ ^2S$) + O($2p^4\ ^3P$) channel, and (iii) the C_{2v} dissociation pathway of the water molecule into $H_2(X\ ^1\Sigma_g^+)$ and O($2p^4\ ^1D$). In the case of cut (i), one of the two O–H bonds and the H–O–H angle were kept fixed at their respective equilibrium values taken from Ref. [139] (R_e = 0.95785 Å and α_e = 104.501 degree, respectively). In the case of the C_{2v}-symmetric cut (ii), the H–O–H angle α was kept fixed at its equilibrium value taken from Ref. [139] (α_e = 104.501 degree). In the case of another C_{2v}-symmetric cut, namely cut (iii), we followed the approximate energy path toward the dissociation of the water molecule into $H_2(X\ ^1\Sigma_g^+)$ and O($2p^4\ ^1D$), determined using the potential function of Ref. [132] and defined by the coordinate Y, which is the distance between the O nucleus and the line connecting the H nuclei, and the properly optimized H–O–H angle α [143,144]. The equilibrium values of Y and α are Y_e = 0.58641 Å and α_e = 104.501 degree. The results of the CR-CC calculations are compared with the results obtained with the conventional

CCSD, CCSD(T), and CCSD(TQ),b methods, which are very well suited for the equilibrium region but are expected to face considerable difficulties when water starts to dissociate into open-shell fragments, and with the results obtained with the MRCI(Q) approach, which is capable of providing an accurate global potential energy surface, including all three dissociation channels listed above.

We also compare the results of various CR-CC and MRCI(Q) calculations with the highly accurate global potential energy surface of water resulting from the use of the ES approach of Varandas [132]. The ES surface of Ref. [132] was obtained by combining and further refining the many-body expansion [172] potential of Murrell and Carter [145] and the polynomial potential form of Polyansky, Jensen, and Tennyson [146]. The ES surface has a spectroscopic or nearly spectroscopic ($\sim 1-10\,\mathrm{cm}^{-1}$) accuracy up to about 19000 cm^{-1} above the global minimum and remains quite accurate at higher energies. With the proper treatment of long-range forces and other suitable refinements, the ES potential of Ref. [132] has an overall double many-body expansion [173–176] quality, making it very useful to study reaction dynamics involving water. In particular, the ES potential of Ref. [132] used in this work, and its multi-sheeted extensions [148, 149], have been exploited in a number of dynamical calculations, including, for example, the successful rate constant and cross-section calculations for the $\mathrm{O}(2p^4\ ^1D) + \mathrm{H}_2(X\ ^1\Sigma_g^+)$ reaction [143, 144]. As mentioned in the Introduction, the $\mathrm{O}(2p^4\ ^1D) + \mathrm{H}_2(X\ ^1\Sigma_g^+) \rightarrow \mathrm{OH}(X\ ^2\Pi) + \mathrm{H}(1s\ ^2S)$ reaction, which takes place on the ground-state potential energy surface of water, is known to play a significant role in combustion and atmospheric chemistry. Two of the above cuts (cut (i) and (iii)) are directly related to this process. In addition to testing the CR-CC (also, MRCI(Q)) methods, by comparing the best CR-CC and MRCI(Q) data with the energies provided by the ES potential function, we suggest ways of improving the ES potential, particularly in intermediate and selected higher-energy regions where precise or well understood spectroscopic data are not always available and where the best CR-CC and MRCI(Q) calculations almost perfectly agree with each other.

All CC and CR-CC calculations were performed using the spin- and symmetry-adapted restricted Hartree-Fock (RHF) determinant as a reference. The MRCI(Q) calculations were performed using the usual multi-determinantal reference obtained in the single-root complete-active-space self-consistent-field (CASSCF) calculations. The active space used in the CASSCF and subsequent MRCI(Q) calculations consisted of six valence orbitals that correlate with the $1s$ shells of the hydrogen atoms and the $2s$ and $2p$ shells of the oxygen atom. In analogy to active orbitals, the lowest-energy molecular orbital ($\sim$1s orbital of oxygen) was optimized in CASSCF calculations, but unlike active orbitals that change occupancies it remained doubly occupied in all reference determinants defining the CASSCF and MRCI(Q) wave functions. Since our *ab initio* results are compared with the spectroscopically accurate ES surface of Ref. [132] and since it is well known that core electrons can significantly contribute to the many-electron correlation effects and affect energy differences between different points on the potential energy surface, including the water potential [139, 147, 177], all electrons were correlated in the CC, CR-CC, and CASSCF-based MRCI(Q) calculations. This distinguishes our calculations from the earlier calculations reported, for example, in Refs. [139, 177], in which the effects of core electrons were added as the *a posteriori* corrections to the results of valence-electron calculations.

The calculations were performed with the aug-cc-pCVXZ basis sets with $X =$ D, T, Q [130, 131], in which additional tight functions are added to the valence basis sets of the aug-cc-pVXZ quality to improve the description of core and core-valence correlation effects. The CCSD, CCSD(T), CCSD(TQ),b, CR-CCSD(T), CR-CCSD(TQ),a, CR-CCSD(TQ),b, CR-CC(2,3),A, CR-CC(2,3) (= CR-CC(2, 3), D), CR-CC(2,3)+Q(a), and CR-CC(2,3)+Q(b) calculations were performed with the system of CC/CR-CC computer codes described in Refs. [24, 34, 45, 118] and incorporated in the GAMESS package [119]. The MRCI(Q) calculations were performed with the MOLPRO package [178]. In addition to the series of MRCI(Q) calculations using the aug-cc-pCVXZ basis sets with $X =$ D, T, Q, we performed the high accuracy MRCI(Q) calculations using the aug-cc-pCV5Z basis, to determine if the aug-cc-pCVQZ results are reasonably well converged. We could not perform such calculations using the CC and CR-CC methods, since the atomic integral package used by GAMESS is restricted to g functions and the aug-cc-pCV5Z basis contains h functions which we did not want to drop in an *ad hoc* manner. Fortunately, we do not observe substantial changes in the results, when going from the aug-cc-pCVQZ to aug-cc-pCV5Z basis sets, which would affect our main conclusions. To facilitate our presentation, we use the simplified notation, in which we refer to the aug-cc-pCVXZ basis set by mentioning the value of it's cardinal number X ($X = 2$ for aug-cc-pCVDZ, $X = 3$ for aug-cc-pCVTZ, $X = 4$ for aug-cc-pCVQZ, and $X = 5$ for aug-cc-pCV5Z). For example, we write "the $X = 3$ basis set" instead of "the aug-cc-pCVTZ basis set." Instead of writing "the CR-CC(2,3) calculations with the the aug-cc-pCVTZ basis set," we simply write "the CR-CC(2,3)/$X = 3$ calculations."

The results of our calculations are summarized in Tables 1–8. Table 1 serves as a reference for the remaining tables, providing the MRCI(Q) and ES energies along the three dissociation pathways considered in our calculations. As one can see, the three potential energy surface cuts probe different energy regions. The single O–H bond dissociation is characterized by the lowest energies (always reported in this work relative to the corresponding energies at the equilibrium geometry of Ref. [139], so that all energies are 0 at the equilibrium geometry), which do not exceed 44000 cm^{-1}. The C_{2v}-symmetric dissociation pathway that leads to the $H_2(X\ ^1\Sigma_g^+)$ and $O(2p^4\ ^1D)$ products goes to higher energies, on the order of 59000 cm^{-1}, and the highest energies, on the order of 81000 cm^{-1}, are reached, when the simultaneous dissociation of both O–H bonds is examined. The results in Table 1 show that the large scale MRCI(Q) calculations with the $X = 5$ basis set agree with the ES potential, to within ~10–200 cm^{-1}, in the $R_e \leq R \leq 2R_e$ and $R \geq 4R_e$ regions of cut (i), $R_e \leq R < 1.5R_e$ and $R \geq 2.5R_e$ regions of cut (ii), and $Y_e \leq Y \leq 1.0$ Å and $Y \geq 1.75$ Å of cut (iii) (R is the O–H separation for the dissociating O–H bond or bonds; the meaning of Y has been explained above). The MRCI(Q)/$X = 4$

Table 1. The ES and MRCI(Q) energies for the three potential energy surface cuts of water examined in this study: (i) the dissociation of a single O–H bond, (ii) the C_{2v}-symmetric dissociation of both O-H bonds, and (iii) the C_{2v} dissociation into $H_2(X\ ^1\Sigma_g^+)$ and $O(2p^4\ ^1D)$. R is an O–H distance defining the dissociating O–H bond(s), Y (in Å) is the distance between O and the line connecting both H nuclei, and α (in degree) is the H–O–H angle. The equilibrium values of R, Y, and α are $R_e = 0.95785$ Å, $Y_e = 0.58641$ Å, and $\alpha_e = 104.501$ degree [139]. All energies E (in cm^{-1}) are reported as $E - E(R_e, \alpha_e)$, where $E(R_e, \alpha_e)$ are the corresponding values of E at the equilibrium geometry. X is a cardinal number defining the aug-cc-pCVXZ basis sets used in the MRCI(Q) calculations. In all MRCI(Q) calculations, all electrons were correlated.

$H_2O(\tilde{X}\ ^1A_1) \rightarrow H(1s\ ^2S) + OH(X\ ^2\Pi)$						
R	α	ES	MRCI(Q)			
			$X = 2$	$X = 3$	$X = 4$	$X = 5$
$1.25R_e$	104.501	7367	6753	7177	7354	7396
$1.50R_e$	104.501	18862	17945	18733	18990	19059
$1.75R_e$	104.501	28857	27478	28510	28838	28928
$2.00R_e$	104.501	35583	34067	35226	35615	35721
$2.50R_e$	104.501	41024	40030	41277	41736	41857
$3.00R_e$	104.501	42816	41521	42811	43282	43407
$4.00R_e$	104.501	43779	41938	43250	43723	43848
$5.00R_e$	104.501	43887	41970	43281	43754	43879
$H_2O(\tilde{X}\ ^1A_1) \rightarrow 2H(1s\ ^2S) + O(2p^4\ ^3P)$						
R	α	ES	MRCI(Q)			
			$X = 2$	$X = 3$	$X = 4$	$X = 5$
$1.25R_e$	104.501	14366	13119	14027	14380	14464
$1.50R_e$	104.501	36152	34506	36188	36706	36843
$1.75R_e$	104.501	54629	52359	54555	55224	55407
$2.00R_e$	104.501	67354	64408	66912	67730	67950
$2.50R_e$	104.501	78170	74341	77126	78113	78374
$3.00R_e$	104.501	80585	76298	79196	80207	80478
$4.00R_e$	104.501	81191	76792	79732	80752	81020
$5.00R_e$	104.501	81219	76840	79776	80795	81062
$H_2O(\tilde{X}\ ^1A_1) \rightarrow H_2(X\ ^1\Sigma_g^+) + O(2p^4\ ^1D)$						
Y	α	ES	MRCI(Q)			
			$X = 2$	$X = 3$	$X = 4$	$X = 5$
0.80	78.808	4998	4501	4809	4981	5023
0.90	69.118	10660	9888	10374	10639	10701
1.00	61.113	17927	16642	17326	17666	17745
1.10	56.072	26589	23953	24887	25287	25383
1.20	53.429	36326	31497	32699	33148	33260
1.30	48.314	43726	38301	39623	40095	40215
1.50	38.432	52137	48794	50171	50662	50786
1.75	29.171	57311	55342	56604	57116	57236
2.00	22.855	58486	56770	57663	58181	58294
4.00	10.589	58696	57534	58058	58590	58694

results are not much different, producing the results that in most cases do not differ from the MRCI(Q)/$X = 5$ results by more than 100–200 cm^{-1} in the higher-energy regions. Thus, with an exception of the region of intermediate R and Y values ($2R_e < R < 4R_e$ for cut (i), $1.5R_e \leq R < 2.5R_e$ for cut (ii), and 1.0 Å $< Y <$ 1.75 Å for cut (iii), the MRCI(Q)/$X = 4$ or MRCI(Q)/$X = 5$ and ES potential functions agree very well. Clearly, they cannot perfectly agree, since MRCI(Q) is not the exact theory. Moreover, our MRCI(Q) calculations are the standard non-relativistic calculations, ignoring relativistic, non-adiabatic, and quantum electrodynamical effects which all contribute to the water potential energy surface [139]. Besides, the minimum energy path defining cut (iii) determined from the ES function is certainly not identical to the similar path that would result from the MRCI(Q) calculations and there may be other small differences in the details of the MRCI(Q)/$X = 4$ or MRCI(Q)/$X = 5$ and ES potentials. On the other hand, the differences between the MRCI(Q)/$X = 4$ or MRCI(Q)/$X = 5$ and ES potential energy surfaces in the afore-mentioned regions of intermediate R and Y values, and intermediate or higher, but not the highest energies, which are as much as 700–800 cm^{-1} for cut (i) in the $R \approx 2.5R_e$ region, 600–800 cm^{-1} for cut (ii) in the $R \approx 1.75R_e$ region, and 3500–3600 cm^{-1} for cut (iii) in the $Y \approx 1.3$ Å region, cannot be explained by the mere neglect of the relativistic, non-adiabatic, and quantum electrodynamical effects. They indicate that either the MRCI(Q) approach is insufficiently accurate or the ES potential needs further refinement. The former is always possible, but the latter explanation would not be a complete surprise either, since we must remember that the ES potential function is constructed by combining the many-body expansion potential of Murrell and Carter [145] and the polynomial potential form of Polyansky, Jensen, and Tennyson [146]. This makes the ES potential very accurate around the minimum, up to about 19000 cm^{-1}, and in the asymptotic regions, including the $H(1s\ ^2S) + OH(X\ ^2\Pi)$, $2H(1s\ ^2S) + O(2p^4\ ^3P)$, and $H_2(X\ ^1\Sigma_g^+) + O(2p^4\ ^1D)$ asymptotes examined in this study, but the accuracy of the ES potential function in the regions of intermediate internuclear separations and energies which connect the spectroscopic and asymptotic regions of the water surface is less certain. A comparison of the MRCI(Q) and best CR-CC data may help us to decide if the ES potential needs further improvements in the regions of intermediate R or Y values and energies that connect the region of the global minimum with the asymptotes.

The various CC and CR-CC results, as compared with the corresponding MRCI(Q) and ES data, are collected in Tables 2–8. The main CC and CR-CC results for the three potential energy surface cuts examined in this work (the CC and CR-CC energies calculated relative to the corresponding energies at the minimum taken from Ref. [139]) are given in Tables 2–4. In order to facilitate the analysis of the data collected in these tables and reading of this section, we give in Tables 5–7 the differences between the CC/CR-CC and the corresponding MRCI(Q) energies for each of the three aug-cc-pCVXZ basis sets used in the CC/CR-CC calculations. In Table 8, we compare the errors in the best CR-CC(2,3)+Q(b)/$X = 4$ results, relative to the

Table 2. The dissociation of a single O–H bond in water (into H(1s ^2S) + OH(X $^2\Pi$); cut(i)). One of the two O–H bonds and the H–O–H angle are kept fixed at their respective equilibrium values taken from Ref. [139] ($R_e = 0.95785$ Å and $\alpha_e = 104.501$ degree, respectively). R is an O–H distance defining the dissociating O–H bond. All energies E (in cm^{-1}) are reported as $E - E(R_e, \alpha_e)$, where $E(R_e, \alpha_e)$ are the corresponding values of E at the equilibrium geometry. X is a cardinal number defining the aug-cc-pCVXZ basis sets used in the calculations. In all CC calculations, all electrons were correlated.

R	CCSD			CCSD(T)			CR-CCSD(T)			CR-CC(2,3),A			CR-CC(2,3)		
	$X=2$	$X=3$	$X=4$	$X=2$	$X=3$	$X=4$	$X=2$	$X=3$	$X=4$	$X=2$	$X=3$	$X=4$	$X=2$	$X=3$	$X=4$
$1.25R_e$	7033	7483	7662	6796	7181	7351	6845	7247	7419	6825	7219	7389	6810	7216	7393
$1.50R_e$	18587	19428	19691	18023	18744	18988	18151	18909	19157	18105	18843	19087	18063	18829	19087
$1.75R_e$	28610	29763	30108	27567	28529	28844	27843	28873	29194	27764	28753	29065	27653	28696	29024
$2.00R_e$	35889	37303	37739	34109	35231	35620	34675	35913	36313	34554	35716	36098	34315	35547	35966
$2.50R_e$	43451	45308	45904	39363	40628	41117	41145	42687	43205	40919	42296	42775	40399	41939	42461
$3.00R_e$	46000	48158	48836	39387	40600	41087	42828	44537	45093	42518	43996	44495	41847	43517	44095
$4.00R_e$	47145	49527	50262	37553	38518	38938	43188	44976	45541	42809	44318	44815	42106	43834	44443
$5.00R_e$	47325	49760	50508	36759	37592	37975	43107	44892	45453	42716	44215	44708	42063	43771	44379

R	CCSD(TQ),b			CR-CCSD(TQ),a			CR-CCSD(TQ),b			CR-CC(2,3)+Q(a)[a]			CR-CC(2,3)+Q(b)[a]		
	$X=2$	$X=3$	$X=4$	$X=2$	$X=3$	$X=4$	$X=2$	$X=3$	$X=4$	$X=2$	$X=3$	$X=4$	$X=2$	$X=3$	$X=4$
$1.25R_e$	6771	7167	7338	6793	7197	7370	6794	7199	7372	6758	7166	7344	6759	7168	7346
$1.50R_e$	17965	18703	18949	18036	18797	19046	18035	18798	19048	17947	18717	18976	17946	18718	18978
$1.75R_e$	27461	28446	28764	27656	28689	29012	27648	28685	29010	27466	28512	28843	27457	28509	28840
$2.00R_e$	33948	35098	35491	34420	35659	36060	34398	35643	36047	34060	35293	35714	34037	35277	35701
$2.50R_e$	39145	40432	40925	40811	42340	42858	40759	42297	42820	40066	41592	42114	40014	41549	42075
$3.00R_e$	39234	40445	40928	42482	44162	44715	42432	44119	44676	41501	43141	43716	41451	43098	43677
$4.00R_e$	37630	38536	38941	42842	44580	45140	42845	44591	45154	41759	43438	44042	41763	43449	44056
$5.00R_e$	36964	37713	38075	42735	44460	45016	42798	44536	45095	41690	43339	43941	41754	43416	44021

[a] CR-CC(2, 3)+Q(x) = CR-CC(2, 3) + [CR-CCSD(TQ), x − CR-CCSD(T)], x = a, b; cf. Eqs. (79) and (80).

Table 3. The C_{2v}-symmetric double dissociation of water (into $2\text{H}(1s\ ^2S) + \text{O}(2p^4\ ^3P)$; cut (ii)). The H–O–H angle is kept fixed at its equilibrium value taken from Ref. [139] ($\alpha_e = 104.501$ degree). R is an O–H distance and $R_e = 0.95785$ Å is the equilibrium value of R [139]. All energies E (in cm^{-1}) are reported as $E - E(R_e, \alpha_e)$, where $E(R_e, \alpha_e)$ are the corresponding values of E at the equilibrium geometry. X is a cardinal number defining the aug-cc-pCVXZ basis sets used in the calculations. In all CC calculations, all electrons were correlated.

R	CCSD			CCSD(T)			CR-CCSD(T)			CR-CC(2,3),A			CR-CC(2,3)		
	$X=2$	$X=3$	$X=4$	$X=2$	$X=3$	$X=4$	$X=2$	$X=3$	$X=4$	$X=2$	$X=3$	$X=4$	$X=2$	$X=3$	$X=4$
$1.25R_e$	13719	14684	15042	13212	14041	14381	13325	14191	14534	13279	14127	14466	13226	14100	14459
$1.50R_e$	35972	37782	38315	34695	36253	36744	35036	36676	37175	34908	36501	36992	34780	36421	36960
$1.75R_e$	55003	57497	58208	52556	54646	55290	53387	55626	56283	53100	55245	55884	52833	55093	55771
$2.00R_e$	68471	71615	72538	64123	66666	67498	66020	68801	69647	65357	67973	68789	64789	67559	68439
$2.50R_e$	79497	84194	85573	67039	70640	71855	75186	79223	80442	72329	75866	77002	71027	74896	76142
$3.00R_e$	79696	85261	86853	56639	59917	61074	74436	79026	80364	69964	73450	74560	68538	72292	73618
$4.00R_e$	78228	84053	85708	43548	45309	46055	72499	77153	78494	67359	70653	71700	66295	69878	71116
$5.00R_e$	77699	83526	85183	39737	40815	41383	71808	76416	77746	66549	69777	70810	65742	69234	70232
R	CCSD(TQ),b			CR-CCSD(TQ),a			CR-CCSD(TQ),b			CR-CC(2,3)+Q(a)[a]			CR-CC(2,3)+Q(b)[a]		
	$X=2$	$X=3$	$X=4$	$X=2$	$X=3$	$X=4$	$X=2$	$X=3$	$X=4$	$X=2$	$X=3$	$X=4$	$X=2$	$X=3$	$X=4$
$1.25R_e$	13157	14010	14351	13208	14079	14424	13212	14085	14430	13109	13988	14349	13113	13993	14355
$1.50R_e$	34552	36146	36641	34749	36395	36897	34752	36403	36908	34493	36139	36682	34496	36148	36693
$1.75R_e$	52314	54437	55086	52907	55144	55804	52892	55142	55806	52352	54611	55293	52338	54608	55295
$2.00R_e$	64074	66549	67364	65481	68219	69059	65469	68208	69053	64250	66976	67851	64238	66965	67845
$2.50R_e$	75356	76491	77198	76337	80006	81153	77686	80953	82021	72177	75679	76853	73526	76626	77722
$3.00R_e$	93104	88747	88276	79038	83153	84401	82900	86370	87477	73140	76418	77654	77003	79635	80730
$4.00R_e$	126095	115610	113773	82241	86600	87911	85748	89689	90907	76037	79325	80533	79544	82415	83529
$5.00R_e$	139384	127204	124945	84989	89436	90773	86028	90109	91361	78923	82253	83260	79962	82926	83848

[a] CR-CC(2, 3)+Q(x) = CR-CC(2, 3) + [CR-CCSD(TQ), x − CR-CCSD(T)], x = a, b; cf. Eqs. (79) and (80).

Table 4. The C_{2v} dissociation of water into $H_2(X\ ^1\Sigma_g^+)$ and $O(2p^4\ ^1D)$ (cut (iii)). The approximate minimum energy path, determined using the potential function of Ref. [132], is defined by the coordinate Y (in Å), which is the distance between O and the line connecting both H nuclei, and the H–O–H angle α (in degree). The equilibrium values of Y and α are $Y_e = 0.58641$ Å and $\alpha_e = 104.501$ degree [139]. All energies E (in cm^{-1}) are reported as $E - E(Y_e, \alpha_e)$, where $E(Y_e, \alpha_e)$ are the corresponding values of E at the equilibrium geometry. X is a cardinal number defining the aug-cc-pCVXZ basis sets used in the calculations. In all CC calculations, all electrons were correlated.

Y	α	CCSD			CCSD(T)			CR-CCSD(T)			CR-CC(2,3),A			CR-CC(2,3)		
		$X=2$	$X=3$	$X=4$	$X=2$	$X=3$	$X=4$	$X=2$	$X=3$	$X=4$	$X=2$	$X=3$	$X=4$	$X=2$	$X=3$	$X=4$
0.80	78.808	4774	5126	5303	4555	4847	5017	4597	4904	5074	4579	4880	5049	4558	4867	5044
0.90	69.118	10376	10928	11201	9977	10432	10691	10056	10536	10797	10022	10492	10752	9992	10468	10744
1.00	61.113	17384	18147	18499	16770	17407	17739	16897	17568	17903	16844	17500	17833	16789	17470	17827
1.10	56.072	24974	25995	26411	24122	24997	25385	24310	25228	25621	24240	25135	25525	24155	25088	25514
1.20	53.429	32808	34103	34568	31700	32836	33268	31969	33158	33597	31887	33039	33471	31771	32963	33438
1.30	48.314	39726	41129	41615	38513	39783	40238	38830	40155	40618	38760	40031	40483	38617	39923	40416
1.50	38.432	49953	51355	51853	48998	50382	50864	49259	50675	51163	49231	50586	51060	49051	50433	50943
1.75	29.171	55899	57130	57642	55485	56838	57358	55596	56941	57462	55595	56915	57430	55444	56787	57331
2.00	22.855	57688	58693	59243	57335	58525	59102	57444	58609	59180	57456	58615	59186	57299	58503	59083
4.00	10.589	59495	60316	60926	58738	59825	60483	59006	60031	60672	59037	60063	60704	58768	59872	60543

Y	α	CCSD(TQ),b			CR-CCSD(TQ),a			CR-CCSD(TQ),b			CR-CC(2,3)+Q(a)[a]			CR-CC(2,3)+Q(b)[a]		
		$X=2$	$X=3$	$X=4$	$X=2$	$X=3$	$X=4$	$X=2$	$X=3$	$X=4$	$X=2$	$X=3$	$X=4$	$X=2$	$X=3$	$X=4$
0.80	78.808	4535	4838	5008	4549	4858	5029	4551	4861	5032	4510	4821	4999	4513	4824	5002
0.90	69.118	9945	10416	10676	9969	10452	10715	9973	10458	10721	9905	10385	10662	9909	10390	10668
1.00	61.113	16727	17385	17717	16767	17443	17779	16772	17450	17787	16659	17345	17703	16664	17352	17711
1.10	56.072	24067	24966	25357	24138	25061	25457	24140	25067	25464	23983	24921	25349	23986	24926	25356
1.20	53.429	31631	32797	33234	31765	32958	33400	31759	32957	33401	31568	32764	33242	31562	32763	33242
1.30	48.314	38454	39757	40218	38655	39979	40444	38633	39964	40430	38442	39747	40242	38420	39732	40229
1.50	38.432	48996	50402	50891	49211	50615	51101	49177	50584	51072	49003	50372	50881	48970	50341	50851
1.75	29.171	55500	56837	57361	55636	56968	57487	55605	56937	57457	55484	56814	57356	55452	56783	57326
2.00	22.855	57312	58468	59047	57498	58653	59225	57463	58618	59191	57354	58548	59128	57318	58512	59094
4.00	10.589	58689	59724	60384	59036	60058	60701	59014	60025	60670	58798	59899	60572	58776	59866	60541

[a] CR-CC(2, 3)+Q(x) = CR-CC(2, 3) + [CR-CCSD(TQ), x − CR-CCSD(T)], x = a, b; cf. Eqs. (79) and (80).

Table 5. The differences between CC/CR-CC energies, calculated relative to their equilibrium values (the CC/CR-CC $[E - E(R_e, \alpha_e)]$ values in Table 3) and the corresponding MRCI(Q) relative energies (the MRCI(Q) $[E - E(R_e, \alpha_e)]$ values in Table 1) for the dissociation of a single O–H bond in water (into $H(1s\ ^2S) + OH(X\ ^2\Pi)$; cut (i)). X is a cardinal number defining the aug-cc-pCVXZ basis sets used in the calculations.

R	CCSD			CCSD(T)			CR-CCSD(T)			CR-CC(2,3),A			CR-CC(2,3)		
	$X=2$	$X=3$	$X=4$	$X=2$	$X=3$	$X=4$	$X=2$	$X=3$	$X=4$	$X=2$	$X=3$	$X=4$	$X=2$	$X=3$	$X=4$
$1.25R_e$	280	306	308	43	4	−3	92	70	65	72	42	35	57	39	39
$1.50R_e$	642	695	701	78	11	−2	206	176	167	160	110	97	118	96	97
$1.75R_e$	1132	1253	1270	89	19	6	365	363	356	286	243	227	175	186	186
$2.00R_e$	1822	2077	2124	42	5	5	608	687	698	487	490	483	248	321	351
$2.50R_e$	3421	4031	4168	−667	−649	−619	1115	1410	1469	889	1019	1039	369	662	725
$3.00R_e$	4479	5347	5554	−2134	−2211	−2195	1307	1726	1811	997	1185	1213	326	706	813
$4.00R_e$	5207	6277	6539	−4385	−4732	−4785	1250	1726	1818	871	1068	1092	168	584	720
$5.00R_e$	5355	6479	6754	−5211	−5689	−5779	1137	1611	1699	746	934	954	93	490	625

R	CCSD(TQ),b			CR-CCSD(TQ),a			CR-CCSD(TQ),b			CR-CC(2,3)+Q(a)[a]			CR-CC(2,3)+Q(b)[a]		
	$X=2$	$X=3$	$X=4$	$X=2$	$X=3$	$X=4$	$X=2$	$X=3$	$X=4$	$X=2$	$X=3$	$X=4$	$X=2$	$X=3$	$X=4$
$1.25R_e$	18	−10	−16	40	20	16	41	22	18	5	−11	−10	6	−9	−8
$1.50R_e$	20	−30	−41	91	64	56	90	65	58	2	−16	−14	1	−15	−12
$1.75R_e$	−17	−64	−74	178	179	174	170	175	172	−12	2	5	−21	−1	2
$2.00R_e$	−119	−128	−124	353	433	445	331	417	432	−7	67	99	−30	51	86
$2.50R_e$	−885	−845	−811	781	1063	1122	729	1020	1084	36	315	378	−16	272	339
$3.00R_e$	−2287	−2366	−2354	961	1351	1433	911	1308	1394	−20	330	434	−70	287	395
$4.00R_e$	−4308	−4714	−4782	904	1330	1417	907	1341	1431	−179	188	319	−175	199	333
$5.00R_e$	−5006	−5568	−5679	765	1179	1262	828	1255	1341	−280	58	187	−216	135	267

[a] $CR-CC(2,3) + Q(x) = CR-CC(2,3) + [CR-CCSD(TQ), x - CR-CCSD(T)]$, x = a, b; cf. Eqs. (79) and (80).

Table 6. The differences between CC/CR-CC energies, calculated relative to their equilibrium values (the CC/CR-CC $[E - E(R_e, \alpha_e)]$ values in Table 3) and the corresponding MRCI(Q) relative energies (the MRCI(Q) $[E - E(R_e, \alpha_e)]$ values in Table 1) for the C_{2v}-symmetric double dissociation of water (into $2H(1s\ ^2S) + O(2p^4\ ^3P)$; cut (ii)). X is a cardinal number defining the aug-cc-pCVXZ basis sets used in the calculations.

R	CCSD			CCSD(T)			CR-CCSD(T)			CR-CC(2,3),A			CR-CC(2,3)		
	$X=2$	$X=3$	$X=4$	$X=2$	$X=3$	$X=4$	$X=2$	$X=3$	$X=4$	$X=2$	$X=3$	$X=4$	$X=2$	$X=3$	$X=4$
$1.25R_e$	600	657	662	93	14	1	206	164	154	160	100	86	107	73	79
$1.50R_e$	1466	1594	1609	189	65	38	530	488	469	402	313	286	274	233	254
$1.75R_e$	2644	2942	2984	197	91	66	1028	1071	1059	741	690	660	474	538	547
$2.00R_e$	4063	4703	4808	−285	−246	−232	1612	1889	1917	949	1061	1059	381	647	709
$2.50R_e$	5156	7068	7460	−7302	−6486	−6258	845	2097	2329	−2012	−1260	−1111	−3314	−2230	−1971
$3.00R_e$	3398	6065	6646	−19659	−19279	−19133	−1862	−170	157	−6334	−5746	−5647	−7760	−6904	−6589
$4.00R_e$	1436	4321	4956	−33244	−34423	−34697	−4293	−2579	−2258	−9433	−9079	−9052	−10497	−9854	−9636
$5.00R_e$	859	3750	4388	−37103	−38961	−39412	−5032	−3360	−3049	−10291	−9999	−9985	−11098	−10542	−10563

R	CCSD(TQ),b			CR-CCSD(TQ),a			CR-CCSD(TQ),b			CR-CC(2,3)+Q(a)[a]			CR-CC(2,3)+Q(b)[a]		
	$X=2$	$X=3$	$X=4$	$X=2$	$X=3$	$X=4$	$X=2$	$X=3$	$X=4$	$X=2$	$X=3$	$X=4$	$X=2$	$X=3$	$X=4$
$1.25R_e$	38	−17	−29	89	52	44	93	58	50	−10	−39	−31	−6	−34	−25
$1.50R_e$	46	−42	−65	243	207	191	246	215	202	−13	−49	−24	−10	−40	−13
$1.75R_e$	−45	−118	−138	548	589	580	533	587	582	−7	56	69	−21	53	71
$2.00R_e$	−334	−363	−366	1073	1307	1329	1061	1296	1323	−158	64	121	−170	53	115
$2.50R_e$	1015	−635	−915	1996	2880	3040	3345	3827	3908	−2164	−1447	−1260	−815	−500	−391
$3.00R_e$	16806	9551	8069	2740	3957	4194	6602	7174	7270	−3158	−2778	−2553	705	439	523
$4.00R_e$	49303	35878	33021	5449	6868	7159	8956	9957	10155	−755	−407	−219	2752	2683	2777
$5.00R_e$	62544	47428	44150	8149	9660	9978	9188	10333	10566	2083	2477	2465	3122	3150	3053

[a] CR-CC(2, 3)+Q(x) = CR-CC(2, 3) + [CR-CCSD(TQ), x − CR-CCSD(T)], x = a, b; cf. Eqs. (79) and (80).

Table 7. The differences between CC/CR-CC energies, calculated relative to their equilibrium values (the CC/CR-CC $[E - E(R_e, \alpha_e)]$ values in Table 4) and the corresponding MRCI(Q) relative energies (the MRCI(Q) $[E - E(R_e, \alpha_e)]$ values in Table 1) for the C_{2v} dissociation of water into $H_2(X\ ^1\Sigma_g^+)$ and $O(2p^4\ ^1D)$ along the approximate minimum energy path determined using the potential function of Ref. [132] (cut (iii)). X is a cardinal number defining the aug-cc-pCVXZ basis sets used in the calculations.

Y	α	CCSD			CCSD(T)			CR-CCSD(T)			CR-CC(2,3),A			CR-CC(2,3)		
		$X=2$	$X=3$	$X=4$	$X=2$	$X=3$	$X=4$	$X=2$	$X=3$	$X=4$	$X=2$	$X=3$	$X=4$	$X=2$	$X=3$	$X=4$
0.80	78.808	273	317	322	54	38	36	96	95	93	78	71	68	57	58	63
0.90	69.118	488	554	562	89	58	52	168	162	158	134	118	113	104	94	105
1.00	61.113	742	821	833	128	81	73	255	242	237	202	174	167	147	144	161
1.10	56.072	1021	1108	1124	169	110	98	357	341	334	287	248	238	202	201	227
1.20	53.429	1311	1404	1420	203	137	120	472	459	449	390	340	323	274	264	290
1.30	48.314	1425	1506	1520	212	160	143	529	532	523	459	408	388	316	300	321
1.50	38.432	1159	1184	1191	204	211	202	465	504	501	437	415	398	257	262	281
1.75	29.171	557	526	526	143	234	242	254	337	346	253	311	314	102	183	215
2.00	22.855	918	1030	1062	565	862	921	674	946	999	686	952	1005	529	840	902
4.00	10.589	1961	2258	2336	1204	1767	1893	1472	1973	2082	1503	2005	2114	1234	1814	1953

Y	α	CCSD(TQ),b			CR-CCSD(TQ),a			CR-CCSD(TQ),b			CR-CC(2,3)+Q(a)[a]			CR-CC(2,3)+Q(b)[a]		
		$X=2$	$X=3$	$X=4$	$X=2$	$X=3$	$X=4$	$X=2$	$X=3$	$X=4$	$X=2$	$X=3$	$X=4$	$X=2$	$X=3$	$X=4$
0.80	78.808	34	29	27	48	49	48	50	52	51	9	12	18	12	15	21
0.90	69.118	57	42	37	81	78	76	85	84	82	17	11	23	21	16	29
1.00	61.113	85	59	51	125	117	113	130	124	121	17	19	37	22	26	45
1.10	56.072	114	79	70	185	174	170	187	180	177	30	34	62	33	39	69
1.20	53.429	134	98	86	268	259	252	262	258	253	71	65	94	65	64	94
1.30	48.314	153	134	123	354	356	349	332	341	335	141	124	147	119	109	134
1.50	38.432	202	231	229	417	444	439	383	413	410	209	201	219	176	170	189
1.75	29.171	158	233	245	294	364	371	263	333	341	142	210	240	110	179	210
2.00	22.855	542	805	866	728	990	1044	693	955	1010	584	885	947	548	849	913
4.00	10.589	1155	1666	1794	1502	2000	2111	1480	1967	2080	1264	1841	1982	1242	1808	1951

[a] CR-CC(2, 3)+Q(x) = CR-CC(2, 3) + [CR-CCSD(TQ), x − CR-CCSD(T)], x = a, b; cf. Eqs. (79) and (80).

Table 8. The differences between the CR-CC(2,3) + Q(b) and MRCI(Q) energies obtained with the aug-cc-pCVQZ ($X = 4$) basis; between the CR-CC(2,3) + Q(b)/aug-cc-pCVQZ and ES energies; and between the MRCI(Q)/aug-cc-pCVXZ ($X = 4, 5$) and ES energies (all energies being calculated relative to their corresponding equilibrium values, as in the earlier tables), for the three cuts of the water potential energy surface: (i) the dissociation of a single O–H bond, (ii) the C_{2v}-symmetric dissociation of both O-H bonds, and (iii) the C_{2v} dissociation into $H_2(X\ ^1\Sigma_g^+)$ and $O(2p^4\ ^1D)$ (see Table 1 for the definitions of R and Y, the corresponding values of the H–O–H angle α, and the equilibrium values of R, Y, and α). All energy differences are in cm^{-1}.

$H_2O(\tilde{X}\ ^1A_1) \rightarrow H(1s\ ^2S) + OH(X\ ^2\Pi)$

R	CR-CC(2,3)+Q(b)/X=4 −MRCI(Q)/X=4	CR-CC(2,3)+Q(b)/X=4 −ES	MRCI(Q)/X=4 −ES	MRCI(Q)/X=5 −ES
$1.25R_e$	−8	−21	−13	29
$1.50R_e$	−12	116	128	197
$1.75R_e$	2	−17	−19	71
$2.00R_e$	86	118	32	138
$2.50R_e$	339	1051	712	833
$3.00R_e$	395	861	466	591
$4.00R_e$	333	277	−56	69
$5.00R_e$	267	134	−133	−8

$H_2O(\tilde{X}\ ^1A_1) \rightarrow 2H(1s\ ^2S) + O(2p^4\ ^3P)$

R	CR-CC(2,3)+Q(b),X=4 −MRCI(Q),X=4	CR-CC(2,3)+Q(b),X=4 −ES	MRCI(Q),X=4 −ES	MRCI(Q),X=5 −ES
$1.25R_e$	−25	−11	14	98
$1.50R_e$	−13	541	554	691
$1.75R_e$	71	666	595	778
$2.00R_e$	115	491	376	596
$2.50R_e$	−391	−448	−57	204
$3.00R_e$	523	145	−378	−107
$4.00R_e$	2777	2338	−439	−171
$5.00R_e$	3053	2629	−424	−157

$H_2O(\tilde{X}\ ^1A_1) \rightarrow H_2(X\ ^1\Sigma_g^+) + O(2p^4\ ^1D)$

Y	CR-CC(2,3)+Q(b),X=4 −MRCI(Q),X=4	CR-CC(2,3)+Q(b),X=4 −ES	MRCI(Q),X=4 −ES	MRCI(Q),X=5 −ES
0.80	21	4	−17	25
0.90	29	8	−21	41
1.00	45	−216	−261	−182
1.10	69	−1233	−1302	−1206
1.20	94	−3084	−3178	−3066
1.30	134	−3497	−3631	−3511
1.50	189	−1286	−1475	−1351
1.75	210	15	−195	−75
2.00	913	608	−305	−192
4.00	1951	1845	−106	−2

MRCI(Q)/$X = 4$ data, with the differences between the CR-CC(2,3)+Q(b)/$X = 4$ and ES energies and the analogous differences between the MRCI(Q)/$X = 4, 5$ and ES energies.

A close inspection of Tables 2–7, particularly Tables 5–7, allows us to appreciate the nature of the challenges the single-reference CC methods are facing when describing global potential energy surfaces along bond breaking coordinates. It also allows us to reemphasize the usefulness of the three dissociation pathways that we chose in this study to test the CC and CR-CC methods, which create different types of bond breaking or bond stretching situations. Indeed, the relatively large differences between the CCSD and MRCI(Q) energies, which exceed ~300–700 cm^{-1} for small stretches of the O-H bond(s) ($R \approx 1.25R_e - 1.5R_e$ in the case of cut (i), $R \approx 1.25R_e$ in the case of cut (ii), and $Y \approx 0.8 - 0.9$ Å in the case of cut (iii)), and huge differences between the CCSD and MRCI(Q) energies, on the order of 4000–7000 cm^{-1}, in the $R \geq 3R_e$ region of cut (i) and $R \approx 2R_e - 3R_e$ region of cut (ii), and more than 1000 cm^{-1} for larger Y values in the case of cut (iii), clearly show that one needs to include higher-than-doubly excited clusters in the CC calculations to obtain reasonable results. Not surprisingly, the CCSD approach is qualitatively correct in the case of cut (i), which corresponds to single-bond breaking (which is, in the zero-order approximation, a two-electron process), producing errors relative to MRCI(Q) which monotonically increase with R, while being completely erratic in the case of the double O–H dissociation defining cut (ii), producing a large hump in the region of intermediate R values. The CCSD approach is also erratic in the case of cut (iii), in which two O–H bonds have to be significantly stretched during the formation of the $H_2(X\ ^1\Sigma_g^+)$ and $O(2p^4\ ^1D)$ products, although the errors relative to MRCI(Q) are not as large in this case as in the other two cuts. One of the reasons is that unlike cuts (i) and (ii), which lead to the fragmentations of the closed-shell water molecule on the singlet ground-state surface into open-shell (doublet or even triplet) products, which introduce very large non-dynamic correlation effects, the minimum energy path that defines cut (iii) leads to the formation of the closed-shell ($H_2(X\ ^1\Sigma_g^+)$) or singlet ($O(2p^4\ ^1D)$) fragments while the O–H bonds are broken. Moreover, the hydrogen product molecule is a two-electron system, which is described exactly by the CCSD approach. These differences between cuts (i) and (ii), on the one hand, and cut (iii), on the other hand, can be seen by examining the largest T_2 cluster amplitudes. The largest spin-free T_2 amplitude, which corresponds to the HOMO $\rightarrow$ LUMO double excitation at $R = 5R_e$ of cut (i) equals, according to the CCSD/$X = 4$ calculations, -0.858278 (this is a single-bond breaking case, so other T_2 amplitudes are much smaller). The two largest T_2 amplitudes, which correspond to the HOMO $\rightarrow$ LUMO and (HOMO $-$ 1) $\rightarrow$ (LUMO $+$ 1) double excitations at $R = 5R_e$ of cut (ii) equal, according to the CCSD/$X = 4$ calculations, -0.774880 and -0.774382, respectively (clearly, there are a few other large T_2 amplitudes in this case, which engage the highest two occupied and lowest two unoccupied orbitals, since this is a double dissociation of water into $2H(1s\ ^2S) + O(2p^4\ ^3P)$ that involves, in the zero-order description, four active orbitals and four electrons). For comparison, the largest T_2

amplitudes along the minimum energy path defining cut (iii) never exceed ≈0.2 and are usually considerably smaller. This explains the observed differences in the performance of the CCSD and various CCSD-based non-iterative CC methods in the calculations for cut (iii), where the overall behavior of the single-reference CC methods, although not perfect, remains quite reasonable, and the other two cuts examined in this work, where the standard CC approximations, including CCSD, CCSD(T), and CCSD(TQ),b, display catastrophic failures.

The above discussion confirms the known fact that one has to go beyond the basic CCSD approximation and account for higher-than-doubly excited clusters to obtain a quantitatively accurate description of the potential energy surface, even in the vicinity of the equilibrium geometry. This is often done with the CCSD(T) approach, which describes the leading effects due to triply excited clusters via non-iterative corrections to the CCSD energy. As shown in our tables, particularly Tables 5–7, the CCSD(T) approach provides excellent results that almost perfectly agree with the results of MRCI(Q) calculations, when stretches of the O–H bonds are small. For example, in the case of cut (i), the differences between the CCSD(T)/$X = 4$ and MRCI(Q)/$X = 4$ energies do not exceed 6 cm^{-1} up to $R = 2R_e$ and for cut (ii) they remain smaller than 38 cm^{-1} up to $R = 1.5R_e$. This is impressive, if we realize that the $R \approx 2R_e$ and $R \approx 1.5R_e$ regions of cuts (i) and (ii), respectively, are characterized by the energies of ≈36000 cm^{-1}. A similarly impressive performance of CCSD(T) is observed for other aug-cc-pCVXZ basis sets, although we should note a rather substantial error increase, relative to MRCI(Q), when the $X = 2$ basis sets is employed (particularly for cuts (i) and (ii); cf. Tables 5 and 6). Interestingly enough, in the case of cut (ii), the CCSD(T) results remain reasonable up to ~67000 cm^{-1} or $R \approx 2R_e$ (unsigned errors relative to MRCI(Q) on the order of 200–300 cm^{-1}). One has to be very careful, however, in interpreting these high accuracies obtained with CCSD(T) for small and, in the case of cut (ii), intermediate stretches of the O–H bonds, particularly that CCSD(T) eventually suffers significant breakdowns and it may not always be easy to predict when one should stop trusting the CCSD(T) approach. Indeed, if we correct the CCSD(T) results for the dominant effects due to T_4 clusters, as is done by the CCSD(TQ),b approach, which can only improve the quality of CC calculations in non-degenerate regions of the potential energy surface, the agreement between the CC and MRCI(Q) results in the $R \leq 2R_e$ regions of cuts (i) and (ii) is no longer as impressive as in the CCSD(T) case. For example, the −2 and 5 cm^{-1} differences between the CCSD(T)/$X = 4$ and MRCI(Q)/$X = 4$ energies obtained for cut (i) at $R = 1.5R_e$ and $2R_e$ increase to −41 and −124 cm^{-1}, respectively, when the CCSD(TQ),b method is employed. The 1 and 38 cm^{-1} differences between the CCSD(T)/$X = 4$ and MRCI(Q)/$X = 4$ energies obtained for cut (ii) at $R = 1.25R_e$ and $1.5R_e$ increase, in absolute value, to 29 and 65 cm^{-1}, when instead of CCSD(T) we use the CCSD(TQ),b method. This means that either the CCSD(T) results are very accurate due to fortuitous cancellation of errors or the MRCI(Q) results that we use as a benchmark are not as accurate as the CCSD(TQ),b results in the region of smaller stretches of the O–H bonds, where CCSD(TQ),b can be safely applied, creating a false impression about superb accuracy of the CCSD(T) approximation. Clearly, both interpretations are possible (the CCSD(T) approach is known to provide the results which are often better than those obtained with the full CCSDT approach, which makes no physical sense whatsoever), but we believe that the CCSD(TQ),b approach is more accurate than MRCI(Q) in the spectroscopic region and moderate stretches of the O–H bond(s), although both methods provide high quality results. The overall superiority of CCSD(TQ),b over MRCI(Q) for moderate stretches of the O–H bond(s) can be seen by comparing the differences between the CCSD(TQ),b/$X = 4$ and ES energies in the $R \leq 2R_e$ regions of cuts (i) and (ii) with the corresponding differences between the MRCI(Q)/$X = 4$ and ES energies. This illustrates the well-known advantage of using the CC methods over MRCI techniques, which are not as effective in accounting for the dynamical correlation effects that dominate electron correlations near the equilibrium geometry as the high-level CC approaches. Interestingly enough, further increase of the basis set makes the agreement between the MRCI(Q) and ES surfaces in the region of smaller stretches of the O–H bond(s), where the ES potential function is nearly spectroscopic, even worse (cf. the MRCI(Q)/$X = 4$ and MRCI(Q)/$X = 5$ results in Table 8). The superiority of the CCSD(TQ),b and related CCSD(TQ$_f$) methods, which account for triply and quadruply excited clusters, over the CCSD(T) approach, which ignores the latter clusters, in applications involving potential energy surfaces near the equilibrium geometry is well-documented as well (cf., e.g. Refs. [12–14, 23, 24, 34, 36, 58, 179–181]) and our calculations confirm this superiority, in spite of the tiny differences between the CCSD(T) and MRCI(Q) energies for cuts (i) and (ii) discussed above.

Before discussing the failures of the CCSD(T) and CCSD(TQ),b methods at larger O–H separations of cuts (i) and (ii) and improvements offered by the CR-CC methods, which are quite apparent when we examine the results shown in Tables 2–7, let us point out that in the case of cut (iii), the behavior of the CCSD(T) and CCSD(TQ),b methods vs. the MRCI(Q) approach is somewhat different, when compared to the other two cuts explored in this work. As already pointed out above, the T_2 (also, T_1) cluster amplitudes along the minimum energy path defining cut (iii) never become large and the corresponding $H_2(X\ ^1\Sigma_g^+)$ and $O(2p^4\ ^1D)$ dissociation products are of the closed-shell ($H_2(X\ ^1\Sigma_g^+)$) or open-shell singlet ($O(2p^4\ ^1D)$) type. Thus, it is not completely surprising to observe the relatively good performance of the standard single-reference CCSD(T) and CCSD(TQ),b approaches in the entire region of Y values shown in our tables (with the exception, perhaps, of the last two points, $Y = 2.0$ and 4.0 Å, although we must remember that these points are located at more than 58000 cm^{-1} above the global minimum, which makes the $\sim$500–1900 cm^{-1} errors relatively small, particularly for the relatively inexpensive single-reference calculations using the RHF reference; cf. the discussion below for the additional remarks). In particular, the differences between the CCSD(T) and MRCI(Q) energies do not exceed 250 cm^{-1} in the entire $Y_e \leq Y \leq 1.75$ Å region, in which energies become as large as $\sim$57000 cm^{-1} (recall that $Y_e = 0.58641$ Å), and are smaller than 100 cm^{-1} when Y does not exceed 1.1 Å (energies below $\sim$27000 cm^{-1}). The CCSD(TQ),b approach reduces these differences even further, showing a nice and systematic behavior of the single-reference CC theory in this case. Clearly, it is interesting to examine if the CR-CC approaches, which are primarily designed to improve the CC results when a system is fragmented into open-shell fragments, can maintain the high accuracies of the CCSD(T) and CCSD(TQ),b calculations and systematic improvements in the results when going from the triples to the quadruples levels of CC theory in the case of the minimum energy path that leads to the closed-shell and singlet, non-closed-shell products.

One issue that we do not address in this study and that may affect the results of all CC and CR-CC calculations in the $Y \geq 2.0$ Å region of cut (iii) (and, perhaps, the best CR-CC results in the $R > 3R_e$ region of cut (ii)) is the issue of the existence of the avoided crossings of ground and excited states of water at larger internuclear separations (see, e.g. Refs. [143, 144, 148, 149] and references therein). For example, for larger values of Y of cut (ii), one can speculate that other dissociation channels may compete with the ground-state $H_2(X\ ^1\Sigma_g^+) + O(2p^4\ ^1D)$ channel, such as $H_2(b\ ^3\Sigma_u^+) + O(2p^4\ ^3P)$, when the H–H bond is somewhat stretched. Thus, it is possible that the CCSD solutions that we found, for example, at $Y = 4.0$ Å, are not necessarily the solutions that correlate with the lowest-energy state of water of the $^1A_1$ symmetry. If this speculation turned out to be true, this would explain a steep increase in the errors of the CCSD(T), CCSD(TQ),b, and all, otherwise very accurate, CR-CC calculations in the $Y \geq 2.0$ Å region of cut (iii), from $\sim$200–400 cm^{-1} at $Y = 1.75$ Å to $\sim$1800–2100 cm^{-1} at $Y = 4.0$ Å. We tried to find other CCSD solutions in the $Y \geq 2.0$ Å region, but we have not been successful. We know, however, that when we perform the calculations of ground and excited states at $Y = 4.0$ Å, using the MMCC-based CR-EOMCCSD(T) approach, in which renormalized triples corrections are added to the energies obtained in the equation-of-motion CCSD calculations [122, 123], we see the appearance of one $^1A_1$ state with a small negative excitation energy, suggesting that the CC and CR-CC energies at $Y = 4.0$ Å reported in Tables 2–4 may not necessarily correlate with the ground-state of water in this region (we have also observed the appearance of the state of $^1B_1$ symmetry at $Y = 4.0$ Å, located in the CR-EOMCCSD(T) calculations $\sim$2500–3000 cm^{-1} below the $^1A_1$ state, but this is a state of different symmetry than the ground state, which has no impact on the ground-state CC/CR-CC calculations). We will examine the issue of avoided crossings and the potential of switching between ground and excited states in solving CC equations at larger internuclear distances in the future, using, for example, the excited-states CR-CC methods, such as CR-EOMCCSD(T), and the excited-state variant of CR-CC(2,3) tested in Refs. [15, 48]. The recent study of the ground and excited states of the ammonia molecule indicates that it is not unusual to observe the switching between the ground and low-lying excited states of the same symmetry in CC calculations in the vicinity of avoided crossings [182].

Much of the above discussion points to the importance of properly balancing various correlation effects and the need to account for the triple as well as quadruple excitations in the high quality calculations of molecular potential energy surfaces. The properly constructed theory should provide an accurate description of triply, quadruply, and, if need be, other higher-order clusters, without the fortuitous cancellation of errors that the CCSD(T) approach often displays, as demonstrated above. Clearly, the good theory should also eliminate the significant failures of the CCSD(T), CCSD(TQ),b, and other similar methods at larger internuclear separations or at least be more robust in this regard. In the case of the water potential examined in this work, these failures are dramatic. As shown in Tables 2–7 (particularly, in Tables 5–7), the unsigned errors in the CCSD(T)/$X = 2 - 4$ energies, relative to the corresponding MRCI(Q)/$X = 2 - 4$ data, range between 619 and 5779 cm^{-1}, when the $R = 2.5R_e - 5R_e$ region of cut (i) is examined, and 6258 and 39412 cm^{-1}, when the $R = 2.5R_e - 5R_e$ region of cut (ii) is explored, and the CCSD(T) energies go considerably below the MRCI(Q) energies. The CCSD(TQ),b approach does not improve the situation at all, producing the 811–5679 cm^{-1} unsigned errors, relative to MRCI(Q), in the $R = 2.5R_e - 5R_e$ region of cut (i) and the 635–62544 cm^{-1} unsigned errors, relative to the corresponding MRCI(Q) data, in the $R = 2.5R_e - 5R_e$ region of cut (ii). Unlike CCSD(T), the CCSD(TQ),b energies are significantly above the corresponding MRCI(Q) energies at larger O–H separations of cut (ii), while being below the MRCI(Q) energies for cut (i). All of this clearly demonstrates the divergent behavior of the standard single-reference CC methods, caused by the large non-dynamic correlation effects (which manifest themselves through large T_2 cluster amplitudes, as described above), the poor description of the wave function by the CCSD approach, on which the CCSD(T) and CCSD(TQ),b methods rely, and the strongly divergent nature of the MBPT series, on which the standard (T) and (TQ) energy corrections are based, in the regions of larger internuclear separations.

As shown in Tables 5–7, the CR-CCSD(T) method, based on the original formulation of the MMCC theory, provides considerable improvements in the CCSD(T) results for the single-bond breaking defining cut (i), reducing, for example, the 4385–4785 cm^{-1} and 5211–5779 cm^{-1} errors in the CCSD(T)/$X = 2 - 4$ energies at $R = 4R_e$ and $5R_e$, relative to the corresponding MRCI(Q)/$X = 2 - 4$ data, to 1250–1818 cm^{-1} and 1137–1699 cm^{-1}, respectively. The CR-CCSD(TQ),a and CR-CCSD(TQ),b methods provide further improvements and a very nice and smooth description of the entire cut (i), with an exception, perhaps, of the last point at $R = 5R_e$, where a small error reduction compared to $R = 4R_e$ may be a signature of the eventual problems somewhere in the $R \gg 5R_e$ region, although we are not sure about it. For example, the differences between the CR-CCSD(TQ),b/$X = 4$ and MRCI(Q)/$X = 4$ energies smoothly increase with R, from 18 and 58 cm^{-1} at $R = 1.25R_e$ and $1.5R_e$, respectively (energies on the order of 7400 and 18900 cm^{-1}), to 432 cm^{-1} at $R = 2R_e$, and 1431 cm^{-1} at $R = 4R_e$, where the energy is almost 44000 cm^{-1}. The situation for the more challenging cut (ii), where both O–H bonds are broken, is, at least to some extent, similar to that observed in the case of cut (i), with the CR-CCSD(TQ),a and CR-CCSD(TQ),b methods eliminating the pathological behavior of CCSD(T) and CCSD(TQ),b at larger stretches of both O–H bonds. Again, we observe a smooth increase of the differences between the CR-CCSD(TQ),a or CR-CCSD(TQ),b and MRCI(Q) energies with R, from 44 (50) and 191 (202) cm^{-1} at $R = 1.25R_e$ and $1.5R_e$, where energies are on the order of

$\sim$14000 and 36000 cm^{-1}, respectively, through 1329 (1323) cm^{-1} at $R = 2R_e$, where the energy exceeds 67000 cm^{-1}, and 9978 (10566) cm^{-1} at $R = 5R_e$, where the energies exceed 81000 cm^{-1}, when the CR-CCSD(TQ),a/$X = 4$ (CR-CCSD(TQ),b/$X = 4$) results are examined. We do not want to claim that these are superb results, but it is quite encouraging to see that the CR-CC methodology is capable of providing significant improvements over the standard CC results, even when the double dissociation of water is examined. There is, of course, a difference between the behavior of the CR-CCSD(T) method in the case of the double O–H dissociation defining cut (ii) and the single-bond breaking defining cut (i). The CR-CCSD(T) approach provides a reasonably smooth description of cut (i), while failing in the case of cut (ii). This is a consequence of ignoring the quadruply excited clusters in the CR-CCSD(T) calculations, which are absolutely critical in cases of double bond breaking (while improving accuracies in the calculations for single-bond breaking). It is interesting to observe, though, the substantial improvements in the poor CCSD(T) results in the $R = 2.5R_e - 5R_e$ region of cut (ii) by the CR-CCSD(T) approach (error reduction in the $X = 4$ calculations at $R = 5R_e$, relative to the corresponding MRCI(Q)/$X = 4$ result, from more than 39000 cm^{-1} in the CCSD(T) case to $\approx$3000 cm^{-1} in the CR-CCSD(T) case). It is also interesting to observe that the CR-CCSD(T), CR-CCSD(TQ),a, and CR-CCSD(TQ),b approaches provide a nice and smooth description of the "easier" cut (iii) as well. In this case, as pointed out above, the conventional CCSD(T) and CCSD(TQ),b approximations work quite well, but it is good to see that the CR-CCSD(T) and CR-CCSD(TQ),x (x = a, b) methods are, more or less, equally effective, with the CR-CCSD(TQ),x approaches providing systematic improvements over the relatively good CR-CCSD(T) results. The $\sim$90–530 cm^{-1} differences between the CR-CCSD(T) and MRCI(Q) energies and the slightly smaller $\sim$50–440 cm^{-1} differences between the CR-CCSD(TQ),x and MRCI(Q) energies in the entire $Y = 0.8 - 1.75$ Å region, where energies grow from about 5000 to more than 57000 cm^{-1} above the global minimum, is clearly an encouraging result, confirming the usefulness of the single-reference CR-CC methods.

The CR-CCSD(T) and CR-CCSD(TQ),x (x = a, b) methods provide substantial improvements in the regions of larger internuclear separations, where the standard CCSD(T) and CCSD(TQ),b approaches fail, but it would be useful to achieve further error reduction in the CR-CC calculations, particularly in the cases of cuts (i) and (ii), which are more challenging for the single-reference CC methods than cut (iii). It would also be useful to improve the results of CR-CC calculations in the vicinity of the equilibrium region, where the CCSD(T) and CCSD(TQ),b methods are somewhat more accurate than the CR-CCSD(T) and CR-CCSD(TQ),x (x = a, b) approaches. As mentioned earlier, the recently proposed CR-CC(2,3) theory not only eliminates the small size extensivity errors from the CR-CCSD(T) results (which in the case of water are negligible, since water molecule is a small, 10-electron system), but it also improves the accuracy of CR-CC calculations by adding various product terms of the $T_1 M_3(2)$, $(T_2+\frac{1}{2}T_1^2)M_3(2)$, etc. (in general, $C_{n-3}(m_A)M_3(2)$) type to the bare $M_3(2)$ terms that are already present in the CR-CCSD(T) triples correction formula (cf., e.g. the discussion at the end of section 2.2). As shown in the earlier studies [45, 46, 48], reporting the discovery and initial tests of the CR-CC(2,3) approach, the CR-CC(2,3) method is essentially as accurate as the full CCSDT approach in the equilibrium and bond breaking region, providing, therefore, the best description of the triples effects that any non-iterative triples CC method can offer. It is, thus, interesting to examine the performance of the CR-CC(2,3) approach using the three cuts of the water potential examined in this work. Since we have already noticed the importance of quadruply excited clusters in improving the results, particularly for the double O–H dissociation defining cut (ii), it is also useful to investigate if correcting the CR-CC(2,3) results for quadruples through the CR-CC(2,3)+Q(a) and CR-CC(2,3)+Q(b) methods defined by Eqs. (79) and (80) gives the desired high accuracies for a wide range of nuclear geometries of the water system explored in this work.

As explained in the previous section, one can consider a few CR-CC(2,3) approximations which differ by the form of the denominator D^{ijk}_{abc}(CCSD), Eq. (69), which enters the relevant desexcitation amplitudes $\tilde{\ell}^{abc}_{ijk}$(CCSD), Eq. (68). Here, we consider only two extreme variants, namely, the most complete variant D, for which we continue to use an acronym CR-CC(2,3) without additional letters, and the simplest variant A, called CR-CC(2,3),A, which is obtained by replacing the denominator D^{ijk}_{abc}(CCSD) by the usual MBPT denominator defining triple excitations, i.e. $(\epsilon_i+\epsilon_j+\epsilon_k-\epsilon_a-\epsilon_b-\epsilon_c)$. As mentioned in section 2.3, the CR-CC(2,3),A approach is equivalent to the CCSD(2)$_T$ method of Ref. [42].

As shown in Tables 2–7, the CR-CC(2,3),A and full CR-CC(2,3) (= CR-CC(2, 3), D) approaches provide improvements in the CR-CCSD(T) results for the single-bond breaking defining cut (i) and the C_{2v} dissociation pathway into $H_2(X\ ^1\Sigma_g^+)$ and $O(2p^4\ ^1D)$ in the entire regions of the corresponding R and Y values. They also improve the description of the double dissociation of water defining cut (ii) by the CR-CCSD(T) method in the $R = R_e - 2.5R_e$ region. For example, in the case of cut (i), the CR-CC(2,3),A approach reduces the 65, 167, 698, 1811, and 1699 cm^{-1} errors in the CR-CCSD(T)/X = 4 results, relative to the corresponding MRCI(Q)/X = 4 data, at $R = 1.25R_e$, $1.5R_e$, $2R_e$, $3R_e$, and $5R_e$ to 35, 97, 483, 1213, and 954 cm^{-1}, respectively. With an exception of $R = 1.25R_e$ and $1.5R_e$, where errors remain almost unchanged, the full CR-CC(2,3) method improves the agreement with the MRCI(Q)/X = 4 energies even further, reducing the 483, 1213, and 954 cm^{-1} errors in the CR-CC(2,3),A energies, relative to MRCI(Q), at $R = 2R_e$, $3R_e$, and $5R_e$ to 351, 813, and 625 cm^{-1}, respectively. We can clearly see the benefits of using the full CR-CC(2,3) approach, where one does not make any additional simplifications in the formula for the denominator D^{ijk}_{abc}(CCSD), Eq. (69), entering the CR-CC(2,3) triples correction. Similar benefits of using the complete expression for D^{ijk}_{abc}(CCSD) are observed, when we compare the results of the CR-CC(2,3),A and full CR-CC(2,3) calculations in the $R = R_e - 2R_e$ region of cut (ii) and for the the entire cut (iii). The 86–1059 cm^{-1} errors, relative to MRCI(Q), in the CR-CC(2,3),A/X = 4 results obtained in the $R = R_e - 2R_e$ region of cut (ii), which are obviously smaller than the 154–1917 cm^{-1} errors obtained with CR-CCSD(T), reduce to 79–709 cm^{-1}, when the full CR-CC(2,3)/X = 4 method is employed. The 68–398 cm^{-1} differences between the CR-CC(2,3),A/X = 4 and MRCI(Q)/X = 4 energies in the Y = 0.8 – 1.75 Å region of cut (iii) reduce to 63–321 cm^{-1}, when the full CR-CC(2,3)/X = 4 method is used. The use of other aug-cc-pCVXZ basis sets does not change any of these accuracy patterns in a substantial manner.

We can conclude that the CR-CC(2,3) approach, with the complete treatment of the D_{abc}^{ijk}(CCSD) denominator, as defined by Eq. (69), provides the overall best results when compared to other non-iterative triples methods examined in this work. It is true that the CR-CC(2,3) approach fails in the $R > 2R_e$ region of cut (ii) and it is also true that the CR-CC(2,3) energies for smaller stretches of the O–H bond(s) appear to be less accurate than the corresponding CCSD(T) energies. We must remember, however, that one needs quadruply excited clusters in the $R > 2R_e$ region of cut (ii), while the perfect agreement between the CCSD(T) and MRCI(Q) data at smaller stretches of the O–H bond(s) is not necessarily a desired behavior. Indeed, the explicit inclusion of quadruples in CC calculations, which can be done in the region of smaller stretches of the O–H bond(s) via the CCSD(TQ),b approach, makes the agreement between the CC and MRCI(Q) data less perfect, as already explained above. These observations agree with the performance of the CR-CC(2,3) approach in a variety of benchmark calculations reported in Refs. [45, 46, 48], where it has been noted that the CR-CC(2,3) energies are very close to the full CCSDT energies, not only when CCSDT works, but also when it fails, as is the case of multiple bond dissociations. In particular, as shown in Refs. [45, 48], the full CCSDT approach completely fails in the $R > 2R_e$ region of the double dissociation of water, analogous to our cut (ii). Thus, the failure of CR-CC(2,3) in the same region is, in a way, a desired result, since one cannot and should not break both O–H bonds in the water molecule without quadruply excited clusters, and the approximate triples methods, such as CR-CC(2,3), should not be better than the full CCSDT approach.

In view of the above discussion, it is very important to examine what happens with the CR-CC(2,3) energies if we augment them by quadruples, as is done in the CR-CC(2,3)+Q(a) and CR-CC(2,3)+Q(b) calculations. As shown in Tables 2–8, the overall agreement between the CR-CC(2,3)+Q(x) ($x = a, b$) and MRCI(Q) data is quite remarkable. For example, in the case of cut (i), corresponding to the dissociation of a single O–H bond, the 39–351 cm^{-1} unsigned differences between the CR-CC(2,3)/X = 4 and MRCI(Q)/X = 4 energies in the $R = R_e - 2R_e$ region, in which these energies increase to more than 35000 cm^{-1}, reduce to 2–86 cm^{-1} when the CR-CC(2,3)+Q(b)/X = 4 approach is employed. The CR-CC(2,3)+Q(a)/X = 4 method provides similar results, although the somewhat more complete CR-CC(2,3)+Q(b) approximation seems better. The maximum error, relative to MRCI(Q), characterizing the CR-CC(2,3)/X = 4 calculations in the entire range of R values, of 813 cm^{-1}, reduces in the CR-CC(2,3)+Q(b)/X = 4 calculations to 395 cm^{-1}. Again, the CR-CC(2,3)+Q(a)/X = 4 approach is almost as effective. The use of other aug-cc-pCVXZ basis sets leads to similar error reductions.

The 79–709 cm^{-1} differences between the CR-CC(2,3)/X = 4 and MRCI(Q)/X=4 energies in the $R = R_e - 2R_e$ region of cut (ii), where both O–H bonds are simultaneously stretched and where the energy goes up to more than 67000 cm^{-1}, reduce (in absolute value) to 24–121 cm^{-1}, when the CR-CC(2,3)+Q(a)/$X = 4$ method is employed, and 13–115 cm^{-1}, when the CR-CC(2,3)+Q(b)/$X = 4$ approach is used. The CR-CC(2,3)+Q(b) approach remains quite accurate up to $R = 3R_e$, where the energy is already larger than 80000 cm^{-1}. The difference between the CR-CC(2,3)+Q(b)/$X = 4$ and MRCI(Q)/$X = 4$ energies is only slightly larger than 500 cm^{-1} in the $R = 3R_e$ region. The CR-CC(2,3)+Q(a) approximation shows the signs of unstable behavior in the $R > 2R_e$ region, with the signed errors relative to MRCI(Q) changing from more than +2000 cm^{-1} to ∼(−3000)–(−2000) cm^{-1}, which is partly due to the fact that we use the first-order MBPT estimates of T_2 cluster amplitudes in defining the (Q) corrections of CR-CC(2,3)+Q(a) (or CR-CCSD(TQ),a) and partly due to the heuristic nature of the CR-CC(2,3)+Q approaches, which should eventually be replaced by the more consistent CR-CC(2,4) theory. The somewhat *ad hoc* nature of the CR-CC(2,3)+Q approximations may also be responsible, at least in part, for the ∼3000 cm^{-1} differences between the CR-CC(2,3)+Q(b) and MRCI(Q) energies in the $R = 4R_e - 5R_e$ region of cut (ii). Again, the accuracy patterns observed in the CR-CC(2,3)+Q(a) and CR-CC(2,3)+Q(b) calculations for cut (ii), when compared to MRCI(Q), are essentially independent of the aug-cc-pCVXZ basis set employed in these calculations.

The CR-CC(2,3)+Q(a) and CR-CC(2,3)+Q(b) results for cut (iii), corresponding to the C_{2v} dissociation pathway into $H_2(X\ ^1\Sigma_g^+)$ and $O(2p^4\ ^1D)$, are very good as well. Both CR-CC(2,3)+Q(a) and CR-CC(2,3)+Q(b) methods reduce the errors resulting from the CR-CC(2,3) calculations and, as a matter of fact, the overall accuracy of the CR-CC(2,3)+Q(b) approach in the $Y = 0.8 - 1.75$ Å region of cut (iii), as judged by the differences with MRCI(Q), is better than the accuracy of the CCSD(TQ),b method, which performs very well in this case. It is quite encouraging to see the relatively small and monotonically increasing 21–210 cm^{-1} differences between the CR-CC(2,3)+Q(b)/$X = 4$ and MRCI(Q)/$X = 4$ energies in the $Y = 0.8 - 1.75$ Å region of cut (iii), where energies go up to ∼57000 cm^{-1} above the minimum.

The overall agreement between the CR-CC(2,3)+Q(b) and MRCI(Q) results for all three potential surface cuts examined in this study is excellent, particularly if we keep in mind the black-box, single-reference nature of the CR-CC(2,3)+Q(b) calculations. The fact that with an exception of the $R = 4R_e - 5R_e$ region of cut (ii) and the $Y = 2.0 - 4.0$ Å region of cut (iii), the CR-CC(2,3)+Q(b) and MRCI(Q) energies agree to within 500 cm^{-1} and, in most cases, to within 100 cm^{-1} or less is a clear demonstration of the large potential offered by the CR-CC theories. The CR-CC(2,3) and CR-CC(2,3)+Q(b) methods seem to be at least as effective as the CCSD(T) and CCSD(TQ),b methods in the vicinity of the equilibrium geometry, where the many-electron correlation effects are primarily of dynamical nature, while providing the accuracy comparable to the MRCI(Q) approach in the higher-energy potential energy surface regions characterized by large non-dynamical correlation effects. In the case of the water molecule, we seem to be able to obtain the relatively small differences between the results of the CR-CC(2,3)+Q(b) and MRCI(Q) calculations, on the order of 100–500 cm^{-1} or less, for the energies as large as 60000–70000 cm^{-1} above the global minimum. This is a remarkable finding, considering the single-reference character and the relatively low cost of all CR-CC calculations. This is also very promising from the point of view of applying the CR-CC methods, particularly the most recent approaches based on the biorthogonal MMCC theory, in calculations aiming at the construction of accurate global potential functions for dynamical studies. The proximity of the CR-CC(2,3)+Q(b) and MRCI(Q) results for the large portion of the global potential energy surface of water and the fact that the CR-CC(2,3)+Q(b) approach remains as accurate as the CCSD(TQ),b method in the vicinity of the equilibrium geometry open up new avenues for constructing highly accurate global potentials, since switching between the CR-CC(2,3)+Q(b) and MRCI(Q) energies which are so similar, should be quite straightforward. Clearly, in the future, one should replace the heuristic CR-CC(2,3)+Q approximations tested in this work by the genuine and properly derived CR-CC(2,4) theory, exploiting the biorthogonal MMCC formalism of Refs. [45,46], as discussed in the earlier sections. The excellent CR-CC(2,3)+Q results for water obtained in this work prompt such development.

Last, but not least, let us address the aforementioned issue of the rather substantial differences between the MRCI(Q)/$X = 4$ or MRCI(Q)/$X = 5$ and ES potential energy surfaces in the regions of the intermediate R and Y values, and intermediate or higher, but not the highest energies, which are as much as 700–800 cm^{-1} for cut (i) in the $R \approx 2.5R_e$ region, 600–800 cm^{-1} for cut (ii) in the $R \approx 1.75R_e$ region, and 3500–3600 cm^{-1} for cut (iii) in the $Y \approx 1.3$ Å region. Such differences cannot be explained by the neglect of the relativistic, non-adiabatic, and quantum electrodynamical effects in MRCI(Q) calculations. They indicate that either the MRCI(Q) approach employing large basis sets is insufficiently accurate or the ES potential needs further refinement in the above regions. Although, as explained earlier, both interpretations are possible, we tend to believe that the accuracy of the ES potential function in the regions of intermediate internuclear separations and energies, which connect the spectroscopic and asymptotic regions, is not as high as the accuracy of the ES surface around the minimum, up to about 19000 cm^{-1}, and in the $\mathrm{H}(1s\ ^2S) + \mathrm{OH}(X\ ^2\Pi)$, $2\mathrm{H}(1s\ ^2S) + \mathrm{O}(2p^4\ ^3P)$, and $\mathrm{H}_2(X\ ^1\Sigma_g^+) + \mathrm{O}(2p^4\ ^1D)$ asymptotic regions. We base our belief on the close proximity of the MRCI(Q) and CR-CC(2,3)+Q(b) energies in the regions of the intermediate R and Y values, where the MRCI(Q)/$X = 4$ or MRCI(Q)/$X = 5$ and ES potentials significantly differ. This is shown in Table 8, where we compare the differences between the best CR-CC(2,3)+Q(b)/$X = 4$ energies and the corresponding MRCI(Q)/$X = 4$ data with the differences between the CR-CC(2,3)+Q(b)/$X = 4$ and ES energies, and the analogous differences between the MRCI(Q)/$X = 4, 5$ and ES energies. As one can see, the 700–800 cm^{-1} differences between the MRCI(Q)/$X = 4, 5$ and ES energies in the $R \approx 2.5R_e$ region of cut (i) are very similar to the $\sim$1000 cm^{-1} difference between the CR-CC(2,3)+Q(b)/$X = 4$ and ES energies in the same region. The 600–800 cm^{-1} differences between the MRCI(Q)/$X = 4, 5$ and ES energies in the $R \approx 1.75R_e$ region of cut (ii) are not much different than the $\sim$700 cm^{-1} difference between the CR-CC(2,3)+Q(b)/$X = 4$ and ES energies in this region.

Finally, the 3500–3600 cm^{-1} differences between the MRCI(Q)/$X = 4, 5$ and ES energies in the $Y \approx 1.3$ Å region of cut (iii) are very similar to the 3000–3500 cm^{-1} differences between the CR-CC(2,3)+Q(b)/$X = 4$ and ES for $Y \approx 1.2 - 1.3$ Å. The large consistency between the MRCI(Q)/$X = 4, 5$ and CR-CC(2,3)+Q(b)/$X = 4$ results in the above regions of the potential energy surface of water makes us believe that the energy values provided by both *ab initio* approaches in these regions are more accurate than those provided by the existing ES potential. This gives us an opportunity to refine the ES global potential in the future by incorporating some MRCI(Q) or CR-CC(2,3)+Q(b) data from the regions of the intermediate R and Y values in the appropriate fitting and ES procedures.

4. SUMMARY AND CONCLUDING REMARKS

In this chapter, we have explored an important issue of the development of black-box single-reference procedures that could be applied to at least some of the most frequent multi-reference situations, such as single and double bond dissociations, by reviewing the recently proposed renormalized CC methods and by reporting test calculations for the potential energy surface of the water molecule. We have focused on a few basic renormalized CC methods, including the older CR-CCSD(T) and CR-CCSD(TQ) approximations [11–14, 24, 33, 34] and the most recent size extensive CR-CC(2,3) and other CR-CC(m_A, m_B) approaches [45, 46]. In the CR-CCSD(T) and CR-CC(2,3) methods, the relatively inexpensive corrections due to triply excited clusters, similar in the computer cost to the triples corrections of the conventional CCSD(T) theory [19], are added to the CCSD energy. In the CR-CCSD(TQ) and CR-CC(2,4) approaches, the CCSD energy is corrected for the dominant effects of triply and quadruply excited clusters in a manner reminiscent of the conventional CCSD(TQ) approximations, such as CCSD(TQ),b [24, 36] or CCSD(TQ$_\mathrm{f}$) [23]. Since the CR-CC(2,4) approach has not been fully developed and implemented yet (we have discussed open issues that one needs to address before proposing the optimum CR-CC(2,4) model), we have considered an approximate form of the CR-CC(2,4) theory, abbreviated as CR-CC(2,3)+Q, in which the CR-CC(2,3) energies are *a posteriori* corrected for the effect of quadruply excited clusters by using the information about quadruples extracted from the CR-CCSD(TQ) calculations.

In addition to discussing specific renormalized CC methods, we have reviewed the MMCC formalism, which is the key concept behind all renormalized CC approaches. In this discussion, we have included the most recent biorthogonal formulation of the MMCC theory employing the left eigenstates of the similarity-transformed Hamiltonian which leads to the CR-CC(2,3) and other CR-CC(m_A, m_B) approaches. We have discussed the similarities and differences between the original MMCC theory of Piecuch and Kowalski, introduced in Refs. [11,24,34], and the biorthogonal MMCC formalism of Piecuch and Włoch, introduced in Ref. [45] and further elaborated on in Ref. [46]. In particular, we have pointed out how the biorthogonal formulation of the MMCC theory enables one to eliminate the overlap denominators, which are present in the original MMCC formalism and which cause small departures from the rigorous size extensivity in the CR-CCSD(T) and CR-CCSD(TQ) calculations, but which are also important to properly renormalize the CCSD(T) and CCSD(TQ) approximations, and how to achieve the size extensive renormalization of the energy corrections due to triples or triples and quadruples through the use of the suitable ansatz for the bra wave function entering the MMCC correction formula.

In order to test the performance and potential benefits of using the single-reference renormalized CC methods, we have compared the results of the CR-CCSD(T), CR-CCSD(TQ), CR-CC(2,3), and CR-CC(2,3)+Q calculations for the three important cuts of the potential energy surface of the water molecule, including the dissociation of one O–H bond, which correlates with the $H(1s\ ^2S) + OH(X\ ^2\Pi)$ asymptote, the simultaneous dissociation of both O–H bonds, which leads to the $2H(1s\ ^2S) + O(2p^4\ ^3P)$ products, and the C_{2v}-symmetric dissociation pathway into $H_2(X\ ^1\Sigma_g^+)$ and $O(2p^4\ ^1D)$, with those obtained in the highly accurate MRCI(Q) calculations and those provided by the spectroscopically accurate ES potential function [132]. We have demonstrated that all renormalized CC methods eliminate or considerably reduce the failures of the conventional CCSD(T) and CCSD(TQ),b approaches in the bond breaking regions of the water potential, while retaining high accuracies of the CCSD(T) and CCSD(TQ),b methods in the vicinity of the equilibrium geometry. The CR-CC(2,3) and CR-CC(2,3)+Q methods are particularly effective in this regard. Unlike CCSD(T), the CR-CC(2,3) approach provides a faithful description of triply excited clusters [45, 46, 48], even in the equilibrium region, where the CCSD(T) approach works well. After correcting the CR-CC(2,3) results for quadruples, as is done in the CR-CC(2,3)+Q(x) (x = a, b) schemes, we obtain the potential energy surfaces of excellent quality. Indeed, as shown in this work, the single-reference, RHF-based, CR-CC(2,3)+Q(b) approach describes the above three cuts of the global potential energy surface of water with accuracies that can only be matched by the high accuracy, CASSCF-based, MRCI(Q) calculations. We find the small differences between the CR-CC(2,3)+Q(b) and MRCI(Q) energies, on the order of 100–500 cm^{-1} or less, for energies as large 60000–70000 cm^{-1}, where the highest possible energies corresponding to the complete atomization of water are on the order of 80000 cm^{-1} and where the existing spectroscopically accurate potentials, such as the ES function, can guarantee very high accuracies up to about 19000 cm^{-1}, to be the most remarkable finding. At the same time, the CR-CC(2,3)+Q(b) approach, which is based on the excellent description of the T_3 cluster contributions by the size extensive CR-CC(2,3) approximation, corrected for T_4 effects, provides a balanced description of triples and quadruples in the bond breaking and equilibrium regions. In the equilibrium region, the accuracy of the CR-CC(2,3)+Q(x) (x = a, b) methods is essentially the same as the accuracy of the CCSD(TQ),b approach, which describes the combined effect of T_3 and T_4 clusters in non-degenerate situations extremely well. Thus, the CR-CC(2,3) method corrected for quadruples enables us to bridge the closed-shell and bond breaking regions of the global potential energy surface of water, while preserving the high accuracy of the CCSD(TQ),b results in the closed-shell regions and matching the high quality of MRCI(Q) results in regions of stretched chemical bonds, where CCSD(TQ),b (and CCSD(T), of course) fails.

The excellent agreement between the CR-CC(2,3)+Q and MRCI(Q) results in regions of intermediate stretches of chemical bonds and higher, but not the highest energies, where the ES surface may be somewhat less accurate, has enabled us to suggest ways of improving the global ES potential function of water that might potentially benefit future reaction dynamics studies. The regions of intermediate stretches of chemical bonds that connect the spectroscopic and asymptotic regions of the water potential energy surface are not as well understood as the spectroscopic and asymptotic regions. Thus, it is difficult to construct the global potential of water without the high accuracy *ab initio* data. The large consistency between the CR-CC(2,3)+Q and MRCI(Q) results in these intermediate regions of the water potential suggest that we should be able to use the CR-CC(2,3)+Q approach and, hopefully, the future CR-CC(2,4) approach to provide the necessary information to improve the ES and other existing global potential functions.

Acknowledgements

One of us (P.P.) would like to thank Professors Souad Lahmar and Jean Maruani for inviting him to give a talk at the Tenth European Workshop on Quantum Systems in Chemistry and Physics (QSCP-X), where some of the results discussed in this contribution were presented. P.P. would like to thank Dr. Stephen Wilson and Pr. Jean Maruani for encouraging him to write this review chapter. He would also like to express his gratitude to Universidade de Coimbra, where this chapter was written, and to the members of the Coimbra Theoretical and Computational Chemistry group, for the warm hospitality during his sabbatical leave. P.P. and M.W. acknowledge several useful discussions with Mr. Jeffrey R. Gour on selected formal aspects of the CR-CC methods discussed in this chapter. This work has been supported by the Chemical Sciences, Geo-sciences and Biosciences Division, Office of Basic Energy Sciences, Office of Science, U.S. Department of Energy (Grant No. DE-FG02-01ER15228; P.P.), the National Science Foundation (Grant No. CHE-0309517; P.P.), and the Fundação para a Ciência e Tecnologia, Portugal (contracts POCI/QUI/60501/2004, POCI/AMB/60261/2004, and REEQ/128/QUI/2005; A.J.C.V.). The calculations reported in this work were performed on the computer systems provided by the High Performance Computing Center at Michigan State University.

References

1. F. Coester, *Nucl. Phys.* **7**, 421, 1958.
2. F. Coester and H. Kümmel, *Nucl. Phys.* **17**, 477, 1960.
3. J. Čížek, *J. Chem. Phys.* **45**, 4256, 1966.
4. J. Čížek, *Adv. Chem. Phys.* **14**, 35, 1969.
5. J. Čížek and J. Paldus, *Int. J. Quantum Chem.* **5**, 359, 1971.

6. R.J. Bartlett, in: *Modern Electronic Structure Theory*, Part I, ed. D.R. Yarkony, World Scientific, Singapore, 1995, pp. 1047–1131.
7. J. Gauss, in: *Encyclopedia of Computational Chemistry*, ed. P.v.R. Schleyer, N.L. Allinger, T. Clark, J. Gasteiger, P.A. Kollman, H.F. Schaefer III, and P.R. Schreiner, Wiley, Chichester, UK, 1998, Vol. 1, pp. 615–636.
8. J. Paldus and X. Li, *Adv. Chem. Phys.* **110**, 1, 1999.
9. T.D. Crawford and H.F. Schaefer III, *Rev. Comp. Chem.* **14**, 33, 2000.
10. J. Paldus, in: *Handbook of Molecular Physics and Quantum Chemistry*, Vol. 2, ed. S. Wilson, Wiley, Chichester, 2003, pp. 272–313.
11. P. Piecuch and K. Kowalski, in: *Computational Chemistry: Reviews of Current Trends*, Vol. 5, ed. J. Leszczyński, World Scientific, Singapore, 2000, pp. 1–104.
12. P. Piecuch, K. Kowalski, I.S.O. Pimienta, and M.J. McGuire, *Int. Rev. Phys. Chem.* **21**, 527, 2002.
13. P. Piecuch, K. Kowalski, P.-D. Fan, and I.S.O. Pimienta, in: *Progress in Theoretical Chemistry and Physics*, Vol. 12, *Advanced Topics in Theoretical Chemical Physics*, ed. J. Maruani, R. Lefebvre, and E. Brändas, Kluwer, Dordrecht, 2003, pp. 119–206.
14. P. Piecuch, K. Kowalski, I.S.O. Pimienta, P.-D. Fan, M. Lodriguito, M.J. McGuire, S.A. Kucharski, T. Kuś, and M. Musiał, *Theor. Chem. Acc.* **112**, 349, 2004.
15. P. Piecuch, M. Włoch, M. Lodriguito, and J.R. Gour, in: *Progress in Theoretical Chemistry and Physics*, Vol. 15, *Recent Advances in the Theory of Chemical and Physical Systems*, ed. J.-P. Julien, J. Maruani, D. Mayou, S. Wilson, and G. Delgado-Barrio, Springer, Berlin, 2006, pp. 45–106.
16. P. Piecuch, M. Włoch, J.R. Gour, D.J. Dean, M. Hjorth-Jensen, and T. Papenbrock, in: *Nuclei and Mesoscopic Physics: Workshop on Nuclei and Mesoscopic Physics WNMP 2004*, AIP Conference Proceedings, Vol. 777, ed. V. Zelevinsky, AIP Press, 2005, pp. 28–45.
17. M. Urban, J. Noga, S.J. Cole, and R.J. Bartlett, *J. Chem. Phys.* **83**, 4041, 1985.
18. P. Piecuch and J. Paldus, *Theor. Chim. Acta* **78**, 65, 1990.
19. K. Raghavachari, G.W. Trucks, J.A. Pople, and M. Head-Gordon, *Chem. Phys. Lett.* **157**, 479, 1989.
20. G.D. Purvis III and R.J. Bartlett, *J. Chem. Phys.* **76**, 1910, 1982.
21. G.E. Scuseria, A.C. Scheiner, T.J. Lee, J.E. Rice, and H.F. Schaefer III, *J. Chem. Phys.* **86**, 2881, 1987.
22. P. Piecuch and J. Paldus, *Int. J. Quantum Chem.* **36**, 429, 1989.
23. S.A. Kucharski and R.J. Bartlett, *J. Chem. Phys.* **108**, 9221, 1998.
24. K. Kowalski and P. Piecuch, *J. Chem. Phys.* **113**, 5644, 2000.
25. W.D. Laidig, P. Saxe, and R.J. Bartlett, *J. Chem. Phys.* **86**, 887, 1987.
26. K.B. Ghose, P. Piecuch, and L. Adamowicz, *J. Chem. Phys.* **103**, 9331, 1995.
27. P. Piecuch, V. Špirko, A.E. Kondo, and J. Paldus, *J. Chem. Phys.* **104**, 4699, 1996.
28. P. Piecuch, S.A. Kucharski, and R.J. Bartlett, *J. Chem. Phys.* **110**, 6103, 1999.
29. P. Piecuch, S.A. Kucharski, and V. Špirko, *J. Chem. Phys.* **111**, 6679, 1999.
30. A. Dutta and C.D. Sherrill, *J. Chem. Phys.* **118**, 1610, 2003.
31. M.L. Abrams and C.D. Sherrill, *J. Chem. Phys.* **121**, 9211, 2004.
32. C.D. Sherrill and P. Piecuch, *J. Chem. Phys.* **122**, 124104, 2005.
33. P. Piecuch, K. Kowalski, I.S.O. Pimienta, and S.A. Kucharski, in: *Low-Lying Potential Energy Surfaces*, ACS Symposium Series, Vol. 828, ed. M.R. Hoffmann and K.G. Dyall, American Chemican Society, Washington, DC, 2002, pp. 31–64.
34. K. Kowalski and P. Piecuch, *J. Chem. Phys.* **113**, 18, 2000.
35. K. Kowalski and P. Piecuch, *Chem. Phys. Lett.* **344**, 165, 2001.
36. P. Piecuch, S.A. Kucharski, and K. Kowalski, *Chem. Phys. Lett.* **344**, 176, 2001.
37. P. Piecuch, S.A. Kucharski, V. Špirko, and K. Kowalski, *J. Chem. Phys.* **115**, 5796, 2001.
38. M.J. McGuire, K. Kowalski, and P. Piecuch, *J. Chem. Phys.* **117**, 3617, 2002.
39. P. Piecuch, K. Kowalski, and I.S.O. Pimienta, *Int. J. Mol. Sci.* **3**, 475, 2002.
40. I.S.O. Pimienta, K. Kowalski, and P. Piecuch, *J. Chem. Phys.* **119**, 2951, 2003.
41. M.J. McGuire, K. Kowalski, P. Piecuch, S.A. Kucharski, and M. Musiał, *J. Phys. Chem. A* **108**, 8878, 2004.
42. S. Hirata, P.-D. Fan, A.A. Auer, M. Nooijen, and P. Piecuch, *J. Chem. Phys.* **121**, 12197, 2004.

43. P.-D. Fan, K. Kowalski, and P. Piecuch, *Mol. Phys.* **103**, 2191, 2005.
44. K. Kowalski and P. Piecuch, *J. Chem. Phys.* **122**, 074107, 2005.
45. P. Piecuch and M. Włoch, *J. Chem. Phys.* **123**, 224105, 2005.
46. P. Piecuch, M. Włoch, J.R. Gour, and A. Kinal, *Chem. Phys. Lett.* **418**, 463, 2005.
47. P. Piecuch, S. Hirata, K. Kowalski, P.-D. Fan, and T.L. Windus, *Int. J. Quantum Chem.* **106**, 79, 2006.
48. M. Włoch, M.D. Lodriguito, P. Piecuch, and J.R. Gour, *Mol. Phys.*, in press (2006).
49. M.D. Lodriguito, K. Kowalski, M. Włoch, and P. Piecuch, *J. Mol. Struct: THEOCHEM*, in press (2006).
50. I. Özkan, A. Kinal, and M. Balci, *J. Phys. Chem. A* **108**, 507, 2004.
51. M.J. McGuire and P. Piecuch, *J. Am. Chem. Soc.* **127**, 2608, 2005.
52. A. Kinal and P. Piecuch, *J. Phys. Chem. A* **110**, 367, 2006.
53. C.J. Cramer, M. Włoch, P. Piecuch, C. Puzzarini, and L. Gagliardi, *J. Phys. Chem. A* **110**, 1991, 2006.
54. S.R. Gwaltney and M. Head-Gordon, *Chem. Phys. Lett.* **323**, 21, 2000.
55. S.R. Gwaltney, C.D. Sherrill, M. Head-Gordon, and A.I. Krylov, *J. Chem. Phys.* **113**, 3548, 2000.
56. S.R. Gwaltney and M. Head-Gordon, *J. Chem. Phys.* **115**, 2014, 2001.
57. S.R. Gwaltney, E.F.C. Byrd, T. Van Voorhis, and M. Head-Gordon, *Chem. Phys. Lett.* **353**, 359, 2002.
58. M. Musiał and R.J. Bartlett, *J. Chem. Phys.* **122**, 224102, 2005.
59. B.O. Roos, P. Linse, P.E.M. Siegbahn, and M.R.A. Blomberg, *Chem. Phys.* **66**, 197, 1982.
60. K. Andersson, P.-Å. Malmqvist, B.O. Roos, A.J. Sadlej, and K. Woliński, *J. Phys. Chem.* **94**, 5483, 1990.
61. K. Andersson, P.-Å. Malmqvist, and B.O. Roos, *J. Chem. Phys.* **96**, 1218, 1992.
62. K. Andersson and B.O. Roos, in: *Modern Electronic Structure Theory*, Vol. 2 of *Advanced Series in Physical Chemistry*, ed. D.R. Yarkony, World Scientific, Singapore, 1995, pp. 55–109.
63. M.F. Rode and H.-J. Werner, *Theor. Chem. Acc.* **114**, 309, 2005.
64. H.-J. Werner and P.J. Knowles, *J. Chem. Phys.* **89**, 5803, 1988.
65. P.J. Knowles and H.-J. Werner, *Chem. Phys. Lett.* **145**, 514, 1988.
66. M.W. Schmidt and M.S. Gordon, *Annu. Rev. Phys. Chem.* **49**, 233, 1998.
67. R.K. Chaudhuri, K.F. Freed, G. Hose, P. Piecuch, K. Kowalski, M. Włoch, S. Chattopadhyay, D. Mukherjee, Z. Rolik, Á. Szabados, G. Tóth, and P.R. Surján, *J. Chem. Phys.* **122**, 134105, 2005.
68. T.H. Schucan and H.A. Weidenmüller, *Ann. Phys. (NY)* **73**, 108, 1972.
69. T.H. Schucan and H.A. Weidenmüller, *Ann. Phys. (NY)* **76**, 483, 1973.
70. A. Hose and U. Kaldor, *J. Phys. B* **12**, 3827, 1979.
71. J.P. Finley, R.K. Chaudhuri, and K.F. Freed, *J. Chem. Phys.* **103**, 4990, 1995.
72. J.P. Finley and K.F. Freed, *J. Chem. Phys.* **102**, 1306, 1995.
73. J.P. Finley, R.K. Chaudhuri, and K.F. Freed, *Phys. Rev. A* **54** 343, 1996.
74. R.K. Chaudhuri and K.F. Freed, *J. Chem. Phys.* **107**, 6699, 1997.
75. S. Zarrabian and J. Paldus, *Int. J. Quantum Chem.* **38**, 761, 1990.
76. J.M. Rintelman, I. Adamovic, S. Varganov, and M.S. Gordon, *J. Chem. Phys.* **122**, 044105, 2005.
77. M.L. Abrams and C.D. Sherrill, *J. Phys. Chem. A* **107**, 5611, 2003.
78. J.S. Sears and C.D. Sherrill, *Mol. Phys.*, **103**, 803, 2005.
79. J. Paldus, in: *Methods in Computational Molecular Physics*, Vol. 293 of *NATO Advanced Study Institute, Series B: Physics*, ed. S. Wilson and G.H.F. Diercksen, Plenum, New York, 1992, pp. 99–194.
80. P. Piecuch and K. Kowalski, *Int. J. Mol. Sci.* **3**, 676, 2002.
81. B. Jeziorski and H.J. Monkhorst, *Phys. Rev. A* **24**, 1668, 1981.
82. X. Li and J. Paldus, *J. Chem. Phys.* **119**, 5320, 2003.
83. X. Li and J. Paldus, *J. Chem. Phys.* **119**, 5334, 2003.
84. X. Li and J. Paldus, *J. Chem. Phys.* **119**, 5346, 2003.
85. X. Li and J. Paldus, *J. Chem. Phys.* **120**, 5890, 2004.
86. J. Paldus and X. Li, *Coll. Czech. Chem. Commun.* **69**, 90, 2004.
87. X. Li and J. Paldus, *Int. J. Quantum Chem.* **99**, 914, 2004.

88. X. Li and J. Paldus, *J. Chem. Phys.* **124**, 034112, 2006.
89. K. Kowalski and P. Piecuch, *J. Mol. Struct.: THEOCHEM* **547**, 191, 2001.
90. K. Kowalski and P. Piecuch, *Mol. Phys.* **102**, 2425, 2004.
91. K. Kowalski and P. Piecuch, *Chem. Phys. Lett.* **334**, 89, 2001.
92. N. Oliphant and L. Adamowicz, *J. Chem. Phys.* **94**, 1229, 1991.
93. N. Oliphant and L. Adamowicz, *J. Chem. Phys.* **96**, 3739, 1992.
94. N. Oliphant and L. Adamowicz, *Int. Rev. Phys. Chem.* **12**, 339, 1993.
95. P. Piecuch, N. Oliphant, and L. Adamowicz, *J. Chem. Phys.* **99**, 1875, 1993.
96. P. Piecuch and L. Adamowicz, *J. Chem. Phys.* **100**, 5792, 1994.
97. P. Piecuch and L. Adamowicz, *Chem. Phys. Lett.* **221**, 121, 1994.
98. P. Piecuch and L. Adamowicz, *J. Chem. Phys.* **102**, 898, 1995.
99. K.B. Ghose and L. Adamowicz, *J. Chem. Phys.* **103,** 9324, 1995.
100. V. Alexandrov, P. Piecuch, and L. Adamowicz, *J. Chem. Phys.* **102**, 3301, 1995.
101. K.B. Ghose, P. Piecuch, S. Pal, and L. Adamowicz, *J. Chem. Phys.* **104**, 6582, 1996.
102. L. Adamowicz, P. Piecuch, and K.B. Ghose, *Mol. Phys.* **94**, 225, 1998.
103. K. Kowalski and P. Piecuch, *J. Chem. Phys.* **113**, 8490, 2000.
104. K. Kowalski and P. Piecuch, *J. Chem. Phys.* **115**, 643, 2001.
105. K. Kowalski and P. Piecuch, *Chem. Phys. Lett.* **347**, 237, 2001.
106. K. Kowalski, S. Hirata, M. Włoch, P. Piecuch, and T.L. Windus, *J. Chem. Phys.* **123**, 074319, 2005.
107. J.R. Gour, P. Piecuch, and M. Włoch, *J. Chem. Phys.* **123**, 134113, 2005.
108. J.R. Gour, P. Piecuch, and M. Włoch, *Int. J. Quantum Chem.*, in press (2006).
109. L. Adamowicz, J.-P. Malrieu, and V.V. Ivanov, *J. Chem. Phys.* **112**, 10075, 2000.
110. V.V. Ivanov and L. Adamowicz, *J. Chem. Phys.* **112**, 9258, 2000.
111. D.I. Lyakh, V.V. Ivanov, and L. Adamowicz, *J. Chem. Phys.* **122**, 024108, 2005.
112. J. Olsen, *J. Chem. Phys.* **113**, 7140, 2000.
113. J.W. Krogh and J. Olsen, *Chem. Phys. Lett.* **344**, 578, 2001.
114. M. Kállay, P.G. Szalay, and P. G. Surján, *J. Chem. Phys.* **117**, 980, 2002.
115. M. Kállay and J. Gauss, *J. Chem. Phys.* **121**, 9257, 2004.
116. L.V. Slipchenko and A.I. Krylov, *J. Chem. Phys.* **123**, 084107, 2005.
117. P.-D. Fan and S. Hirata, *J. Chem. Phys.* **124**, 104108, 2006.
118. P. Piecuch, S.A. Kucharski, K. Kowalski, and M. Musiał, *Comp. Phys. Commun.* **149**, 71, 2002.
119. M.W. Schmidt, K.K. Baldridge, J.A. Boatz, S.T. Elbert, M.S. Gordon, J.H. Jensen, S. Koseki, N. Matsunaga, K.A. Nguyen, S.J. Su, T.L. Windus, M. Dupuis, and J.A. Montgomery, *J. Comput. Chem.* **14**, 1347, 1993.
120. K. Kowalski and P. Piecuch, *J. Chem. Phys.* **115**, 2966, 2001.
121. K. Kowalski and P. Piecuch, *J. Chem. Phys.* **116**, 7411, 2002.
122. K. Kowalski and P. Piecuch, *J. Chem. Phys.* **120**, 1715, 2004.
123. M. Włoch, J.R. Gour, K. Kowalski, and P. Piecuch, *J. Chem. Phys.* **122**, 214107, 2005.
124. D.M. Chipman, *J. Chem. Phys.* **124**, 044305, 2006.
125. M.I.M. Sarker, C.S. Kim, and C.H. Choi, *Chem. Phys. Lett.* **411**, 297, 2005.
126. P.V. Avramov, I. Adamovic, K.M. Ho, C.Z. Wang, W.C. Lu, and M.S. Gordon, *J. Phys. Chem. A* **109**, 6294, 2005.
127. K. Kowalski, *J. Chem. Phys.* **123**, 014102, 2005.
128. J. Noga and R.J. Bartlett, *J. Chem. Phys.* **86**, 7041, 1987; **89**, 3401, 1988 (Erratum).
129. G.E. Scuseria and H.F. Schaefer III, *Chem. Phys. Lett.* **152**, 382, 1988.
130. T.H. Dunning, Jr., *J. Chem. Phys.* **90**, 1007 1989; D.E. Woon and T.H. Dunning, Jr., *J. Chem. Phys.* **103**, 4572 1995; R.A. Kendall, T.H. Dunning, Jr., and R.J. Harrison, *J. Chem. Phys.* **96**, 6769, 1992.
131. Basis sets were obtained from the Extensible Computational Chemistry Environment Basis Set Database, Version 02/25/04, as developed and distributed by the Molecular Science Computing Facility, Environmental and Molecular Sciences Laboratory which is part of the Pacific Northwest Laboratory, P.O. Box 999, Richland, Washington 99352, USA, and funded by the U.S. Department of Energy. The Pacific Northwest Laboratory is a multi-program laboratory operated by Battelle Memorial Institute for the U.S. Department of Energy under contract DE-AC06-76RLO 1830. Contact Karen Schuchardt for further information.

132. A.J.C. Varandas, *J. Chem. Phys.* **105**, 3524, 1996.
133. L. Wallace, P. Bernath, W. Livingstone, K. Hinkle, J. Busler, B. Gour, and K. Zhang, *Science* **268**, 1155, 1995.
134. T. Oka, *Science* **277**, 328, 1997.
135. R.P. Wayne, *Chemistry of Atmospheres*, Oxford University Press, Oxford, 2000, pp. 50–58.
136. A.G. Császár, G. Tarczay, M.L. Leininger, O.L. Polyansky, J. Tennyson, and W.D. Allen, in: *Spectroscopy from Space*, NATO ASI Series C, ed. J. Demaison and K. Sarka, Kluwer, Dordrecht, 2001, pp. 317–339.
137. R.N. Tolchenov, J. Tennyson, J.W. Brault, A.A.D. Canas, and R. Schermaul, *J. Mol. Spectrosc.* **215**, 269, 2002.
138. R.N. Tolchenov, J. Tennyson, S.V. Shirin, N.F. Zobov, O.L. Polyansky, and A.N. Maurellis, *J. Mol. Spectrosc.* **221**, 99, 2003.
139. O.L. Polyansky, A.G. Csaszar, S.V. Shirin, N.F. Zobov, P. Bartletta, J. Tennyson, D.W. Schwenke, and P.J. Knowles, *Science* **299**, 539, 2003.
140. R.N. Tolchenov, O. Naumenko, N.F. Zobov, S.V. Shirin, O.L. Polyansky, J. Tennyson, M. Carleer, P.-F. Coheur, S. Fally, A. Jenouvrier, and A.C. Vandaele, *J. Mol. Spectrosc.* **233**, 68, 2005.
141. P.F. Coheur, P.F. Bernath, M. Carleer, R. Colin, O.L. Polyansky, N.F. Zobov, S.V. Shirin, R.J. Barber, and J. Tennyson, *J. Chem. Phys.* **122**, 074307, 2005.
142. R.N. Tolchenov and J. Tennyson, *J. Mol. Spectrosc.* **231**, 23, 2005.
143. A.J.C. Varandas, A.I. Voronin, A. Riganelli, and P.J.S.B. Caridade, *Chem. Phys. Lett.* **278**, 325, 1997.
144. A.J.C. Varandas, A.I. Voronin, P.J.S.B. Caridade, and A. Riganelli, *Chem. Phys. Lett.* **331**, 331, 2000.
145. J.N. Murrell and S. Carter, *J. Phys. Chem.* **88**, 4887, 1984.
146. O.L. Polyansky, P. Jensen, and J. Tennyson, *J. Chem. Phys.* **101**, 7651, 1994.
147. H. Partridge and D.W. Schwenke, *J. Chem. Phys.* **106**, 4618, 1997.
148. A.J.C. Varandas, *J. Chem. Phys.* **107**, 867, 1997.
149. A.J.C. Varandas, A.I. Voronin, and P.J.S.B. Caridade, *J. Chem. Phys.* **108**, 7623, 1998.
150. K. Emrich, *Nucl. Phys. A* **351**, 379, 1981.
151. J. Geertsen, M. Rittby, and R.J. Bartlett, *Chem. Phys. Lett.* **164**, 57, 1989.
152. J.F. Stanton and R.J. Bartlett, *J. Chem. Phys.* **98**, 7029, 1993.
153. P. Piecuch and R.J. Bartlett, *Adv. Quantum Chem.* **34**, 295, 1999.
154. E.A. Salter, G.W. Trucks, and R.J. Bartlett, *J. Chem. Phys.* **90**, 1752, 1989.
155. L.V. Kantorovich and V.I. Krylov, *Approximate Methods of Higher Analysis*, Interscience, New York, 1958, pp. 150.
156. K. Jankowski, J. Paldus, and P. Piecuch, *Theor. Chim. Acta* **80**, 223, 1991.
157. X. Li and J. Paldus, *J. Chem. Phys.* **115**, 5759, 2001.
158. X. Li and J. Paldus, *J. Chem. Phys.* **115**, 5774, 2001.
159. X. Li and J. Paldus, *J. Chem. Phys.* **117**, 1941, 2002.
160. X. Li and J. Paldus, *J. Chem. Phys.* **118**, 2470, 2003.
161. J. Paldus, J. Čížek, and M. Takahashi, *Phys. Rev. A* **30**, 2193, 1984.
162. P. Piecuch, R. Toboła, and J. Paldus, *Phys. Rev. A* **54**, 1210, 1996.
163. J. Paldus and J. Planelles, *Theor. Chim. Acta* **89**, 13, 1994.
164. G. Peris, J. Planelles, and J. Paldus, *Int. J. Quantum Chem.* **62**, 137, 1997.
165. L. Stolarczyk, *Chem. Phys. Lett.* **217**, 1, 1994.
166. X. Li and J. Paldus, *J. Chem. Phys.* **107**, 6257, 1997.
167. X. Li and J. Paldus, *J. Chem. Phys.* **108**, 637, 1998.
168. X. Li and J. Paldus, *Chem. Phys. Lett.* **286**, 145, 1998.
169. M. Włoch, M.D. Lodriguito, J.R. Gour, and P. Piecuch, in preparation.
170. J.F. Stanton, *Chem. Phys. Lett.* **281**, 130, 1997.
171. S. Hirata, M. Nooijen, I. Grabowski, and R.J. Bartlett, *J. Chem. Phys.* **114**, 3919, 2001; **115**, 3967, 2001 (Erratum).

172. J.N. Murrell, S. Carter, S.C. Farantos, P. Huxley, and A.J.C. Varandas, *Molecular Potential Energy Functions*, Wiley, Chichester, 1984.
173. A.J.C. Varandas, *Adv. Chem. Phys.* **74**, 255, 1988.
174. A.J.C. Varandas, *Chem. Phys. Lett.* **194**, 333, 1992.
175. A.J.C. Varandas and A.I. Voronin, *Mol. Phys.* **95**, 497, 1995.
176. A.J.C. Varandas, in: *Conical Intersections: Electronic Structure, Dynamics, and Spectroscopy*, Vol. 15 of *Advanced Series in Physical Chemistry*, ed. W. Domcke, D.R. Yarkony, and H. Köppel, World Scientific, Singapore, 2004, pp. 205–270.
177. E.F. Valeev, W.D. Allen, H.F. Schaefer III, and A.G. Császár, *J. Chem. Phys.* **114**, 2875, 2001.
178. MOLPRO, version 2002.6; a package of *ab initio* programs; H.-J. Werner, P.J. Knowles, R. Lindh, M. Schütz, P. Celani, T. Korona, F.R. Manby, G. Rauhut, R.D. Amos, A. Bernhardsson, A. Berning, D.L. Cooper, M.J.O. Deegan, A.J. Dobbyn, F. Eckert, C. Hampel, G. Hetzer, A.W. Lloyd, S.J. McNicholas, W. Meyer, M.E. Mura, A. Nicklass, P. Palmieri, R. Pitzer, U. Schumann, H. Stoll, A.J. Stone, R. Tarroni, and T. Thorsteinsson; see http://www.molpro.net (Birmingham, UK, 2003).
179. R.L. DeKock, M.J. McGuire, P. Piecuch, W.D. Allen, H.F. Schaefer III, K. Kowalski, S.A. Kucharski, M. Musiał, A.R. Bonner, S.A. Spronk, D.B. Lawson, and S.L. Laursen, *J. Phys. Chem. A* **108**, 2893, 2004.
180. S.A. Kucharski and R.J. Bartlett, *J. Chem. Phys.* **110**, 8233, 1999.
181. S.A. Kucharski and R.J. Bartlett, *Chem. Phys. Lett.* **302**, 295, 1999.
182. S. Nangia, D.G. Truhlar, M.J. McGuire, and P. Piecuch, *J. Phys. Chem. A* **109**, 11643, 2005.

HYPERSPHERICAL AND RELATED TYPES OF COORDINATES FOR THE DYNAMICAL TREATMENT OF THREE-BODY SYSTEMS

MIRCO RAGNI, ANA CARLA PEIXOTO BITENCOURT,
AND VINCENZO AQUILANTI

Dipartimento di Chimica dell'Università di Perugia, 06123, Perugia, Italy

Abstract We present an explicit list of relevant formulae connecting the various coordinate sets for the representation of the potential energy surface of triatomic systems. The connections are made to those coordinates which give the potential energy surface dependence on the internuclear distances. Reference will also be made to computer programs which are made available on the Internet. Applications are indicated for molecular and chemical physics.

1. INTRODUCTION

In this paper we consider some coordinate sets used for the treatment in classical and quantum mechanics of the motion of three particles in space. The alternative sets of coordinate systems for the three-body problem have been studied extensively [1–5] and a good choice of the coordinate systems is of crucial importance. Key references for the basic theory are [1–4, 6], where also history is sketched and credits are given.

In the laboratory frame the motion of the three particles depends on nine variables, three of which define the position of the center-of-mass. Other three coordinates are needed to describe the rotation of the system in the space and therefore the internal motion is described by the three remaining coordinates. For example, in molecular dynamics the potential energy surface in general is calculated and presented using geometrical coordinates, such the interparticle distances, or two "bond" distances and an angle. But it is convenient and necessary to use different coordinate systems to describe and understand the dynamics of the particles, because of the rotational terms which appear in the full Hamiltonian. In this context, we will present the transformation equations from the interparticle distances to coordinate sets of the hyperspherical and related types, successful in the treatment of the dynamics.

S. Lahmar et al. (eds.), Topics in the Theory of Chemical and Physical Systems, 123–146.

Early basic formulations [7] and applications were restricted to physically collinear (mathematically two-dimensional) problems [8] and semiclassical approaches [9], nonadiabatic effects [10] and resonances [11] were studied.

A historical account of the development of orthogonal coordinates for elementary chemical reactions has been given by one of the protagonists [12]. The early hyperspherical treatment for the helium atom as a tree-body quantum-mechanical problem [13, 14], reviewed in Morse and Feshbach's treatise [15], was taken up in two basic papers by Fock [16]. They essentially used the parametrization referred to as asymmetrical in the following. Further important work in atomic physics [17,18] used the same hyperspherical parametrization, as did the investigations by Delves [19] on the breakdown of systems of many particles.

Gallina *et al.* [20] introduced the hyperspherical symmetrical parametrization in a particle-physics context, as did Zickendraht later [21, 22]. At the same time, F.T. Smith [23] gave the definitions of internal coordinates following Fock's work already mentioned [16], Clapp [24, 25] and others and established, for the symmetrical and asymmetrical parametrization, the basic properties and the notation we follow. Since then, applications have been extensive, especially for bound states. For example, the symmetrical coordinates have often been used in atomic [26], nuclear [27] and molecular [28–31] physics. This paper accounts for modern applications, with particular reference to the field of reaction dynamics, in view of the prominent role played by these coordinates for dealing with rearrangement problems.

Exploiting a four-dimensional rotation group analysis, the transformation between harmonic expansions in the two coordinates systems was given explicitly [32], as well as the most general representation in terms of Jacobi functions [2]. In practice, however, the two representations are in one form or another those being used in all applications and specifically in recent treatments of the elementary chemical reactions as a three-body problem [11, 33–36]. For example, Eqs. (29)–(31) and Eqs. (47)–(49) permitted to establish [37] the explicit connection between coordinates for entrance and exit channels to be used in sudden approximation treatments of chemical reactions [38].

The full three-body problem in the physical three-dimensional space required development of hyperspherical harmonic expansions [39]. Crucial for further progress was the introduction of discrete analogues for the latter [40–43], based on hyperangular momentum theory [44, 45] and leading to the efficient hyperquantization algorithm [46–49]. For other hyperspherical approaches to reaction dynamics, see [50–63].

The content of the paper is as follows. In section 2 we revisit formulas for the various coordinate sets; in section 3 we list relationships with interatomic distances; in section 4 the computer implementation for the use of these formulas is described. An overview of past and perspective applications concludes the paper (section 5).

All programs can be downloaded from the web site http://www.chm.unipg.it/chimgen/mb/theo2/home/pagine/ricerca/cc3.html.

2. DEFINITION OF SETS AND RELATIONS

Here, we present the various sets, starting from that commonly used where a molecule is seen as a central atom, B, two bonds and an internal angle. We consider a system of three particles (A, B, and C), with masses m_A, m_B, and m_C, respectively. Distances will be indicated as r_{AB}, r_{BC}, and r_{AC}.

2.1. Two distances and one angle

2.1.1. Vectors with center at B

In this coordinate set, we use two vectors and an angle, which can be defined considering two vectors, connecting the particle, centered on one particle. Assuming B as center at Figure 1, the vectors with center at B are

(1) $$|\mathbf{r}_{AB}| = r_{AB}$$

(2) $$|\mathbf{r}_{BC}| = r_{BC}$$

(3) $$\cos\gamma = \frac{r_{AB}^2 + r_{BC}^2 - r_{AC}^2}{2r_{BC}r_{AB}}.$$

2.1.2. Jacobi vectors

To proceed, we choose a particular configuration, indicated by the suffix α, where the atom A is assumed to impinge on the molecule BC. Other two choices are clearly possible. Quantities independent of this choice are called "kinematic invariants".

In the three-body problem we can write down two Jacobi vectors, one ($\mathbf{x}_\alpha$) is the interparticle distance between two particles and the other ($\mathbf{X}_\alpha$) connects their center-of-mass to the third particle. So, the choice of the Jacobi vectors is not unique [1]. Here we will consider $\mathbf{x}_\alpha$ as the vector from the particle B to the particle C, and $\mathbf{X}_\alpha$ as the vector from the particle A to the center-of-mass of the BC couple (see Figure 2).

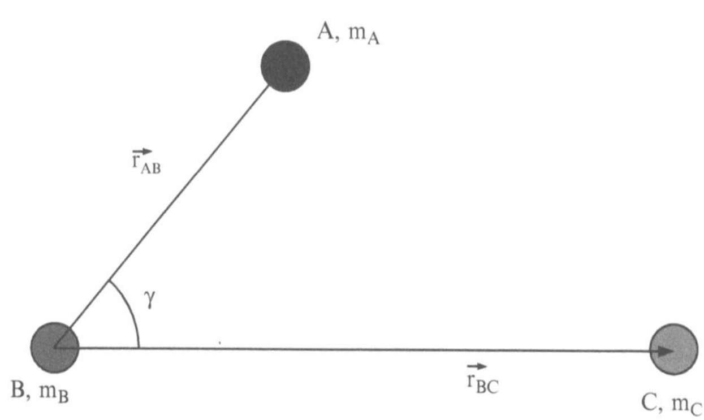

Figure 1. Vectors with center at B.

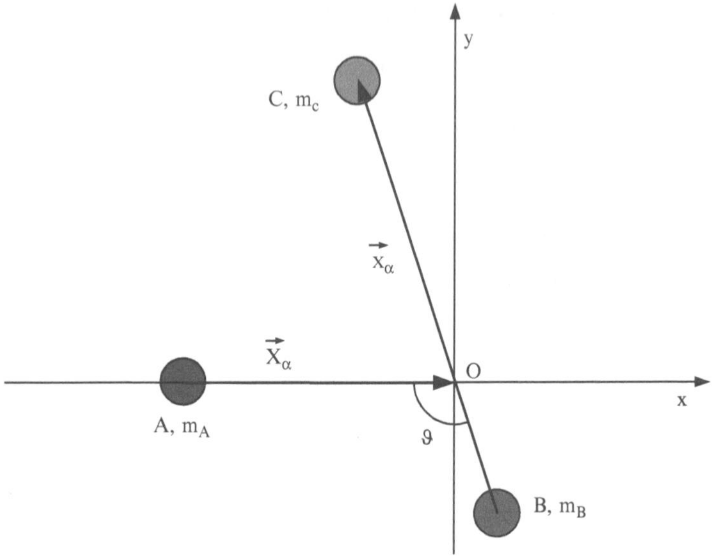

Figure 2. The Jacobi vectors.

Then, from the interparticle distances we have:

$$(4)\qquad |\mathbf{X}_\alpha| = \left[\frac{m_B}{m_B+m_C}r_{AB}^2 + \frac{m_C}{m_B+m_C}r_{AC}^2 - \frac{m_Bm_C}{(m_B+m_C)^2}r_{BC}^2\right]^{\frac{1}{2}}$$

$$(5)\qquad |\mathbf{x}_\alpha| = r_{BC}$$

$$(6)\qquad \cos\vartheta = \frac{r_{AC}^2 - r_{AB}^2 - \frac{m_B-m_C}{m_B+m_C}r_{BC}^2}{r_{BC}\left[\frac{4m_B}{m_B+m_C}r_{AB}^2 + \frac{4m_C}{m_B+m_C}r_{AC}^2 - \frac{4m_Bm_C}{(m_B+m_C)^2}r_{BC}^2\right]^{\frac{1}{2}}}$$

When the denominator of (6) vanishes, $\cos\vartheta$ is zero. The interparticle distances from the Jacobi vectors can be calculated by:

$$(7)\qquad r_{BC} = |\mathbf{x}_\alpha|$$

$$(8)\qquad r_{AB} = \left[\frac{m_C^2}{(m_B+m_C)^2}|\mathbf{x}_\alpha|^2 - \frac{2m_C}{m_B+m_C}|\mathbf{x}_\alpha||\mathbf{X}_\alpha|\cos\vartheta + |\mathbf{X}_\alpha|^2\right]^{\frac{1}{2}}$$

$$(9)\qquad r_{AC} = \left[\frac{m_B^2}{(m_B+m_C)^2}|\mathbf{x}_\alpha|^2 + \frac{2m_B}{m_B+m_C}|\mathbf{x}_\alpha||\mathbf{X}_\alpha|\cos\vartheta + |\mathbf{X}_\alpha|^2\right]^{\frac{1}{2}}$$

2.1.3. The mass-scaled Jacobi vectors

Under proper mass-scaling the three possible Jacobi vectors sets can be related to each other by a planar rotation by an angle which depends only on the masses of A,

B, and C particles and is an extension of the so-called skewing angle concept [3]. For each set of the two Jacobi vectors the mass-scaling can be written

$$(10) \qquad \mathbf{r}_\alpha = \left[\frac{\mu_{BC}}{\mu}\right]^{\frac{1}{2}} \mathbf{x}_\alpha \quad \textit{and} \quad \mathbf{R}_\alpha = \left[\frac{\mu}{\mu_{BC}}\right]^{\frac{1}{2}} \mathbf{X}_\alpha$$

where μ_{BC} is the two-body reduced mass for the BC couple and

$$(11) \qquad \mu = \sqrt{m_A m_B m_C/(m_A + m_B + m_C)}\,.$$

is the three-body reduced mass. Equivalently, we can write

$$(12) \qquad \mathbf{r}_\alpha = \left[\frac{\mu_{BC}}{\mu_{A,BC}}\right]^{\frac{1}{4}} \mathbf{x}_\alpha \quad \textit{and} \quad \mathbf{R}_\alpha = \left[\frac{\mu_{A,BC}}{\mu_{BC}}\right]^{\frac{1}{4}} \mathbf{X}_\alpha$$

where

$$(13) \qquad \mu_{A,BC} = \frac{m_A(m_B + m_C)}{m_A + m_B + m_C}\,.$$

2.1.4. The Radau–Smith vectors

Besides this coordinate sets, other sets of orthogonal vectors have been considered in the literature. Kinematic Rotations by mass-dependent matrices allows to relate different particle couplings in the Jacobi scheme, and to build up alternative systems such as those based on the Radau–Smith vectors and hyperspherical coordinates [1,3]. The Radau–Smith vectors $\mathbf{RS}_1$, $\mathbf{RS}_2$ and the angle $\cos\vartheta_{RS}$ $(0 \le \vartheta_{RS} \le \pi)$, showed in Figure 3 (the D point is defined by $\overline{OD}^2 = \overline{OE} \times \overline{OA}$, where O is the center-of-mass of the BC couple, E is the center-of-mass of the three particles and A is the position of the A particle), can be calculated from the Jacobi vectors $\mathbf{x}_\alpha$ and $\mathbf{X}_\alpha$ using:

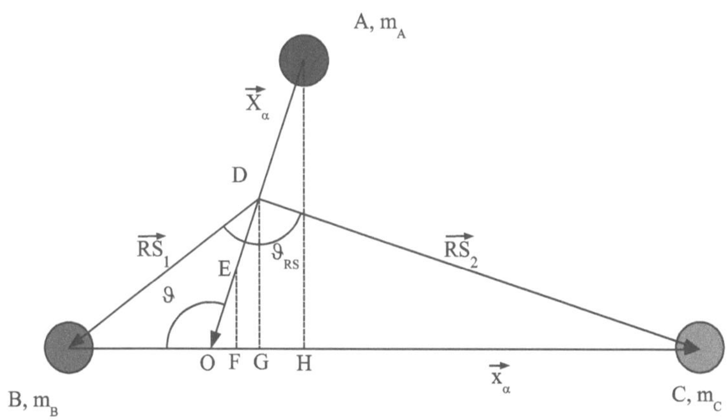

Figure 3. The Radau Smith vectors.

$$|\mathbf{RS}_1| = \left[\frac{m_C^2}{(m_B+m_C)^2}|\mathbf{x}_\alpha|^2 - \frac{2m_C}{m_B+m_C}\sqrt{\frac{m_A}{m_T}}|\mathbf{x}_\alpha||\mathbf{X}_\alpha|\cos\vartheta + \frac{m_A}{m_T}|\mathbf{X}_\alpha|^2\right]^{\frac{1}{2}} \quad (14)$$

$$|\mathbf{RS}_2| = \left[\frac{m_B^2}{(m_B+m_C)^2}|\mathbf{x}_\alpha|^2 + \frac{2m_B}{m_B+m_C}\sqrt{\frac{m_A}{m_T}}|\mathbf{x}_\alpha||\mathbf{X}_\alpha|\cos\vartheta + \frac{m_A}{m_T}|\mathbf{X}_\alpha|^2\right]^{\frac{1}{2}} \quad (15)$$

$$\cos\vartheta_{RS} = \frac{|\mathbf{RS}_1|^2 + |\mathbf{RS}_2|^2 - |\mathbf{x}_\alpha|^2}{-2|\mathbf{RS}_1||\mathbf{RS}_2|} \quad (16)$$

where m_T indicates the total mass of the system:

$$m_T = m_A + m_B + m_C \quad (17)$$

The Jacobi vectors from the Radau–Smith vectors can be calculated by:

$$|\mathbf{x}_\alpha| = \left[|\mathbf{RS}_1|^2 + |\mathbf{RS}_2|^2 - 2|\mathbf{RS}_1||\mathbf{RS}_2|\cos\vartheta_{RS}\right]^{\frac{1}{2}} \quad (18)$$

$$|\mathbf{X}_\alpha| = \left[\frac{m_B m_T}{m_A(m_B+m_C)}|\mathbf{RS}_1|^2 + \frac{m_C m_T}{m_A(m_B+m_C)}|\mathbf{RS}_2|^2 + \frac{m_B m_C m_T}{m_A(m_B+m_C)^2}|\mathbf{x}_\alpha|^2\right]^{\frac{1}{2}} \quad (19)$$

$$\cos\vartheta = \frac{|\mathbf{RS}_1|^2 - \frac{m_C^2}{(m_B+m_C)^2}|\mathbf{x}_\alpha|^2 - \frac{m_A}{m_T}|\mathbf{X}_\alpha|^2}{-2\frac{m_C}{m_B+m_C}\sqrt{\frac{m_A}{m_T}}|\mathbf{x}_\alpha||\mathbf{X}_\alpha|} \quad (20)$$

where $|\mathbf{x}_\alpha|$ cannot vanish because it is the distance between B and C; instead $|\mathbf{X}_\alpha|$ can vanish: in this case the particle A is in the barycenter of B and C, here $\cos\vartheta$ is taken as zero.

2.2. Hyperspherical and related coordinates

2.2.1. Basic invariants and the hyperradius

The basic idea of the hyperspherical approach is the introduction of the ρ variable, which plays the role of a radius of a hypersphere. In hyperspherical coordinates systems the hyperradius is a critical quantity and is found to be:

$$\rho^2 = |\mathbf{R}_\alpha|^2 + |\mathbf{r}_\alpha|^2 \quad (21)$$

and enjoys the important property of being independent on the particular choice of A, B, and C, i.e. of being kinematically invariant.

Let's define three coordinates in terms of mass-scaled Jacobi vectors:

$$w_1 = |\mathbf{r}_\alpha|^2 - |\mathbf{R}_\alpha|^2 \tag{22}$$

$$w_2 = 2\,(\mathbf{r}_\alpha \cdot \mathbf{R}_\alpha) \tag{23}$$

$$w_3 = 2\,|(\mathbf{r}_\alpha \times \mathbf{R}_\alpha)| \tag{24}$$

with $-\infty < w_1, w_2 < \infty$, and $0 \leq w_3 < \infty$. It can be shown that w_3 is kinematically invariant, while w_1 and w_2 are not. These have very interesting proprieties, but are seldom used. It is convenient to define:

$$\xi_\alpha = w_2/\rho \tag{25}$$

$$\eta \;= w_3/\rho \tag{26}$$

$$\zeta_\alpha = w_1/\rho \tag{27}$$

where only η is kinematic invariant and $-\infty < \xi\alpha, \zeta\alpha < \infty$, and $0 < \eta < \infty$. In function of these sets one can define the hyperradius as

$$\rho^2 = \left[w_1^2 + w_2^2 + w_3^2\right]^{\frac{1}{2}} = \xi_\alpha^2 + \eta^2 + \zeta_\alpha^2 \tag{28}$$

2.2.2. Asymmetric hyperspherical coordinates

The (hyper)-angles $\vartheta\alpha$ and χ_α can be defined by as a polar representation of Eq. (25)-(27):

$$\xi\alpha = \rho \sin(2\chi\alpha)\cos(\vartheta\alpha) \tag{29}$$

$$\eta \;= \rho \sin(2\chi\alpha)\sin(\vartheta\alpha) \tag{30}$$

$$\zeta\alpha = \rho \cos(2\chi\alpha) \tag{31}$$

Therefore the asymmetric parametrization [2] can be expressed in terms of two coordinates referred to an internal system: the angle $\vartheta\alpha$ is the same encountered before as that formed by the two Jacobi vectors and the angle χ_α is related to their ratio:

$$\tan \chi_\alpha = \frac{|\mathbf{R}_\alpha|}{|\mathbf{r}_\alpha|} \tag{32}$$

The asymmetric hyperspherical coordinates from the mass-scaled Jacobi vectors can be calculated by:

$$|\mathbf{R}_\alpha| = \sqrt{\frac{\rho^2 \tan \chi_\alpha^2}{1 + \tan \chi_\alpha^2}} \tag{33}$$

$$|\mathbf{r}_\alpha| \;= \sqrt{\frac{\rho^2}{1 + \tan \chi_\alpha^2}} \tag{34}$$

2.2.3. Symmetric hyperspherical coordinates

The variables $\xi\alpha$, η and $\zeta\alpha$ of the Eq. (25)-(27) serve to define the so-called symmetric hyperspherical coordinates.

$$(35)\qquad \xi\alpha = \rho\cos(2\Theta)\sin(2\Phi\alpha)$$

$$(36)\qquad \eta = \rho\sin(2\Theta)$$

$$(37)\qquad \zeta\alpha = \rho\cos(2\Theta)\cos(2\Phi\alpha)$$

where $0 \leq \Theta \leq \pi/4$ and $0 \leq \Phi\alpha < \pi$. The dependence of Φ on change of labelling (kinematic rotation) is trivial. The suffix α will be implied in the following but neglected to simplify the notation.

The symmetric parametrization can be achieved by taking as internal reference system the one that diagonalizes the inertia tensor, placing the principal axis in correspondence with that of maximal inertia [6, 64]. The symmetric hyperspherical coordinates can be calculated from the asymmetric hyperspherical coordinates:

$$(38)\qquad \sin(2\Theta) = \frac{2\tan\chi}{1+\tan^2\chi}\sin\vartheta$$

$$(39)\qquad \cos(2\Phi) = \frac{1-\tan^2\chi}{1+\tan^2\chi}\frac{1}{\cos(2\Theta)}$$

$$(40)\qquad \sin(2\Phi) = \frac{2\tan\chi}{1+\tan^2\chi}\frac{\cos\vartheta}{\cos(2\Theta)}$$

When $\cos(2\Theta)$ vanishes Φ is undefined. The asymmetric hyperspherical coordinates from the symmetric hyperspherical coordinates can be calculated by:

$$(41)\qquad \tan\chi = \sqrt{\frac{1-\cos(2\Theta)\cos(2\Phi)}{1+\cos(2\Theta)\cos(2\Phi)}}$$

$$(42)\qquad \cos\vartheta = \cos(2\Theta)\sin(2\Phi)\frac{1+\tan^2\chi}{2\tan\chi}$$

Coherently with the definition of asymmetric hyperspherical coordinates, $\cos\vartheta$ is zero when $\tan\chi_\alpha$ vanishes.

2.2.4. Other relationships

It can be useful to know the relationship between these new coordinate sets. So here we present some inverse formulae. For example, to pass from the ξ, η, and ζ set to symmetric hyperspherical coordinates one can use:

$$(43)\qquad \rho = \left[\xi^2+\eta^2+\zeta^2\right]^{\frac{1}{2}}$$

$$(44)\qquad \cos(2\Theta) = \left[\frac{\xi^2+\zeta^2}{\xi^2+\eta^2+\zeta^2}\right]^{\frac{1}{2}}$$

$$(45)\qquad \cos(2\Phi) = \frac{\zeta}{\rho\cos(2\Theta)}$$

$$(46)\qquad \sin(2\Phi) = \frac{\xi}{\rho\cos(2\Theta)}$$

As in Eqs. (41) and (42), when $\cos(2\Theta)$ vanishes the angle Φ is undefined.

The variables w_1, w_2, and w_3 using Eqs. (25)–(27), and Eqs. (35)–(37) are given by following relations [5]:

$$(47) \qquad w_1 = \rho^2 \cos(2\Theta)\cos(2\Phi)$$

$$(48) \qquad w_2 = \rho^2 \cos(2\Theta)\sin(2\Phi)$$

$$(49) \qquad w_3 = \rho^2 \sin(2\Theta)$$

And the inverses are:

$$(50) \qquad \rho = \left[w_1^2 + w_2^2 + w_3^2\right]^{\frac{1}{4}}$$

$$(51) \qquad \cos(2\Theta) = \left[\frac{w_1^2 + w_2^2}{w_1^2 + w_2^2 + w_3^2}\right]^{\frac{1}{2}}$$

$$(52) \qquad \cos(2\Phi) = \frac{w_1}{\rho^2 \cos(2\Theta)}$$

$$(53) \qquad \sin(2\Phi) = \frac{w_2}{\rho^2 \cos(2\Theta)}$$

where the angle Φ is undefined when $\cos(2\Theta)$ vanishes.

2.2.5. *Elliptic coordinates*

In the elliptic coordinates set, introduced in [2] (see also [5]), the three variables are indicated with ρ, ψ, and ω. Similar coordinates have been used in other types of problems, especially in connection with momentum space techniques [65] and Sturmian basis sets [66] of interest in quantum chemistry. In [5] we also describe relationship with a set which Tolstikhin and coworkers [67–75] presented and utilized. The variable ρ is again the hyperradius and can be calculated with Eq. (28). The last two variables are two angles and have ranges $0 \le \psi \le \pi$ and $-\pi/2 \le \omega \le \pi/2$. To give more flexibility to the elliptic set, there are two parameters k and k':

$$(54) \qquad k^2 + k'^2 = 1$$

where $0 \le k, k' \le 1$. Sometimes, for a process A + BC → AB + C, k is equal to the the cosine of the angle [5]:

$$(55) \qquad \gamma = \tan^{-1}\sqrt{\frac{m_B m_T}{m_A m_C}}$$

This set is defined [5] by

$$(56) \qquad w_1 = \rho^2 \cos\psi \sqrt{1 - k'^2 \cos^2 \omega}$$

$$(57) \qquad w_2 = \rho^2 \sin\psi \sin\omega$$

$$(58) \qquad w_3 = \rho^2 \cos\omega \sqrt{1 - k^2 \cos^2 \psi}$$

and the inverse, when k and k' are different from zero, are:

$$(59) \qquad \cos^2\omega = \frac{k'^2 w_3^2 + \rho^4 - k^2 w_1^2 - \sqrt{(k'^2 w_3^2 + \rho^4 - k^2 w_1^2)^2 - 4\rho^4 k'^2 w_3^2}}{2\rho^4 k'^2}$$

$$(60)\qquad \cos^2\psi = \frac{k^2w_1^2+\rho^4-k'^2w_3^2-\sqrt{(k^2w_1^2+\rho^4-k'^2w_3^2)^2-4\rho^4k^2w_1^2}}{2\rho^4k^2}$$

where $\cos\psi$ and $\sin\omega$ have the same sign of w_1 and w_2 respectively. Notice that the values under the roots of the Eqs. (59) and (60) are identical.

The relations valid for $k=0$ ($k'=1$) are:

$$(61)\qquad w_1 = \rho^2\cos\psi\sin\omega$$

$$(62)\qquad w_2 = \rho^2\sin\psi\sin\omega$$

$$(63)\qquad w_3 = \rho^2\cos\omega$$

and the inverses are:

$$(64)\qquad \cos\omega = \frac{w_3}{\rho^2}$$

$$(65)\qquad \cos\psi = \frac{w_1}{\rho^2\sin\omega}$$

From Eq. (62) we deduce that $\sin\omega$ has the sign of w_2, and when $\sin\omega$ vanishes the angle ψ is undefined.

Another case is obtained with $k=1$ ($k'=0$) and the relationships are:

$$(66)\qquad w_1 = \rho^2\cos\psi$$

$$(67)\qquad w_2 = \rho^2\sin\omega\sin\psi$$

$$(68)\qquad w_3 = \rho^2\cos\omega\sin\psi$$

and the inverses are:

$$(69)\qquad \cos\psi = \frac{w_1}{\rho^2}$$

$$(70)\qquad \sin\omega = \frac{w_2}{\rho^2\sin\psi}$$

When $\sin\psi$ vanishes the angle ω is undefined.

2.2.6. *Other notations*

As far as applications are concerned, a list of alternative notations used by recent authors follows. Regarding the choice of the arbitrary mass and the consequent mass-scaling, our definitions (section 2.1.3) lead to a hyperradius ρ whose unit is length, and is used by most. Babamov and Marcus [76] impose a unitary three body mass, with the consequence that theie hyperradius is $(\mu_{BC}\mu_{A,BC})^{\frac{1}{2}}\rho$. In their numerous applications, Manz, Römelt, and co-workers [33] do not scale $\mathbf{r}_{BC}$, and their hyperradius is $(\mu/\mu_{BC})^{\frac{1}{2}}\rho$.

Symbols and scalings for hyperangles may lead to confusion. As discussed in section 2, Eqs. (29)–(31), often angles parameterizing the (center-of-mass frame) internal coordinates are twice those which parameterize the (laboratory-frame) six-dimensional vector. In considering what we called Smith's kinetic plane [3] for the representation of the potential-energy surfaces, Kuppermann [77] proposed to double the $\chi\alpha$ angle. Actually, Kuppermann in this pioneering and recent work [35] uses twice the angle

$$\eta\alpha = \arctan \frac{|\mathbf{x}\alpha|}{|\mathbf{X}\alpha|} \tag{71}$$

i.e. $\eta\alpha = \pi/2 + \chi\alpha$, (see also ref. [37] and [38]). This eliminates the superfluous repetitions due to an inversion-plus-reflection symmetry in the kinetic plane: a sequence like ABC is indistinguishable from CBA in $2D$ and $3D$, even if not in $1D$, but in any case the potential-energy surface is invariant to such a symmetry operation (see ref. [2] for a full discussion). Kuppermann's doubling of the angle of the asymmetrical hyperspherical parametrization appears to have stimulated Johnson [78] to redefine the angles of the symmetrical parametrization: he used an angle $\theta\alpha = \pi/2 - 2\Theta\alpha$ and an angle $\phi\alpha = \pi\alpha/2 - 2\Phi\alpha$. He also gave useful explicit quantum [79] and classical [80] Hamiltonians, used since then by many [81, 82]. Pack [83], in what he defines **APH** coordinates, uses the symbol χ for the kinematic angle $\Phi\alpha$, and θ as Johnson instead of Θ (he also suggests an alternative orientation of the three atoms from that of Smith, a choice unimportant for exact calculations but of perspective interest for some approximations). Other representations are also simply related: for example, Linderberg and co-workers [84, 85], following Mead [86], use coordinates which, in the present notation, are ρ, $\xi\alpha/\rho$, and $\zeta\alpha$.

For his the principal inertial axis formulation, Eckart (and the recent applications by Robert and Baudon [87–89]) defined r_1 and r_2 corresponding to Q and q. De Celles and Darling [90] used ξ and η, respectively.

3. RELATIONS WITH THE INTERPARTICLE DISTANCES

In this section we present the formulas to obtain the different sets, introduced in the previous section, directly from the interparticle distances as well as their inverses.

3.1. Radau–Smith vectors

The interparticle distances are related with Radau–Smith vectors by:

$$r_{BC} = \left[|\mathbf{RS}_1|^2 + |\mathbf{RS}_2|^2 - 2|\mathbf{RS}_1||\mathbf{RS}_2| \cos\vartheta_{RS}\right]^{\frac{1}{2}} \tag{72}$$

$$r_{AB} = \left[C_{RS2}|\mathbf{RS}_1|^2 + C_{RS3}|\mathbf{RS}_2|^2 + C_{RS4}|\mathbf{RS}_1||\mathbf{RS}_2| \cos\vartheta_{RS}\right]^{\frac{1}{2}} \tag{73}$$

$$r_{AC} = \left[C_{RS5}|\mathbf{RS}_1|^2 + C_{RS6}|\mathbf{RS}_2|^2 + C_{RS7}|\mathbf{RS}_1||\mathbf{RS}_2| \cos\vartheta_{RS}\right]^{\frac{1}{2}} \tag{74}$$

and the inverses are:

$$|\mathbf{RS}_1| = \left[C_{DI2} r_{BC}^2 + C_{DI3} r_{AB}^2 + C_{DI4} r_{AC}^2\right]^{\frac{1}{2}} \tag{75}$$

$$(76) \qquad |\mathbf{RS}_2| = \left[C_{DI5}r_{BC}^2 + C_{DI6}r_{AB}^2 + C_{DI7}r_{AC}^2\right]^{\frac{1}{2}}$$

$$(77) \qquad \cos\vartheta_{RS} = \frac{|\mathbf{RS}_1|^2 + |\mathbf{RS}_2|^2 - r_{BC}^2}{2|\mathbf{RS}_1||\mathbf{RS}_2|}$$

where the coefficients are the following:

$$C_{RS2} = \left[\frac{m_C}{m_B+m_C} + \frac{m_B}{m_B+m_C}\sqrt{\frac{m_T}{m_A}}\right]^2$$

$$C_{RS3} = \frac{m_C^2}{(m_B+m_C)^2}\left[\sqrt{\frac{m_T}{m_A}} - 1\right]^2$$

$$C_{RS4} = 2\left[\frac{-m_C^2}{(m_B+m_C)^2} + \frac{m_C^2 - m_Bm_C}{(m_B+m_C)^2}\sqrt{\frac{m_T}{m_A}} + \frac{m_Bm_Cm_T}{(m_B+m_C)^2m_A}\right]$$

$$C_{RS5} = \frac{m_B^2}{(m_B+m_C)^2}\left[\sqrt{\frac{m_T}{m_A}} - 1\right]^2$$

$$C_{RS6} = \left[\frac{m_B}{m_B+m_C} + \frac{m_C}{m_B+m_C}\sqrt{\frac{m_T}{m_A}}\right]^2$$

$$C_{RS7} = 2\left[\frac{-m_B^2}{(m_B+m_C)^2} + \frac{m_B^2 - m_Bm_C}{(m_B+m_C)^2}\sqrt{\frac{m_T}{m_A}} + \frac{m_Bm_Cm_T}{(m_B+m_C)^2m_A}\right]$$

$$C_{DI2} = \frac{m_C^2}{(m_B+m_C)^2} + \frac{m_Bm_C - m_C^2}{(m_B+m_C)^2}\sqrt{\frac{m_A}{m_T}} - \frac{m_Am_Bm_C}{(m_B+m_C)^2m_T}$$

$$C_{DI3} = \frac{m_C}{m_B+m_C}\sqrt{\frac{m_A}{m_T}} + \frac{m_Am_B}{(m_B+m_C)m_T}$$

$$C_{DI4} = \frac{-m_C}{m_B+m_C}\sqrt{\frac{m_A}{m_T}} + \frac{m_Am_C}{(m_B+m_C)m_T}$$

$$C_{DI5} = \frac{m_B^2}{(m_B+m_C)^2} + \frac{m_Bm_C - m_B^2}{(m_B+m_C)^2}\sqrt{\frac{m_A}{m_T}} - \frac{m_Am_Bm_C}{(m_B+m_C)^2m_T}$$

$$C_{DI6} = \frac{-m_B}{m_B+m_C}\sqrt{\frac{m_A}{m_T}} + \frac{m_Am_B}{(m_B+m_C)m_T}$$

$$C_{DI7} = \frac{m_B}{m_B+m_C}\sqrt{\frac{m_A}{m_T}} + \frac{m_Am_C}{(m_B+m_C)m_T}$$

3.2. Hyperspherical and related coordinates

To simplify the notation of the following equations we define these parameters:

$$(78) \qquad d = \left[\frac{m_A(m_B+m_C)^2}{m_Bm_Cm_T}\right]^{\frac{1}{4}}$$

(79) $$E = \frac{d^2 m_C}{m_B + m_C} = \sqrt{\frac{m_A m_C}{m_B m_T}}$$

(80) $$F = \frac{d^2 m_B}{m_B + m_C} = \sqrt{\frac{m_A m_B}{m_C m_T}}$$

3.2.1. Asymmetric hyperspherical coordinates

The interparticle distances are related with asymmetric hyperspherical coordinates by the following relations:

(81) $$r_{BC} = \frac{\rho\, d}{\sqrt{1+\tan^2\chi}}$$

(82) $$r_{AB} = \frac{\rho d}{\sqrt{1+\tan^2\chi}}\left[E^2 + \tan^2\chi - 2E\cos\vartheta\tan\chi\right]^{\frac{1}{2}}$$

(83) $$r_{AC} = \frac{\rho d}{\sqrt{1+\tan^2\chi}}\left[F^2 + \tan^2\chi + 2F\cos\vartheta\tan\chi\right]^{\frac{1}{2}}$$

and the inverses are:

(84) $$\rho^2 = \sqrt{\frac{m_A m_B}{m_C m_T}} r_{AB}^2 + \sqrt{\frac{m_A m_C}{m_B m_T}} r_{AC}^2 + \sqrt{\frac{m_B m_C}{m_A m_T}} r_{BC}^2$$

(85) $$\tan\chi_\alpha = \left[\frac{m_A(m_B+m_C)}{m_C m_T}\frac{r_{AB}^2}{r_{BC}^2} + \frac{m_A(m_B+m_C)}{m_B m_T}\frac{r_{AC}^2}{r_{BC}^2} - \frac{m_A}{m_T}\right]^{\frac{1}{2}}$$

(86) $$\cos\vartheta = \frac{r_{AC}^2 - r_{AB}^2 - \frac{m_B - m_C}{m_B + m_C} r_{BC}^2}{r_{BC}\left[\frac{4m_B}{m_B+m_C} r_{AB}^2 + \frac{4m_C}{m_B+m_C} r_{AC}^2 - \frac{4m_B m_C}{(m_B+m_C)^2} r_{BC}^2\right]^{\frac{1}{2}}}$$

When the denominator of Eq. (86) is zero, $\cos\vartheta$ vanishes.

3.2.2. Symmetric hyperspherical coordinates

The interparticle distances are related with symmetric hyperspherical coordinates by:

(87) $$r_{BC} = \frac{\rho d}{\sqrt{2}}\left[1 + \cos(2\Theta)\cos(2\Phi)\right]^{\frac{1}{2}}$$

(88) $$r_{AB} = \frac{\rho}{d\sqrt{2}}\Big[E^2 + 1 + \cos(2\Theta)\cos(2\Phi)(E^2 - 1) - \cos(2\Theta)\sin(2\Phi)2E\Big]^{\frac{1}{2}}$$

(89) $$r_{AC} = \frac{\rho}{d\sqrt{2}}\Big[F^2 + 1 + \cos(2\Theta)\cos(2\Phi)(F^2 - 1) + \cos(2\Theta)\sin(2\Phi)2F\Big]^{\frac{1}{2}}$$

The symmetric hyperspherical coordinates from the interparticle distances, where ρ is given by the Eq. (84), can be written as:

$$\sin(2\Theta) = \frac{\left[2(r_{AB}^2 r_{BC}^2 + r_{AC}^2 r_{BC}^2 + r_{AB}^2 r_{AC}^2) - r_{AB}^4 - r_{AC}^4 - r_{BC}^4\right]^{\frac{1}{2}}}{\sqrt{\frac{m_A m_B}{m_C m_T}} r_{AB}^2 + \sqrt{\frac{m_A m_C}{m_B m_T}} r_{AC}^2 + \sqrt{\frac{m_B m_C}{m_A m_T}} r_{BC}^2} \tag{90}$$

$$\cos(2\Phi) = \left[\frac{2r_{BC}^2}{\rho^2 d^2} - 1\right] / \cos(2\Theta) \tag{91}$$

When $\cos(2\Theta)$ vanishes in Eq. (91), Φ is not defined.

3.2.3. Other relationships

The interparticle distances are related with ζ, ξ, and η coordinates by:

$$r_{BC} = \frac{d}{\sqrt{2}}\left[\eta^2 + \xi^2 + \zeta^2 + (\eta^2 + \xi^2 + \zeta^2)^{\frac{1}{2}}\zeta\right]^{\frac{1}{2}} \tag{92}$$

$$r_{AB} = \frac{1}{d\sqrt{2}}\Big[(E^2 + 1)(\eta^2 + \xi^2 + \zeta^2) + [\zeta(E^2 - 1) - 2\xi E](\eta^2 + \xi^2 + \zeta^2)^{\frac{1}{2}}\Big]^{\frac{1}{2}} \tag{93}$$

$$r_{AC} = \frac{1}{d\sqrt{2}}\Big[(F^2 + 1)(\eta^2 + \xi^2 + \zeta^2) + [\zeta(F^2 - 1) + 2\xi F](\eta^2 + \xi^2 + \zeta^2)^{\frac{1}{2}}\Big]^{\frac{1}{2}} \tag{94}$$

and the inverse are:

$$\zeta = \frac{2r_{BC}^2}{\rho d^2} - \rho \tag{95}$$

$$\xi = \frac{r_{AC}^2 - r_{AB}^2 - \frac{m_B - m_C}{m_B + m_C} r_{BC}^2}{\rho} \tag{96}$$

$$\eta = \left[\rho^2 - \zeta^2 - \xi^2\right]^{\frac{1}{2}}. \tag{97}$$

The interparticle distances are related with w_1, w_2, w_3 coordinates by:

$$r_{BC} = \frac{d}{\sqrt{2}}\left[(w_1^2 + w_2^2 + w_3^2)^{\frac{1}{2}} + w_1\right]^{\frac{1}{2}} \tag{98}$$

$$r_{AB} = \frac{1}{d\sqrt{2}}\left[(E^2 + 1)(w_1^2 + w_2^2 + w_3^2)^{\frac{1}{2}} + w_1(E^2 - 1) - 2w_2 E\right]^{\frac{1}{2}} \tag{99}$$

$$r_{AC} = \frac{1}{d\sqrt{2}}\left[(F^2 + 1)(w_1^2 + w_2^2 + w_3^2)^{\frac{1}{2}} + w_1(F^2 - 1) + 2w_2 F\right]^{\frac{1}{2}} \tag{100}$$

and the inverses are:

(101) $$w_1 = \left[\frac{1+EF}{d^2}\right] r_{BC}^2 - F\, r_{AB}^2 - E\, r_{AC}^2$$

(102) $$w_2 = r_{AC}^2 - r_{AB}^2 - \left[\frac{m_B - m_C}{m_B + m_C}\right] r_{BC}^2$$

(103) $$w_3 = \left[\rho^4 - w_1^2 - w_2^2\right]^{\frac{1}{2}}$$

Where ρ is defined at Eq. (84).

3.2.4. Elliptic coordinates

The interparticle distances are related with elliptic coordinates by:

(104) $$r_{BC} = \frac{\rho d}{\sqrt{2}}\left[1 + \cos\psi\sqrt{1 - k'^2\cos^2\omega}\right]^{\frac{1}{2}}$$

$$r_{AB} = \frac{\rho}{d\sqrt{2}}\Big[E^2 + 1 + \cos\psi\sqrt{1 - k'^2\cos^2\omega}(E^2 - 1)$$

(105) $$- 2E\sin\psi\sin\omega\Big]^{\frac{1}{2}}$$

$$r_{AC} = \frac{\rho}{d\sqrt{2}}\Big[F^2 + 1 + \cos\psi\sqrt{1 - k'^2\cos^2\omega}(F^2 - 1)$$

(106) $$+ 2F\sin\psi\sin\omega\Big]^{\frac{1}{2}}$$

When $k' = 0$ the elliptic coordinates become:

(107) $$r_{BC} = \frac{\rho d}{\sqrt{2}}\left[1 + cos\psi\right]^{\frac{1}{2}}$$

(108) $$r_{AB} = \frac{\rho}{d\sqrt{2}}\left[E^2 + 1 + (E^2 - 1)\cos\psi - 2E\sin\psi\sin\omega\right]^{\frac{1}{2}}$$

(109) $$r_{AC} = \frac{\rho}{d\sqrt{2}}\left[F^2 + 1 + (F^2 - 1)\cos\psi + 2F\sin\psi\sin\omega\right]^{\frac{1}{2}}$$

and when $k' = 1$ the elliptic coordinates become:

(110) $$r_{BC} = \frac{\rho d}{\sqrt{2}}\left[1 + \cos\psi\sin\omega\right]^{\frac{1}{2}}$$

(111) $$r_{AB} = \frac{\rho}{d\sqrt{2}}\left[E^2 + 1 + (E^2 - 1)\cos\psi\sin\omega - 2E\sin\psi\sin\omega\right]^{\frac{1}{2}}$$

(112) $$r_{AC} = \frac{\rho}{d\sqrt{2}}\left[F^2 + 1 + (F^2 - 1)\cos\psi\sin\omega + 2F\sin\psi\sin\omega\right]^{\frac{1}{2}}.$$

Here we present the inverse relationships of the elliptic coordinates from the interparticle distances. The hyperradius can be calculated from the interparticle distances using Eq. (84). To calculate the angle ψ and ω it is necessary to consider the value of k'. When k' is different of 0 and 1 the two angles can be calculate using the inverses of Eqs. (104)–(106):

$$\cos^2\omega = \frac{-\delta - \sqrt{\delta^2 + 4k'^2(\beta^2 + \alpha^2 - 1)}}{2k'^2} \tag{113}$$

$$\cos\psi = \frac{\alpha}{\sqrt{1 - k'^2\cos^2\omega}} \tag{114}$$

$$\sin\psi = \sqrt{1 - \cos^2\psi} \tag{115}$$

$$\sin\omega = \frac{\beta}{\sin\psi} \tag{116}$$

where

$$\alpha = \cos\psi\sqrt{1 - k'^2\cos^2\omega} = \frac{2r_{BC}^2}{\rho^2 d^2} - 1 \tag{117}$$

$$\beta = \sin\psi\sin\omega = \frac{d^2}{E}\left[\frac{E^2+1}{2d^2} + \frac{E^2-1}{2d^2}\alpha - \frac{r_{AB}^2}{\rho^2}\right] \tag{118}$$

$$\delta = \alpha^2 + k'^2(\beta^2 - 1) - 1 \tag{119}$$

In Eq. (115) $0 \leq \sin\psi \leq 1$ because $0 \leq \psi \leq \pi$. When $k' = 0$ the valid relations are:

$$\cos\psi = \frac{2r_{BC}^2}{\rho^2 d^2} - 1 \tag{120}$$

$$\sin\omega = \frac{d^2}{E\sin\psi}\left[\frac{E^2+1}{2d^2} + \frac{E^2-1}{2d^2}\cos\psi - \frac{r_{AB}^2}{\rho^2}\right] \tag{121}$$

The angle ω is undefined when $\sin\psi$ vanishes. The relation for $k' = 1$ are:

$$\cos\psi\sin\omega = \frac{2r_{BC}^2}{\rho^2 d^2} - 1 \tag{122}$$

$$\sin\psi\sin\omega = \frac{d^2}{E}\left[\frac{E^2+1}{2d^2} + \frac{E^2-1}{2d^2}\cos\psi\sin\omega - \frac{r_{AB}^2}{\rho^2}\right] \tag{123}$$

$$\sin\omega = \pm\left[\cos^2\psi\sin^2\omega + \sin^2\psi\sin^2\omega\right]^{\frac{1}{2}} \tag{124}$$

When $\sin\omega$ is zero the angle ψ is undefined, otherwise its sign is given by the sign of $\sin\psi\sin\omega$.

4. IMPLEMENTATION

We have implemented all the equations described previously in three programming languages (C, FORTRAN and, JAVA). Also we have constructed two useful programs. The first program, called *Zmatrix*, permits to describe the configuration of the three particles from one of the sets above to the standard of the Z Matrix implemented in common quantum chemical programs (e.g. Gaussian, Gamess, and Molpro). The second program implements the conversion routines in a Interface Programming. This program, called *Pcc3*, allows us to understand the behavior of the set. This was made using the Graphical User Interfaces (GUI) of the Operative System (OS), in this way simplifying the introduction of the coordinates in the program. A goal of this implementation is that one can see on the screen the position of the three particles. So we can familiarize with the nature of the coordinate sets and follow the evolution of the system when the coordinates are varying.

4.1. Routines

The conversion equations, in section 2 and 3, are functions of some coefficients that depend from the masses of the three particles. Since in the routines these coefficients are calculated separately from the coordinates, we do not need to calculate them each time we want a conversion. This also permitted to write conversion routines for general purposes and very simple calling procedures.

4.2. Programs

The wide possibilities of conversion routines have allowed us to write two much faster programs.

4.2.1. Zmatrix

The first program works with the command line, with an appropriate input from file and/or keyboard it acquires the coordinates of the point in the pre-chosen set. Then it converts these coordinates into the Z matrix. This comes out on a file. This program can do the inverse calculation too. It takes a file that describes the position of the three particles with the Z matrix and it calculates its representation in the chosen set.

4.2.2. Pcc3

Another program has been developed (Pcc3). With this program it is very easy to understand the behavior of the coordinates. In fact it shows what happens to the particles when one varies the coordinates and/or the masses. Inserting the data and making them vary it is possible to visualize the mutual position of three particles. To print those three particles on the plane of the screen we need six degrees of freedom. Three of these six are the interparticle distances and we have introduced the other three degrees according to the Figures 2 and 4.

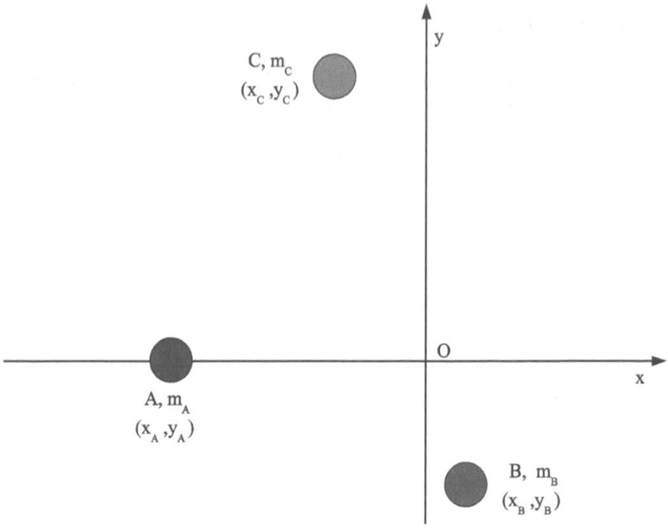

Figure 4. Cartesian coordinates.

Here we present the equations that permit to calculate the Cartesian coordinates directly from the distances:

$$x_A = -\left[\frac{m_B}{m_B+m_C}r_{AB}^2 + \frac{m_C}{m_B+m_C}r_{AC}^2 - \frac{m_B m_C}{(m_B+m_B)^2}r_{BC}^2\right]^{\frac{1}{2}}$$

$$y_A = 0$$

$$x_B = \frac{r_{AB}^2 - r_{AC}^2 + \frac{m_B-m_C}{m_B+m_C}r_{BC}^2}{\left[\frac{4m_B(m_B+m_C)}{m_C^2}r_{AB}^2 + \frac{4(m_B+m_C)}{m_C}r_{AC}^2 - \frac{4m_B}{m_C}r_{BC}^2\right]^{\frac{1}{2}}}$$

$$y_B = -\left[\frac{-r_{AC}^4 + 2r_{AC}^2 r_{AB}^2 + 2r_{AC}^2 r_{BC}^2 - r_{AB}^4 + 2r_{AB}^2 r_{BC}^2 - r_{BC}^4}{\frac{4m_B(m_B+m_C)}{m_C^2}r_{AB}^2 + \frac{4(m_B+m_C)}{m_C}r_{AC}^2 - \frac{4m_B}{m_C}r_{BC}^2}\right]^{\frac{1}{2}}$$

$$x_C = \frac{r_{AC}^2 - r_{AB}^2 - \frac{m_B-m_C}{m_B+m_C}r_{BC}^2}{\left[\frac{4(m_B+m_C)}{m_B}r_{AB}^2 + \frac{4m_C(m_C+m_C)}{m_B^2}r_{AC}^2 - \frac{4m_C}{m_B}r_{BC}^2\right]^{\frac{1}{2}}}$$

$$y_C = \left[\frac{-r_{AC}^4 + 2r_{AC}^2 r_{AB}^2 + 2r_{AC}^2 r_{BC}^2 - r_{AB}^4 + 2r_{AB}^2 r_{BC}^2 - r_{BC}^4}{\frac{4(m_B+m_C)}{m_B}r_{AB}^2 + \frac{4m_C(m_C+m_C)}{m_B^2}r_{AC}^2 - \frac{4m_C}{m_B}r_{BC}^2}\right]^{\frac{1}{2}}$$

The two denominators vanish when the particle A is in the center-of-mass of particles B and C. In this case $(x_A, y_A) = (0, 0)$, $(x_B, y_B) = (0, -r_{AB})$ and $(x_C, y_C) = (0, r_{AC})$.

We released a version for "Windows SO" that can be downloaded from the website http://www.chm.unipg.it/chimgen/mb/theo2/home/pagine/ricerca/cc3.html. Also it is possible to run a JAVA version of the Pcc3 program directly online.

5. APPLICATIONS AND PERSPECTIVES

The hyperspherical and related coordinates which have been considered in this work have served for the visualization of critical features of potential energy surfaces [91, 92], crucial for the understanding of reactivity (role of the ridge [93] and the kinetic paths [94]). In [95], the PES for the $O + H_2$ reaction was studied. A discrete hyperspherical harmonics representation is presented in [96] for proton transfer in malonaldehyde.

For a review of the use of hyperspherical harmonics as orbitals in quantum chemistry, (see [97]). Applications to bound state problems have mainly regarded nuclear physics, and are outside the scope of this article. The hyperquantization algorithm had been successfully applied to the prototype ion–molecule reaction $He + H_2^+ \rightarrow HeH^+ + H$ [98, 99] and atom–molecule reaction $F + H_2 \rightarrow HF + H$ [100, 101]. For the latter, resonances were characterized [102, 103] and benchmark state-to-state differential cross sections and rate constants [104, 105] were given.

Progress towards extension to four or more bodies is to be recorded [106–111]. See also [112–116]. Recently, application of the hyperspherical view to many body systems had been made to the study of classical dynamics of atomic clusters [117–121]

Acknowledgements

We thank S. Cavalli, D. De Fazio and G. Grossi for discussions, and the Italian Ministero per l'Universitá e la Ricerca (MIUR) for the PRIN and FIRB Grants.

References

1. V. Aquilanti and S. Cavalli, Coordinates for molecular dynamics: orthogonal local systems. *J. Chem: Phys.*, 85:1355–1361, 1986.
2. V. Aquilanti, S. Cavalli, and G. Grossi, Hyperspherical coordinates for molecular dynamics by the method of trees and the mapping of potential energy surfaces for triatomic systems. *J. Chem. Phys.*, 86:1362–1375, 1986.
3. V. Aquilanti, S. Cavalli, G. Grossi, and R.W. Anderson, Representation in hyperspherical and related coordinates of the potential-energy surface for triatomic reactions. *J. Chem. Soc. Faraday Trans*, 86(s):1681–1687, 1990.
4. V. Aquilanti, G. Capecchi, and S. Cavalli, Hyperspherical coordinates for chemical reaction dynamics. *Adv. Quant. Chem.*, 36:341–361, 1999.
5. V. Aquilanti and S. Tonzani, Three-body problem in quantum mechanics: Hyperspherical elliptic coordinates and harmonic basis sets. *J. Chem. Phys.*, 120(9):4073, 2004.
6. R.G. Littlejohn, K.A. Mitchell, V. Aquilanti, and S. Cavalli, Body frames and frame singularities for three-atom systems. *Phys. Rev. A*, 58:3715–3717, 1998.

7. V. Aquilanti, G. Grossi, and A. Laganá, Hyperspherical diabatic and adiabatic representations for chemical reactions. *Chem. Phys. Lett*, 93:174–178, 1982.
8. V. Aquilanti, S. Cavalli, and A. Laganá, Hyperspherical adiabatic description of interference effects and resonances in collinear chemical reactions. *Chem. Phys. Lett*, 93:179–183, 1982.
9. V. Aquilanti, S. Cavalli, G. Grossi, and A. Laganá, A semiclassical approach to the dynamics of chemical reactions within the hyperspherical formalism. *J. Mol. Struct.*, 93:319–323, 1983.
10. V. Aquilanti, S. Cavalli, G. Grossi, and A. Laganá, Nonadiabatic effects in the hyperspherical description of elementary chemical reactions. *J. Mol. Struct.*, 107:95–99, 1984.
11. V. Aquilanti, Resonances in reactions: a semiclassical view. *The Theory of Chemical Reaction Dynamics, D. C. Clary,* ed. Reidel, Dordrecht, 383–413, 1986.
12. J.O. Hirschfelder, My adventures in theoretical chemistry. *Annu. Rev. Phys. Chem.*, 34:1–30, 1983.
13. T.H. Gronwall, *Ann. Math.*, 33:279, 1932.
14. T.H. Gronwall, The helium wave equation. *Phys. Rev.*, 51:655–660, 1937.
15. P.M. Morse and H. Feshbach, *Methods of Theoretical Physics*. McGraw-Hill, New York, 1953.
16. V. Fock, *K. Norske Vidensk. selsksk Forh.*, 41:138–145, 1958.
17. J.H. Macek, Properties of autoionizing states of He. *J. Phys. B*, 1:831–843, 1968.
18. C.D. Lin, Properties of high-lying doubly excited states of H^-. *Phys. Rev. A*, 25:1535–1545, 1982.
19. L.M. Delves, Tertiary and general-order collisions (II). *Nucl. Phys.*, 20:275–308, 1960.
20. L. Bianchi V. Gallina, P. Nata and G. Viano, *Nuovo Cimento*, 24:835, 1962.
21. W. Zickendraht, Construction of a complete orthogonal system for the quantum-mechanical three-body problem. *Ann. Phys.*, 35:18–41, 1965.
22. W. Zickendraht, Configuration-space approach to three-particle scattering. *Phys. Rev.*, 159: 1448–1455, 1967.
23. F.T. Smith, A symmetric representation for three-body problems. I. Motion in a plane. *J. Math. Phys.*, 3:735–748, 1962.
24. R.E. Clapp, The binding energy of the triton. *Phys. Rev.*, 76:873–874, 1949.
25. R.E. Clapp, A complete orthogonal expansion for the nuclear three-body problem. *Ann. Phys.*, 13:187–236, 1961.
26. H. Klar, A model for triatomic large amplitude vibrations. Energy levels for the water molecule. *Phys. Rev. A*, 15:1452–1458, 1977.
27. Yu A. Simonov, *Soviet J. Nucl. Phys.*, 3:461, 1966.
28. A. Riganelli, F.V. Prudente, and A.J.C. Varandas, Evaluation of vibrational partition functions for polyatomic systems: quantum versus classical methods for H_2O and $ArCN$. *Phys. Chem. Chem. Phys.*, 2:4121–4129, 2000.
29. J.J. Soares Neto and F.V. Prudente, A novel finite element method implementation for calculating bound states of triatomic systems: Application to the water molecule. *Theor. Chim. Acta*, 89: 415–427, 1994.
30. R. Wallace, *Chem. Phys.*, 34:93, 1979.
31. R. Wallace, A model for triatomic large amplitude vibrations. Energy levels for the water molecule. *Chem. Phys. Lett.*, 67:442–443, 1979.
32. V. Aquilanti, G. Grossi, and A. Laganá, On hyperspherical mapping and harmonic expansions for potential energy surfaces. *J. Chem. Phys.*, 76:1587–1588, 1982.
33. J. Manz, Molecular dynamics along hyperspherical coordinates. *Comm. Atom. Mol. Phys.*, 17:91–113, 1985.
34. J.M. Launay and B. Lepetit, Three-dimensional quantum study of the reaction $H + FH(vj) \rightarrow HF(v'j') + H$ in hyperspherical coordinates. *Chem. Phys. Lett.*, 144:346–352, 1988.
35. P.G. Hipes and A. Kuppermann, Lifetime analysis of high-energy resonances in three-dimensional reactive scattering. *Chem. Phys. Lett.*, 133:1–7, 1987.
36. G.C. Schatz, Quantum reactive scattering using hyperspherical coordinates: Results for H+H_2 and Cl+HCl. *Chem. Phys. Lett.*, 150:92–98, 1988.
37. G. Grossi, Angular parametrizations in the hyperspherical description of elementary chemical reactions. *J. Chem. Phys.*, 81:3355–3356, 1984.

38. H. Nakamura, A. Ohsaki, and M. Baer, New implementation to approximate quantum mechanical treatment of atom-diatom chemical reactions. *J. Phys. Chem.*, 90:6176–6184, 1986.
39. V. Aquilanti, G. Grossi, A. Laganá, E. Pelikan, and H. Klar, A decoupling scheme for a three-body problem treated by expansions into hyperspherical harmonics: the hydrogen molecular ion. *Lett. Nuovo Cim.*, 41:541–544, 1984.
40. V. Aquilanti and G. Grossi, Discrete representations by artificial quantization in the quantum mechanics of anisotropic interactions. *Lett. Nuovo Cim.*, 42:157–162, 1985.
41. V. Aquilanti, S. Cavalli, G. Grossi, and R.W. Anderson, Stereodirected states in molecular dynamics: A discrete basis representation for the quantum mechanical scattering matrix. *J. Phys. Chem.*, 95:8184–8193, 1991.
42. V. Aquilanti, S. Cavalli, and G. Grossi, Discrete analogs of spherical harmonics and their use in quantum mechanics: The hyperquantization algorithm. *Theor. Chim. Acta*, 79:283–296, 1991.
43. V. Aquilanti, S. Cavalli, and D. De Fazio, Angular and hyperangular momentum coupling coefficients as hahn polynomials. *J. Phys. Chem.*, 99:15694–15698, 1995.
44. V. Aquilanti, S. Cavalli, C. Coletti, D. De Fazio, and G. Grossi, Hyperangular momentum: Applications to atomic and molecular science. *New Methods in Quantum Theory,* eds. C.A. Tsipis, V.S. Popov, D.R. Herschbach, J.S. Avery, Kluwer, pages 233–250, 1996.
45. V. Aquilanti and G. Capecchi, Harmonic analysis and discrete polynomials from semiclassical angular momentum theory to the hyperquantization algorithm. *Theor. Chem. Accounts*, 104:183–188, 2000.
46. V. Aquilanti, S. Cavalli, and D. De Fazio, Hyperquantization algorithm: I. theory for triatomic systems. *J. Chem. Phys.*, 109:3792–3804, 1998.
47. V. Aquilanti, S. Cavalli, D. De Fazio, A. Volpi, A. Aguilar, X. Gimenez, and J. Maria Lucas, Hyperquantization algorithm: II. Implementation for the $F+H_2$ reaction dynamics including open-shell and spin-orbit interaction. *J. Chem. Phys.*, 109:3805–3818, 1998.
48. V. Aquilanti, S. Cavalli, A. Volpi, and D. De Fazio, The a + bc reaction by the hyperquantization algorithm: the symmetric hyperspherical parametrization for $J>0$. *Adv. Quant. Chem.*, 39:103–121, 2001.
49. D. De Fazio, S. Cavalli, and V. Aquilanti, Orthogonal polynomials of a discrete variable as expansion basis sets in quantum mechanics. The hyperquantization algorithm. *Int. J. Quant. Chem.*, 93:91–111, 2003.
50. K. Museth and A. Kuppermann, Asymptotic analysis of state-to-state tetraatomic reactions using row-orthonormal hyperspherical coordinates. *J. Chem. Phys.*, 115:8285–8297, 2001.
51. A. Kuppermann, Reactive scattering with row-orthonormal hyperspherical coordinates. 2. Transformation properties and hamiltonian for tetraatomic systems. *J. Phys. Chem.*, 101:6368–6383, 1997.
52. A. Kuppermann, Reactive scattering with row-orthonormal hyperspherical coordinates. 1. Transformation properties and hamiltonian for triatomic systems (vol 100, pg 2635, 1996). *J. Phys. Chem.*, 100:11202–11202, 1996.
53. A. Kuppermann, Reactive scattering with row-orthonormal hyperspherical coordinates. 1. Transformation properties and hamiltonian for triatomic systems. *J. Phys. Chem.*, 100:2621–2636, 1996(Erratum 100:11202–11202, 1996).
54. F.D. Colavecchia, F. Mrugala, G.A. Parker, and R.T. Pack, Accurate quantum calculations on three-body collisions in recombination and collision-induced dissociation. ii. the smooth variable discretization enhanced renormalized numerov propagator. *J. Chem. Phys.*, 118:10387–10398, 2003.
55. G.A. Parker, R.B. Walker, B.K. Kendrick, and R.T. Pack, Accurate quantum calculations on three-body collisions in recombination and collision-induced dissociation. I. converged probabilities for the $H+Ne_2$ system. *J. Chem. Phys.*, 117:6083–6102, 2002.
56. G.A. Parker, A. Laganá, S. Crocchianti, and R.T. Pack, A detailed 3-dimensional quantum study of the Li+FH reaction. *J. Chem. Phys.*, 102:1238–1250, 1995.
57. C.Y. Yang, S.J. Klippenstein, J.D. Kress, and A. Laganá, Comparison of transition-state theory with quantum scattering-theory for the reaction Li+HF $\rightarrow$ LiF+H. *J. Chem. Phys.*, 100:4917–4924, 1994.
58. A. Laganá, R.T. Pack, and G.A. Parker, Li+FH reactive cross-sections from $J = 0$ accurate quantum reactivity. *J. Chem. Phys.*, 99:2269–2270, 1993.

59. G.A. Parker and R.T. Pack, Quantum reactive scattering in 3 dimensions using hyperspherical (APH) coordinates. 6. Analytic basis method for surface functions. *J. Chem. Phys.*, 98:6883–6896, 1993.
60. G.A. Parker, R.T. Pack, and A. Laganá, Accurate 3d-quantum reactive probabilities of $li + fh$. *Chem. Phys. Lett.*, 202:75–81, 1993.
61. J.D. Kress, Z. Bacic Z, G.A Parker, and R.T Pack, Quantum reactive scattering in 3 dimensions using hyperspherical (aph) coordinates. 5. comparison between 2 accurate potential energy surfaces for $H + H_2$ and $D + H_2$. *J. Phys. Chem.*, 94:8055–8058, 1990.
62. J.D. Kress, R.T. Pack, and G.A. Parker, Accurate 3-dimensional quantum scattering calculations for $F{+}H_2 \rightarrow HF{+}H$ with total angular-momentum J = 1. *Chem. Phys. Lett.*, 170:306–310, 1990.
63. Z. Bacic, J.D. Kress, G.A. Parker, and R.T. Pack, Quantum reactive scattering in 3 dimensions using hyperspherical (aph) coordinates. 4. discrete variable representation (dvr) basis functions and the analysis of accurate results for $f + h_2$. *J. Chem. Phys.*, 92:2344–2361, 1990.
64. F.T. Smith, Generalized angular momentum in many-body collisions. *Phys. Rev.*, 120:1058–1069, 1960.
65. D. Delande, PhD thesis, Université Pierre et Marie Curie, Paris, France, 1988.
66. V. Aquilanti, A. Caligiana, S. Cavalli, and C. Coletti, Hydrogenic orbitals in momentum space and hyperspherical harmonics elliptic sturmian basis sets. *Int. J. Quant. Chem.*, 92:212–228, 2003.
67. O.I. Tolstikhin and H. Nakamura, Hyperspherical elliptic coordinates for the theory of light atom transfer reactions in atom-diatom collisions. *J. Chem. Phys.*, 108:8899–8921, 1998.
68. O.I. Tolstikhin, V. N. Ostrovsky, and H. Nakamura, Cumulative reaction probability without absorbing potentials. *Phys. Rev. Lett.*, 80:41–44, 1998.
69. O.I. Tolstikhin, I. Yu. Tolstikhina, and C. Namba, Interference effects in the decay of resonance states in three-body coulomb systems. *Phys. Rev. A*, 60:4673–4692, 1999.
70. O.I. Tolstikhin and C. Namba, Hyperspherical calculations of low-energy rearrangement processes in $dt\mu$. *Phys. Rev. A*, 60:5111–5114, 1999.
71. K. Nobusada, O.I. Tolstikhin, and H. Nakamura, Quantum mechanical elucidation of reaction mechanisms of heavy-light-heavy systems: Role of potential ridge. *J. Chem. Phys.*, 108:8922–8930, 1998.
72. K. Nobusada, O.I. Tolstikhin, and H. Nakamura, Quantum reaction dynamics of heavy-light-heavy systems: Reduction of the number of potential curves and transitions at avoided crossings. *J. Phys. Chem. A*, 102:9445–9453, 1998.
73. K. Nobusada, O.I. Tolstikhin, and H. Nakamura, Quantum reaction dynamics of Cl + HCl → HCl + Cl: vibrationally non-adiabatic reactions. *J. Mol. Struct.*, Theochem 461-2:137–144, 1999.
74. O.I. Tolstikhin and M. Matsuzawa, Hyperspherical elliptic harmonics and their relation to the heun equation. *Phys. Rev. A*, 63:032510/1–032510/8, 2001.
75. O.I. Tolstikhin and M. Matsuzawa, Exploring the separability of the three-body coulomb problem in hyperspherical elliptic coordinates. *Phys. Rev. A*, 63:062705/1–062705/23, 2001.
76. V.K. Babamov and R.A. Marcus, Dynamics of hydrogen atom and proton transfer reactions. Symmetric case. *J. Chem. Phys.*, 74:1790–1798, 1981.
77. A. Kuppermann, A useful mapping of triatomic potential energy surfaces. *Chem. Phys. Lett.*, 32:374–375, 1975.
78. B.R. Johnson, On hyperspherical coordinates and mapping the internal configurations of a three body system. *Chem. Phys.*, 73:5051–5058, 1980.
79. B.R. Johnson, The classical dynamics of three particles in hyperspherical coordinates. *J. Chem. Phys.*, 79:1906–1915, 1983.
80. B.R. Johnson, The quantum dynamics of three particles in hyperspherical coordinates. *J. Chem. Phys.*, 79:1916–1925, 1983.
81. L. Wolniewicz and J. Hinze, Atom-diatomic molecular reactive scattering: Investigation of the hyperangular integration. *J. Chem. Phys.*, 85:2012–2018, 1986.
82. J.T. Muckerman, R.D. Gilbert, and G.D. Billing, A classical path approach to reactive scattering. i. use of hyperspherical coordinates. *J. Chem. Phys.*, 88:4779–4787, 1988.
83. R.T. Pack, Coordinates for an optimum CS approximation in reactive scattering. *Chem. Phys. Lett.*, 108:333–338, 1984.

84. M. Mishra and J. Linderberg, Hyperspherical representations of triatomic energy surfaces. *Mol. Phys.*, 50:91, 1983.
85. J. Linderberg and B. Vessal, Reactive scattering in hyperspherical coordinates. *Int. J. Quant. Chem.*, 31:65, 1987.
86. C.A. Mead, Superposition of reactive and nonreactive scattering amplitudes in the presence of a conical intersection. *J. Chem. Phys.*, 72:3839–3840, 1980.
87. J. Robert and J. Baudon, A molecular description of molecular collisions. *J. Phys. B*, 19:171–184, 1986.
88. J. Robert and J. Baudon, *J. Phys. (Paris)*, 47:631, 1986.
89. J. Robert and J. Baudon, *Europhys. Lett.*, 2:363, 1986.
90. M. De Celles and B.T. Darling, *J. Mol. Spectrosc.*, 29:66, 1969.
91. V. Aquilanti, A. Laganá, and R.D. Levine, On the all channels representation of the potential energy surface for reactive collisions. *Chem. Phys. Lett.*, 158:87–94, 1989.
92. V. Aquilanti, S. Cavalli, and G. Grossi, Dynamics on reactive potential energy surfaces: the hyperspherical view. *Advances in Molecular Vibrations and Collision Dynamics,* ed. J.M. Bowman, JAI Press, Greenwhich (Conn), 2A:147–181, 1993.
93. V. Aquilanti, S. Cavalli, and G. Grossi, On the ridge effect in mode transitions: semiclassical analysis of the quantum pendulum. *Chem. Phys. Lett.*, 110:43–48, 1984.
94. V. Aquilanti and S. Cavalli, Hyperspherical analysis of kinetic paths for elementary chemical reactions and their angular momentum dependence. *Chem. Phys. Lett.*, 141:309–314, 1987.
95. V. Aquilanti, S. Cavalli, G. Grossi, V. Pellizzari, M. Rosi, A. Sgamellotti, and F. Tarantelli, Potential energy surfaces in hyperspherical coordinate: abinitio kinetic paths for the $O(3P)$ + H_2 reaction. *Chem. Phys. Lett.*, 162:179–184, 1989.
96. V. Aquilanti, G. Capecchi, S. Cavalli, C. Adamo, and V. Barone, Representation of potential energy surfaces by discrete polynomials: proton transfer in malonaldehyde. *Phys. Chem. Chem. Phys.*, 2:4095–4103, 2000.
97. V. Aquilanti, S. Cavalli, C. Coletti, D. Di Domenico, and G. Grossi, Hyperspherical harmonics as sturmian orbitals in momentum space: a systematic approach to the few-body coulomb problem. *Int. Rev. in Phys. Chem.*, 20:673–709, 2001.
98. V. Aquilanti, G. Capecchi, S. Cavalli, D. De Fazio, P. Palmieri, C. Puzzarini, A. Aguilar, X. Gimnez, and J.M. Lucas, He + H_2^+ reaction: a dynamical test on potential energy surfaces for a system exhibiting a pronounced resonance pattern. *Chem. Phys. Lett.*, 318:619–628, 2000.
99. P. Palmieri, C. Puzzarini, V. Aquilanti, G. Capecchi, S. Cavalli, D. De Fazio, Aguilar, X. Gimenez, and J.M. Lucas, Ab initio dynamics of the He + $H_2^+ \rightarrow HeH^+$ + H reaction: a new potential energy surfaces and quantum mechanical cross sections. *Mol. Phys.*, 98:1835– 1849, 2000,
100. V. Aquilanti, S. Cavalli, D. De Fazio, A. Volpi, A. Aguilar, X. Gimnez, and J.M. Lucas, Exact reaction dynamics by the hyperquantization algorithm: integral and differential cross section for F + H_2, including long-range and spin-orbit effects. *Physical Chem. Chem. Phys.*, 4:401–415, 2002.
101. V. Aquilanti, A. Beddoni, A. Lombardi, and R. Littlejohn, Hyperspherical harmonics for polyatomic systems: basis sets for kinematic rotations. *Int. J. Quant. Chem.*, 89:277–291, 2002.
102. V. Aquilanti, S. Cavalli, A. Simoni, A. Aguilar, J.M. Lucas, and D. De Fazio, Lifetime of reactive scattering resonances: Q-matrix analysis and angular momentum ddependence for the F + H_2 reaction by the hyperquantization algorithm. *J. Chem. Phys.*, 121:11675–11690, 2004.
103. V. Aquilanti, S. Cavalli, D. De Fazio, A. Simoni, and T.V. Tscherbul, Direct evaluation of the lifetime matrix by the hyperquantization algorithm: Narrow resonances in the F + H_2 reaction dynamics and their splitting for nonzero angular momentum. *J. Chem. Phys.*, 123(054314):1–15, 2005.
104. V. Aquilanti, S. Cavalli, D. De Fazio, A. Volpi, A. Aguilar, J.M. Lucas, and X. Gimnez, Reactivity enhanced by under-barrier tunneling and resonances: the F + $H_2 \rightarrow$ HF + H reaction. *Chem. Phys. Lett.*, 371:504–509, 2003.
105. V. Aquilanti, S. Cavalli, D. De Fazio, A. Volpi, A. Aguilar, and J.M. Lucas, Benchmark rate constants by the hyperquantization algorithm. the F + H_2 reaction for various potential energy surfaces: features of the entrance channel and of the transition state, and low temperature reactivity. *Chem. Phys.*, 308:237–253, 2005.

106. V. Aquilanti, L. Bonnet, and S. Cavalli, Kinematic rotations for four-center reactions: Mapping tetra-atomic potential energy surfaces on the kinetic sphere. *Mol. Phys.*, 89:1–12, 1996.
107. V. Aquilanti and S. Cavalli, The quantum-mechanical hamiltonian for tetra-atomic systems in symmetric hyperspherical coordinates. *J. Chem. Soc. Faraday Trans.*, 93:801–809, 1997.
108. R.G. Littlejohn, K.A. Mitchell, M. Reinsch, V. Aquilanti, and S. Cavalli, Internal spaces, kinematic rotations and body frames for four-atom systems. *Phys. Rev.*, 58(A):3718–3738, 1998.
109. R.G. Littlejohn, K.A. Mitchell, and V. Aquilanti, Quantum dynamics of kinematic invariants in tetra- and polyatomic systems. *Phys Chem. Chem. Phys.*, 1:1259–1264, 1999.
110. V. Aquilanti, A. Beddoni, S. Cavalli, A. Lombardi, and R. Littlejohn, Collective hyperspherical coordinates for polyatomic molecules and clusters. *Mol. Phys.*, 98:1763–1770, 2000.
111. V. Aquilanti, A. Lombardi, and R.G. Littlejohn, Hyperspherical harmonics for polyatomic systems: basis set for collective motions. *Theor. Chem. Accounts*, 111:400–406, 2004.
112. D.S. Wang and A. Kuppermann, Use of symbolic algebra in the calculation of hyperspherical harmonics. *Int. J. Quant. Chem.*, 106:152–166, 2006.
113. A. Kuppermann, Hyperspherical harmonics for tetraatomic systems. 2. the weak interaction region. *J. Phys. Chem. A*, 108:8894–8904, 2004.
114. A. Kuppermann, Quantum reaction dynamics and hyperspherical harmonics. *Isr. J. Chem.*, 43:229–241, 2003.
115. D.S. Wang and A. Kuppermann, Hyperspherical harmonics for triatomic systems. *J. Phys. Chem. A*, 107:7290–7310, 2003.
116. D.S. Wang and A. Kuppermann, Hyperspherical harmonics for tetraatomic systems. *J. Chem. Phys.*, 115:9184–9208, 2001.
117. V. Aquilanti, A. Lombardi, and E. Yurtsever, Global view of classical clusters: the hyperspherical approach to structure and dynamics. *Phys. Chem. Chem. Phys.*, 4:5040–5051, 2002.
118. V. Aquilanti, A. Lombardi, M.B. Sevryuk, and E. Yurtsever, Phase-space invariants as indicators of the critical behavior of nanoaggregates. *Phys. Rev. Lett.*, 93(4):113–402, 2004.
119. V. Aquilanti, A. Lombardi, and M.B. Sevryuk, Phase-space invariants for aggregates of particles: hyperangular momenta and partitions of the classical kinetic energy. *J. Chem. Phys.*, 121:5579–5589, 2004.
120. V. Aquilanti, E. Carmona-Novillo, E. Garcia, A. Lombardi, M.B. Sevryuk, and E. Yurtsever, Invariant energy partitions in chemical reactions and cluster dynamics simulations. *Computational Materials Science*, 35:187–191, 2005.
121. M.B. Sevryuk, A. Lombardi, and V. Aquilanti, Hyperangular momenta and energy partitions in multi-dimensional many-particle classical mechanics: the invariance approach to cluster dynamics. *Phys. Rev. A*, 72:033201, 2005.

ON THE TIME-DEPENDENT SOLUTIONS OF THE SCHRÖDINGER EQUATION

ALEJANDRO PALMA AND I. PEDRAZA

Instituto de Física (BUAP), Apartado Postal J-48, Puebla, Pue. 72570, México

Abstract Recently, there has been some interest in finding exact solutions to the time-dependent Schrödinger equation, specifically in the case of a time-dependent linear potential, though surprisingly in all those works very cumbersome methods are used. In the present report we want to emphasize that there exists another method, quite general and simple, to solve such kind of problems. The method was proposed several years ago and it is based on the so called Wei–Norman theorem.

Recently, there has been some interest in the solutions of the Schrödinger equation for the time-dependent linear potential [1–6]. Most of the authors use the method of the Lewis–Riesenfeld invariant while Feng [6] used a a space time transformation method. In this work we want to emphasize that there is yet another method, simpler and straightforward, based on the Wei–Norman theorem [7]. A particular version of this method has been used by Rau *et al.* [8] to analyze the same problem and later on applied to the quantum Liouville–Bloch equation [9]. Curiously, in this publication [9], they do not give credit to the work of Wei and Norman although they do in the first one [8]. Our approach is quite different from that of Rau *et al.* since we avoid guessing the solution (ansatz) and, instead, a closed algebra is defined by adding some operators to the original problem. Although this would seem to complicate the problem, it happens to be just the opposite: the problem can be solved straightforward and, from the very beginning, the coefficients of the new operators are set equal to zero, thus leading to the solution we are looking for.

As an example of how this method works, let us consider the equation [8]

$$(1) \qquad i\frac{\partial \psi}{\partial t} = \left\{ -\frac{1}{2}\,\frac{\partial^2}{\partial x^2} + E_0 x \sin wt \right\} \psi,$$

S. Lahmar et al. (eds.), Topics in the Theory of Chemical and Physical Systems, 147–150.

which can be written as

$$(2)\qquad i\frac{\partial\psi}{\partial t}=\left\{\sum_{i=1}^{4}a_iH_i\right\}\psi$$

where $a_1 = 0,\ H_1 = 1;\ \ a_2 = E_0\sin wt,\ H_2 = x;\ \ a_3 = 0,\ H_3 = \frac{\partial}{\partial x};\ \ a_4 = -\frac{1}{2},\ H_4 = \frac{\partial^2}{\partial x^2}$. It is very simple to show that $\pounds = \{H_1, H_2, H_3, H_4\}$ is a solvable Lie Algebra [7] since $\pounds'' = \{0\}$, so that the problem can be solved by quadratures.

Actually, solving the above equation is equivalent to do so for the evolution operator

$$(3)\qquad i\frac{\partial U}{\partial t}=\dot{g}_1H_1U+\dot{g}_2H_2U+\dot{g}_3UH_3+\dot{g}_4UH_4,$$

where $\Psi(x,t) = U(t)\Psi(x,0) = e^{g_1H_1}e^{g_2H_2}e^{g_3H_3}e^{g_4H_4}\varphi(x)$ and the upper dots denote differentiation with respect to t, and $\Psi(x,0) = \varphi(x)$ is the solution of the time independent Schrödinger equation.

Application of some well–known operator algebra techniques, leads to a set of four linear equations:

$$(4a)\qquad i\dot{g}_4=-\frac{1}{2}$$

$$(4b)\qquad i\dot{g}_2=E_0\sin wt$$

$$(4c)\qquad \dot{g}_3-2\dot{g}_4g_2=0$$

$$(4d)\qquad \dot{g}_1-\dot{g}_3g_2+\dot{g}_4g_2^2=0$$

which can be easily integrated to give the desired solution, that is, the one reported by Rau and Unnikrishnan [8]. The case with $E = E_0\cos wt$ can be treated in a similar fashion, and also those analyzed in references [1–6]. The advantage of using properly the Wei–Norman method is that we can know in advance whether the problem is soluble or not, as it is shown to be the case here, because it corresponds to a solvable Lie algebra.

Another important feature of the Wei–Norman method is that the solution is global [10], i.e. it is valid in the whole domain of variable t, restricted only to time-dependent equations where solutions are of the exponential type.

Let us consider a more general equation, the one which is called the reduced velocity gauge or the Airy–Gordon–Volkov wave equation [11].

$$(5)\qquad i\frac{\partial\psi}{\partial t}=\left\{-\frac{1}{2m}\frac{\partial^2}{\partial x^2}+i\frac{A_0\cos wt}{m}\frac{\partial}{\partial x}+V-Fx\right\}\psi,$$

which can be transformed to an equivalent one for the evolution operator, as we did in the previous case:

$$(6)\qquad \frac{dU}{dt}=\left\{\frac{i}{2m}\frac{\partial^2}{\partial x^2}+\frac{A_0\cos wt}{m}\frac{\partial}{\partial x}+iFx-iV\right\}U(t)=H(t)U(t).$$

Introducing now the definitions

(7a) $$a_1(t) = \frac{i}{2m}$$

(7b) $$a_2(t) = \frac{A_0 \cos wt}{m}$$

(7c) $$a_3(t) = iF$$

(7d) $$a_4(t) = iV$$

(7e) $$H_1 = \frac{\partial^2}{\partial x^2}$$

(7f) $$H_2 = \frac{\partial}{\partial x}$$

(7g) $$H_3 = x$$

(7h) $$H_4 = I$$

This set of four operators forms a solvable Lie algebra, as we pointed out above, and the proposed Eq. (5) must have an elementary solution. In order to find it, we propose again:

(8) $$\Psi(x_1 t) = U(t)\Psi(x,0) = e^{g_1 H_1} e^{g_2 H_2} e^{g_3 H_3} e^{g_4 H_4} \varphi(x)$$

Repeating the procedure outlined above we obtain a set of four *linear* differential equations:

(9a) $$\dot{g}_1 = a_1$$

(9b) $$\dot{g}_2 + 2\dot{g}_3 g_1 = a_2$$

(9c) $$\dot{g}_3 = a_3$$

(9d) $$\dot{g}_4 + \dot{g}_3 g_2 = a_4$$

which can be easily integrated thus obtaining:

(10a) $$g_1(t) = \frac{it}{2m}$$

(10b) $$g_2(t) = \frac{A_0 \sin wt}{mw} + \frac{F}{2m} t^2$$

(10c) $$g_3(t) = iFt$$

(10d) $$g_4(t) = iVt + i\frac{A_0 F \cos wt}{mw^2} - i\frac{F^2}{6m} t^3 - i\frac{A_0 F}{mw^2}$$

It is important to point out that this solution for the evolution operator $U(t)$ is not unique, since it depends on the order in which we arrange the elements of the corresponding Lie algebra, we can obtain a unique solution by using the BCH formula [12]:

$$(11) \qquad \Psi(x,t) = e^{-iEt} e^{i\frac{FA_0 \cos wt}{mw^2}} A_i \left[-(2mF)^{1/3} \left(x + \frac{A_0 \sin wt}{mw} + \frac{E-V}{F} \right) \right]$$

where A_i is the Airy function.

There are several other similar cases, called length gauge, velocity gauge, Kramers–Henneberger frame, which are particular cases of Eq. (2) with appropriate coefficient $a_i's$. All of them are related to the same Lie algebra but with different Hamiltonians and their solution has been reported elsewhere [13].

We have shown in this work the power and elegance of Lie algebraic methods in the solution of differential equations, which, when used properly, lead easily to the desired solution. Application of this method to the case of the Caldirola–Kanai Hamiltonian is in progress and will be published elsewhere.

Acknowledgements

One of us (AP) acknowledges stimulating and fruitful discussions with R. Lefebvre. This work was supported by CONACYT (Consejo Nacional de Ciencia y Tecnología, México), under project SEP-2004-C01-47090, and by VIEP (Vicerrectoría de Investigación y Estudios de Posgrado, Benemérita Universidad Autónoma de Puebla).

‡ palma@sirio.ifuap.buap.mx
Visiting Professor at Instituto Nacional de Astrofísica, Óptica y Electrónica (INAOE)

References

1. P.G. Luan and C.S. Tang, ArXiv: quant-ph/0309174 v2 (14 Sept. 2004).
2. J.Q. Shen, ibid. A 0311002 v1 (1 Nov. 2003).
3. I. Guedes, *Phys. Rev.* **68**, 16102, 2003; **63**, 034102, 2001.
4. H. Bekkar, F. Benamira, and M. Maamache, *Phys Rev. A* **68**, 016101, 2003.
5. J. Bauer, *Phys. Rev. A* **65**, 036101, 2002.
6. M. Feng, *Phys. Rev. A* **64**, 034101, 2001.
7. J. Wei and E. Norman, *J. Math. Phys.* **4**, 575, 1963.
8. A.R.P. Rau and K. Unnikrishnan, *Phys. Lett. A* **222**, 304, 1996.
9. A.R.P. Rau, *Phys. Rev. Lett.* **81**, 4785, 1998.
10. J. Wei and E. Norman, *Proc. Am. Math. Soc.*, **15**, 327–334, 1964.
11. R. Lefebvre, private communication; see also: *Int. J. Quantum Chem.* **80** 110, 2000.
12. B. Mielnick and J. Plebanski, *Ann. Inst. H. Poincaré*, XII(3), 215–254, 1970.
13. I. Pedraza, Master thesis, BUAP, 2005.

Part II
Interactions and Clusters

AN IMPROVED 6-D POTENTIAL ENERGY SURFACE FOR AMMONIA

S. RASHEV[1], D.C. MOULE[2], AND S.T. DJAMBOVA[3]

[1]*Institute of Solid State Physics, Bulgarian Academy of Sciences, 72 Tsarigradsko Chaussee, 1784 Sofia, Bulgaria*
[2]*Department of Chemistry, Brock University, St. Catharines, ON, L2S3A1, Canada*
[3]*Physics Department, UACEG, 1 Hr. Smirnensky Bd, 1164 Sofia, Bulgaria*

Abstract In this work we discuss a fully symmetrized vibrational calculation designed for studying the vibrational level structure and the ground electronic state 6-D potential energy surface (PES) of ammonia. The PES of ammonia was modeled in a simple analytical form, as a Taylor series expansion in terms of the molecular vibrational coordinates. Calculations on the energy levels of ammonia $^{14}NH_3$ at the higher excess vibrational energies (up to about 7000 cm^{-1}) are presented; and compared to the experimental data. The values of the most important force constants of the ammonia PES were determined from a fitting of the calculated and experimentally measured data.

1. INTRODUCTION

The potential energy surface (PES) of ammonia has been studied repeatedly by many authors ([1–11], and references therein), and continues to be an object of active theoretical interest. Most authors start their analysis with an abinitio (or semi-empirical) calculation of the PES and then perform an additional refinement to achieve an agreement between the calculated and experimental vibrational frequencies. Lately, the discrete variable representation has received particular attention and is currently one of the preferred methods [3, 7, 8, 10–12].

Full-scale vibrational calculations on polyatomic molecules (including all vibrational degrees of freedom) represent a difficult problem, especially at the higher vibrational excitation energies, due to the steeply growing vibrational level densities. In our recent work [13] we described our fully symmetrized procedure and algorithm, especially designed for large-scale vibrational calculations on ammonia, and

S. Lahmar et al. (eds.), Topics in the Theory of Chemical and Physical Systems, 153–160.

in general, XH_3 molecules. The specific symmetric top symmetry of these molecules has been exploited by constructing an unrestricted, fully symmetrized, and separable (in product form) vibrational basis set, allowing for the reduction of the available level density to the submanifold of levels that belong to a single symmetry species. This approach enables the reliable exploration of highly excited vibrational levels in symmetric top molecules at reduced computational cost. For the purposes of the calculation, the ammonia PES was modeled in a simple analytical form, depending parametrically on a limited number of the most important harmonic and anharmonic force constants. The ammonia PES is expanded in a Taylor series around the totally symmetric planar configuration, in the way typical for a semirigid molecule, of D_{3h} symmetry. The flexibility is then taken into account by adopting a specific analytical form for the double well potential along the large amplitude out-of-plane vibration.

In this work we elaborate and extend our model description of the ammonia 6-D PES [13], in order to achieve a better reproduction of the molecular vibrational level structure at the higher vibrational energies. Our major aim is to propose a simple and transparent analytical expression for the PES of the electronic ground state of ammonia, which allows a straightforward assessment of the major intramolecular interactions and the determination of the complex rovibronic level structure [1, 15–18]. Making use of our model-vibrational Hamiltonian and method, which were specifically designed for symmetric top molecules, we have performed 6-D vibrational calculations using a least squares fitting routine to adjust the calculated vibrational frequencies to experimentally measured data. In this manner, we have been able to determine the values of a number of harmonic and anharmonic force constants as well as of other parameters which characterize the PES of ammonia.

2. VIBRATIONAL COORDINATES, HAMILTONIAN AND BASIS SET

We start with the six conventional curvilinear vibrational coordinates for ammonia [7, 14]: three (N–H) bond stretches r_i, and three interbond angle (H–N–H) distortions θ_i, $(i = 1, 2, 3)$. The molecular vibrational Hamiltonian and, in particular, our analytical PES are built with reference to the molecular totally symmetrical planar configuration. A basic feature of this formalism is its full D_{3h} symmetrization. For this purpose we define symmetrized curvilinear vibrational coordinates in a somewhat unconventional, complex form:

$$\begin{aligned} q_1(A_1') &= (r_1 + r_2 + r_3)/\sqrt{3}, \\ q_{3a}(E_a') &= (r_1 + G^* r_2 + G r_3)/\sqrt{3}, \\ q_{3b}(E_b') &= (r_1 + G r_2 + G^* r_3)/\sqrt{3}, \\ q_{4a}(E_a') &= R_0(\varphi_1 + G^*\varphi_2 + G\varphi_3)/\sqrt{3}, \\ q_{4b}(E_b') &= R_0(\varphi_1 + G\varphi_2 + G^*\varphi_3)/\sqrt{3}, \\ q_2'(A_1') &= R_0(\varphi_1 + \varphi_2 + \varphi_3)/\sqrt{3}, \end{aligned} \tag{1}$$

where R_0 is the N–H bond length at planar configuration and $G = e^{2i\pi/3}$, $G^* = e^{-2i\pi/3}$. They correspond to the symmetric stretch (q_1), the doubly degenerate antisymmetric stretch (q_{3a}, q_{3b}), and the antisymmetric bend (q_{4a}, q_{4b}). Instead of the symmetric bend coordinate q_2', we shall employ $q_2 = R_0\theta$ (that has the required symmetry A''), where θ is the out-of-plane angle between one N–H bond and the plane of two remaining bonds, for a symmetrical pyramidal configuration. Here q_2 is a function of q_2' defined by the relation:

$$\cos(q_2/R_0) = -\cos(2\pi/3 - q_2'/R_0\sqrt{3})/\cos(2\pi/6 - q_2'/R_0\sqrt{12}).$$

The zeroth-order vibrational Hamiltonian H_0 has been defined as the sum of three major parts [13]: a local mode (LM) Hamiltonian $H_0{}^{\mathrm{NH}}(r_1, r_2.r_3)$, involving three identical (N–H stretch) Morse oscillators, a 2D harmonic oscillator Hamiltonian $H_0{}^{(4)}(q_{4a}, q_{4b})$, describing the asymmetric bend (mode #4) in terms of the pair of symmetrized coordinates q_{4a}, q_{4b} and the inversion Hamiltonian $H_0{}^{(2)}(q_2)$. The eigenfunctions of the zeroth Hamiltonian H_0, i.e. the basis functions for the present vibrational treatment, are obtained as products of the eigenfunctions for each one of these three parts, each one in fully symmetrized form. The symmetrized eigenfunctions of the LM Hamiltonian $H_0{}^{\mathrm{NH}}(r_1, r_2.r_3)$ have been obtained in complex form, as simple linear combinations over the products of three Morse eigenfunctions, as discussed in detail in our preceding work [13]. They are designated as $|L;S(\mathrm{f});\mathrm{p}_1\mathrm{k}_2\mathrm{l}_3)$, where $\mathrm{p}_1, \mathrm{k}_2, \mathrm{l}_3$ are occupation (excitation) quantum numbers for Morse oscillators #1,2,3 respectively, S is the symmetry, L (taking values 1,3, or 6) is the number of terms in the linear combination, and f is 1 or i (imaginary unit). The eigenfunctions (n_{4a}, n_{4b}) (where $n_{4a}, n_{4b} = 0, 1, 2, \ldots$) of $H_0{}^{(4)}(q_{4a}, q_{4b})$ are directly obtained in complex symmetrized form [13].

The inversion double well potential in ammonia is a difficult problem that has been the object of numerous theoretical studies over the years [2–12]. We chose to model this potential as the superposition of a harmonic part [modeling its overall shape to the zeroth Hamiltonian $H_0{}^{(2)}(q_2)$], and two Gaussian terms (modeling the barrier), supplemented by three small corrections to the overall shape, of fourth, sixth, and eighth powers in (q_2):

$$\begin{aligned} U(q_2) &= 0.5F_{2.2}q_2^{\,2}/(hc) + B\exp(-dq_2^{\,2}/K_2^{\,2}) \\ &\quad + B_1\exp(-d_1q_2^{\,2}/K_2^{\,2}) + (F_4q_2^{\,4} + F_6q_2^{\,6} + F_8q_2^{\,8})/(hc)[\mathrm{cm}^{-1}], \end{aligned} \tag{2}$$

The kinetic energy dependence on the out-of-plane angle θ, can be derived in the form [14]:

$$\begin{aligned} G_{\theta\theta}^{(2)} &= \frac{\mu_H}{R_0^2}\left[1 + \frac{2}{\cos^2\theta} - \frac{32\tan^2\theta}{16 - [\cos^2\theta - \cos\theta\sqrt{\cos^2\theta + 8}]^2}\right] \\ &\quad + 3\frac{\mu_N}{R_0^2}\frac{4 - \cos^2\theta + \cos\theta\sqrt{\cos^2\theta + 8}}{4 + \cos^2\theta - \cos\theta\sqrt{\cos^2\theta + 8}}. \end{aligned} \tag{3}$$

At $\theta = 0$ (planar configuration), this expression is reduced to: $G_{00}^{(2)} = (3\mu_{\mathrm{H}} + 9\mu_{\mathrm{N}})/R_0^2$. Here μ_{H} and μ_{N} are the inverse masses of the H and N–atoms, respectively, $K_2 = \sqrt{\hbar\sqrt{G_{00}{}^{(2)}F_{2,2}}}$, h is Planck's constant, c the light velocity, and $B, B_1[\mathrm{cm}^{-1}], d, d_1$ [dimensionless], F_4, F_6, F_8 [aJ and Angström] are variable parameters. The basis functions for the treatment of this mode serve as the eigenfunctions $|n_2\rangle$ of the harmonic oscillator zeroth-order Hamiltonian $H_0{}^{(2)}$, with frequency $\omega_2 = \sqrt{F_{2,2}G_{00}{}^{(2)}}/(2\omega c)[\mathrm{cm}^{-1}]$, with symmetries A' and A'' for n_2 even or uneven, respectively.

As was pointed out above, a basic feature of this approach is the completely symmetrized, separable and unrestricted basis set employed: $|i\rangle = |L; S(\mathrm{f}); p_1k_2l_3\rangle \times (n_{4\mathrm{a}}, n_{4\mathrm{b}}) \times |n_2\rangle$. Besides $|i\rangle$, each factor in this product is symmetrized (belongs to well defined symmetry species of $D_{3\mathrm{h}}$) and, in addition, all (excitation) quantum numbers involved can take arbitrarily high values: $p, k, l, n_{4\mathrm{a}}, n_{4\mathrm{b}}, n_2 = 0, 1, 2, \ldots$ In order to account for the various (anharmonic) interactions among basis states $|i\rangle$, we employ interaction Hamiltonian H^{int} terms of orders from 2 to 8. These interaction terms can be either kinetic or potential. The kinetic terms are obtained as G-matrix derivatives of appropriate orders, in the symmetrized curvilinear coordinates, as discussed in our previous work [13]. Only for the inversion mode has an explicit q_2-dependent kinetic energy expression (Eq. 3) been employed. The following potential terms have been included in the calculations, involving the relevant force constants that correspond to a Taylor series expansion of the PES at the planar totally symmetric reference configuration. The only potential terms that are not of this type are those modeling the large amplitude inversion motion (Eq. 2):

Zeroth potential:

$$\begin{aligned} U_0(r_1, r_2, r_3, q_2, q_{4\mathrm{a}}, q_{4\mathrm{b}}) &= D[1-\exp(-a_{\mathrm{r}}r_1)]^2 + \\ & D[1-\exp(-a_{\mathrm{r}}r_2)]^2 + D[1-\exp(-a_{\mathrm{r}}r_3)]^2 + \\ & F_{4,4}q_{4\mathrm{a}}q_{4\mathrm{b}} + 0.5F_{2,2}q_2{}^2, \end{aligned} \tag{4}$$

where: $D[1-\exp(-a_{\mathrm{r}}r)]^2 = f_{\mathrm{rr}}r^2/2 + f_{\mathrm{rrr}}r^3/6 + \ldots$.

Interaction potential:

$$\begin{aligned} U^{\mathrm{int}}(r_1, r_2, r_3, q_2, q_{4\mathrm{a}}, q_{4\mathrm{b}}) &= f_{1,2}(r_1r_2 + r_2r_3 + r_3r_1) + f_{1,2,3}r_1r_2r_3 \\ &+ f_{1,1,2}[r_1{}^2(r_2+r_3) + r_2{}^2(r_1+r_3) + r_3{}^2(r_1+r_2)] \\ &+ f_{1,1,2,2}(r_1{}^2r_2{}^2 + r_2{}^2r_3{}^2 + r_3{}^2r_1{}^2) \\ &+ f_{1,1,2,3}(r_1{}^2r_2r_3 + r_2{}^2r_3r_1 + r_3{}^2r_1r_2) \\ &+ f_{1,1,1,2}[r_1{}^3(r_2+r_3) + r_2{}^3(r_1+r_3) + r_3{}^3(r_2+r_1)] \\ &+ f_{1,1,2,2,3,3}r_1{}^2r_2{}^2r_3{}^2 + f_{1,1,1,2,2,2}(r_1{}^3r_2{}^3 + r_2{}^3r_3{}^3 + r_3{}^3r_1{}^3) \\ &+ f_{1,1,1,1,2,3}(r_1{}^4r_2r_3 + r_2{}^4r_3r_1 + r_3{}^4r_1r_2) \\ &+ f_{1,1,1,1,1,2}[r_1{}^5(r_2+r_3) + r_2{}^5(r_3+r_1) + r_3{}^5(r_1+r_2)] \\ &+ f_{1,1,1,2,2,3}[r_1{}^3(r_2{}^2r_3 + r_3{}^2r_2) + r_2{}^3(r_3{}^2r_1 \end{aligned}$$

$$+ r_1{}^2 r_3) + r_3{}^3(r_2{}^2 r_1 + r_1{}^2 r_2)] + f_{1,1,1,1,2,2}[r_1{}^4(r_2{}^2 + r_3{}^2)$$
$$+ r_2{}^4(r_3{}^2 + r_1{}^2) + r_3{}^4(r_1{}^2 + r_2{}^2)] + F_{3,4}(q_{3a}q_{4b} + q_{3b}q_{4a})$$
$$+ F_{3,4,4}(q_{3a}q_{4a}{}^2 + q_{3b}q_{4b}{}^2) + F_{1,4,4}q_1 q_{4a}q_{4b} + F_{1,2,2}q_1 q_2{}^2$$
$$+ F_{2,2,4,4}q_2{}^2(q_{4a}q_{4b}) + F_{2,2,2,2,4,4}q_2{}^4(q_{4a}q_{4b})$$
$$+ F_{2,2,4,4,4,4}q_2{}^2(q_{4a}q_{4b})^2 + F_{1,1,1,1,2,2}q_1{}^4 q_2{}^2$$
$$+ F_{1,1,2,2,2,2}q_1{}^2 q_2{}^4 + F_{1,1,1,1,1,1,2,2}q_1{}^6 q_2{}^2$$
$$+ F_{1,1,1,1,2,2,2,2}q_1{}^4 q_2{}^4 + F_{1,1,2,2,2,2,2,2}q_1{}^2 q_2{}^6$$
$$+ F_{2,2,3,4}q_2{}^2(q_{3a}q_{4b} + q_{3b}q_{4a}) + F_{4,4,4,4}(q_{4a}q_{4b})^2$$
$$+ F_{4,4,4,4,4,4}(q_{4a}q_{4b})^3 + (hc)[B\exp(-dq_2{}^2/K_2^2)$$
$$+ B_1\exp(-d_1 q_2{}^2/K_2{}^2)] + F_2{}^{(4)}q_2{}^4 + F_2{}^{(6)}q_2{}^6 + F_2{}^{(8)}q_2{}^8.$$

Here, U_0 is the potential part of H_0, while U^{int} pertains to H^{int}. All Hamiltonian interaction terms [both kinetic and potential, except $G_{\theta\theta}{}^{(2)}$ (Eq. 3)] were expressed, in terms of raising and lowering operators, in explicit totally symmetric form, thus allowing an analytical calculation of the required matrix elements $\langle i|\, H^{\text{int}}\, |k\rangle$ for the Hamiltonian matrix.

3. VIBRATIONAL CALCULATIONS. RESULTS AND DISCUSSION

Six-dimension (6-D) calculations of the vibrational energy-level structure of ammonia were carried out, using a nonperturbative vibrational procedure including the following stages: (i) artificial intelligence (AI) search for the selection of an active space (AS) of N basis vectors $|i\rangle$ that are involved in substantial couplings among each other, all of them belonging to one and the same symmetry species of the molecular group D_{3h}. The algorithm and criteria for the search have been described earlier [13]. During this search, the relevant $N \times N$ Hamiltonian matrix $H_{\text{i,k}}$ was calculated; (ii) next, the resulting $H_{\text{i,k}}$ matrix was first Lanczos tridiagonalized and then diagonalized for the eigenvalues, i.e. the molecular vibrational energy levels of a given symmetry species, in an energy range around an appropriately chosen initial basis state $|0\rangle$ (for starting the AI search). However, the search algorithm selects also basis vectors $|i\rangle$ that are energetically located at some distance from $|0\rangle$ if they can contribute to the vibrational mixing pattern around $|0\rangle$. We employed a nonlinear least squares fitting routine, to adjust the calculated frequencies to a set of experimentally measured and reliably assigned fundamental, overtone and combination vibrational levels of ammonia $^{14}NH_3$, in the range up to about 7000 cm^{-1}, by varying the values of the force constants defined above (Eqs. 4, 5). We note that, using this vibrational calculation procedure, it is possible to compute very highly excited vibrational states in ammonia; however such results are not reported here, since the vibrational levels are congested and it was difficult to assign reliably the calculated vibrational frequencies. The results from calculations on higher vibrational levels in ammonia will be the object of a forthcoming detailed publication.

Table 1 contains the vibrational frequencies of ammonia calculated in this work together with experimentally measured values. Table 2 summarizes the values of all the force constants and parameters characterizing the ammonia PES (Eqs. 4, 5) corresponding to the vibrational frequencies of Table 1. The parameters of the calculation algorithm (see Ref. 13 for details) were fixed at values ensuring conversion of the calculated frequencies, $\sim 0.1\ \mathrm{cm}^{-1}$.

As it is seen on Table 1, the vibrational frequencies of ammonia are adequately reproduced by the PES given in Eqs. 4, 5, and in Table 2, both for the fundamentals, the highly excited inversion levels, and the overtones and combinations of all vibrational modes. From a comparison with our preceding work [13] it is seen that some of the force constants for ammonia are different from our previous work. This is

Table 1. Calculated vibrational frequencies for ammonia $^{14}NH_3$, compared to experimentally measured data.

Our notation	$n_2 = 1$	$N_2 = 2$	$n_2 = 3$	$n_2 = 4$	$n_2 = 5$	$n_2 = 6$
NM	$0(A_1'')$	$\nu_2(A_1')$	$\nu_2(A_1'')$	$2\nu_2(A_1')$	$2\nu_2(A_1'')$	$3\nu_2(A_1')$
Calculated	0.56	931.2	967.8	1597. 8	1882.8	2383.0
Experim.[Ref.]	0.79[5]	932.43[5]	968.12[5]	1597.47[15]	1882.18[15]	2384.15[16]

$n_2 = 7$	$n_2 = 8$	$n_{4a} = 1$	$n_{4a} = n_{4b} = 1$	$N_{4a} = 2$	$n_{4a} = n_2 = 1$	$n_{4a} = 1, n_2 = 2$
$3\nu_2(A_1'')$	$4\nu_2(A_1')$	$\nu_4(E')$	$2\nu_4(A_1')$	$2\nu_4(E')$	$\nu_4(E'')$	$\nu_2 + \nu_4(E')$
2895.8	3461.8	1626.3	3216.5	3240.9	1626.5	2538.2
2895.52[16]	3462[17]	1626.28[15]	3216.1[18]	3240.44[18]	1627.3[15]	2540.43[18]

$n_{4a}=1, n_2=3$	$n_{4a,4b}=1, n_2=2$	$n_{4a,4b}=1, n_2=3$	$\lvert 3;\, A_1';\, 1_1)$	$\lvert 3;\, E_a';1_1)$	$\lvert 3;\, A_1';1_1), n_2=1$
$\nu_2 + \nu_4(E'')$	$\nu_2 + 2\nu_4(A_1')$	$\nu_2 + 2\nu_4(A_1'')$	$\nu_1(A_1')$	$\nu_3(E_a')$	$\nu_1(A_1'')$
2583.0	4117.3	4172.2	3336.0	3443.2	3336.7
2586.02[18]	4115.62[18]	4173.25[18]	3336.08[17]	3443.68[17]	3337.10[17]

$\lvert 3;\, A_1';1_1), n_2 = 2$	$\lvert 3;\, A_1';\, 1_1), n_2 = 3$	$\lvert 3;\, E_a';1_1)$	$\lvert 3;\, E_a';\, 1_1), n_2 = 1$	$\lvert 3;\, E_a';\, 1_1), n_2 = 2$
$\nu_1 + \nu_2(A_1')$	$\nu_1 + \nu_2(A_1'')$	$\nu_3(E')$	$\nu_3(E'')$	$\nu_2 + \nu_3(E')$
4292.3	4320.5	3443.0	3443.4	4413.2
4294.51[18]	4320.93[18]	3443.68[17]	3443.99[17]	4416.91[18]

$\lvert 3;\, E_a';1_1), n_2 = 3$	$\lvert 3;\, A_1';\, 2_1)$	$\lvert 3;\, E_a';2_1)$	$\lvert 3;\, A_1';1_11_1)$	$\lvert 3;\, E_a';\, 1_11_1)$
$\nu_2 + \nu_3(E'')$	$2\nu_1(A_1')$	$\nu_1 + \nu_3(E')$	$2{\nu_3}^0(A_1')$	$2{\nu_3}^2(E')$
4431.6	6606.1	6609.1	6796.2	6850
4435.40[18]	6606.0[1]	6608.83[1]	6795.31[1]	6850.20[1]

Table 2. Force constants and parameters characterizing the ammonia PES. The description and units for the force constants are given in Eqs. 4, 5.

f_{rr}	f_{rrr}	$f_{1,2}$	$f_{1,1,2}$	$f_{1,2,3}$	$f_{1,1,2,2}$	$F_{1,1,2,3}$	$f_{1,1,1,2}$
7.0811	−42.537	−0.0938	0.8174	0.3526	−24.627	41.7442	−6.0178

$f_{1,1,1,2,2,2}$	$f_{1,1,2,2,3,3}$	$f_{1,1,1,1,2,3}$	$f_{1,1,1,1,1,2}$	$f_{1,1,1,2,2,3}$	$f_{1,1,1,1,2,2}$
886.59	38.9629	172.344	−123.064	−168.873	134.365

$F_{4,4}$	$F_{4,4,4,4}$	$F_{4,4,4,4,4,4}$	$F_{3,4}$	$F_{3,4,4}$
0.21943	−0.00927	0.276474	0.00872	−0.002437

$F_{1,4,4}$	$F_{1,1,4,4}$
−0.005703	−0.06185

$F_{1,2,2}$	$F_{3,2,2,4}$	$F_{1,1,2,2}$	$F_{2,2,4,4}$	$F_{2,2,2,2,4,4}$	$F_{2,2,4,4,4,4}$
−0.00374	0.03747	−0.22782	0.024768	−0.012765	0.029736

$F_{1,1,1,1,2,2}$	$F_{1,1,2,2,2,2}$	$F_{1,1,1,1,1,1,2,2}$	$F_{1,1,1,1,2,2,2,2}$	$F_{1,1,2,2,2,2,2,2}$
0.228428	−0.007123	−0.487485	0.7552235	0.188535

$F_{2,2}$	$F_2^{(4)}$	$F_2^{(6)}$	$F_2^{(8)}$	B	d
0.07457	0.0054	0.00324	−0.000882	15986.3	0.05456

partially due to the improved accuracy of the present determination in reproducing a large number of higher excited vibrational levels, including a larger number of (anharmonic) force constants, in the description of the molecular PES, as well as to the removal of some errors connected with the description of the inversion mode in ammonia.

4. CONCLUSIONS

In this study we have further elaborated and improved our fully symmetrized vibrational Hamiltonian description of ammonia, which was presented in recent work [13]. Our main objective was to develop the molecular 6-D PES in simple analytical form, as a Taylor series expansion in terms of the conventionally defined symmetrized force constants (harmonic and anharmonic) of various orders. Based on this PES, we have performed 6-D vibrational calculations allowing for a satisfactory reproduction of the experimental frequencies of both fundamental, overtone, and combination levels of ammonia $^{14}NH_3$, up to about $7000\,cm^{-1}$ of vibrational excitation energy. From this fit the values of the molecular force constants of various orders were determined. The values, especially for the higher-order anharmonic constants, should not be considered as final: they will be subject to further refinement as we introduce higher excited vibrational levels in the treatment.

Acknowledgements

Financial support for this work is gratefully acknowledged from the Bulgarian Ministry of Education and Scientific Research, Grant Φ–1415, and the National Science and Engineering Reseach Council of Canada.

References

1. G. Herzberg, *Molecular Spectra and Molecular Structure II. Infrared and Raman Spectra of Polyatomic Molecules*, Van Nostrand, Princeton, 1945.
2. P.R. Bunker and P. Jensen, *Molecular Symmetry and Spectroscopy*, NRC, Ottawa, 1998.
3. H. Lin, W. Thiel, S.N. Yurchenko, M. Carajal, and P. Jensen, *J. Chem. Phys.* **117**, 11265, 2002.
4. D. Rush and K. Wiberg, *J. Phys. Chem.* A **101**, 3143, 1997.
5. V. Spirko, *J. Mol. Spectrosc.* **101**, 30, 1983.
6. C. Leonard, N.C. Handy, S. Carter, and J.M. Bowman, *Spectrochim. Acta A* **58**, 825, 2002.
7. T. Rajamaki, A. Miani and L. Halonen, *J. Chem. Phys.* **118**, 6358, 2003.
8. F. Gatti, C. Jung, C. Leforestier, and X. Chapuisat, *J. Chem. Phys.* **111**, 7236, 1999.
9. J.M.L. Martin, T.J. Lee, and P.R. Taylor, *J. Chem. Phys.* **97**, 8361, 1992.
10. D. Lauvergnat and A. Nauts, *Chem. Phys.* **305**, 105, 2004.
11. D. Luckhaus, *J. Chem. Phys.* **113**, 1329, 2000.
12. S.N. Yurchenko, J. Zheng, H. Lin, P. Jensen, and W. Thiel, *J. Chem. Phys.* **123**, 134308, 2005.
13. S. Rashev and D.C. Moule, *J. Mol. Spectrosc.* **235**, 93, 2006.
14. E.B. Wilson, J.C. Decius, and P.C. Cross, *Molecular Vibrations*, McGraw-Hill, New York, 1955.
15. C. Kottaz, I. Kleiner, G. Tarrago, L.R. Brown, J.S. Margolis, R.L. Poynter, H.M. Pickett, T. Fouchet, P. Drossart, and E. Lellouch, *J. Mol. Spectrosc.* **203**, 285, 2000.
16. I. Kleiner, G. Tarrago and L.R. Brown, *J. Mol. Spectrosc.* **193**, 46, 1999.
17. I. Kleiner, L.R. Brown, G. Tarrago, Q.-L. Kou, N. Picque, G. Guelachvili, V. Dana, and J.-Y. Mandin, *J. Mol. Spectrosc.* **173**, 120, 1995.
18. V. Spirko and W. Kraemer, *J. Mol. Spectrosc.* **133**, 331, 1989.

A REVIEW ON GOLD–AMMONIA BONDING PATTERNS

E.S. KRYACHKO[1,2] AND F. REMACLE[1]

[1] *Department of Chemistry, Blg. B6c, University of Liege, Sart-Tilman, 4000 Liege 1, Belgium*
[2] *Bogoliubov Institute for Theoretical Physics, 03143-Kiev, Ukraine*

Bonding Patterns that Neutral and Charged Gold Clusters form with Small Ammonia Clusters and which implement Quantum Logic Gates

Abstract We demonstrate that the charge states of the gold atom and of the triangular gold cluster drastically influence their reactive properties and bonding patterns with ammonia clusters. These bonding patterns are a multifaceted phenomenon that exhibits different characteristics. We show how the specific bonding patterns of the landscapes of the potential energy surfaces of $Au_{1,3}{}^{Z}$-$(NH_3)_{1\leq n\leq 3}$ can be used for implementing two single-qubit quantum gates.

Keywords: gold, gold cluster, ammonia oligomer, anchoring bond, nonconventional hydrogen bond, proton acceptor ability, quantum gate, logic gate, Hadamard gate.

1. INTRODUCTION

The investigation of the gold–nitrogen (Au–N) bond, up to now, has mostly been confined to either the gold atom and its cation (i.e. $Au_1{}^{0,+1}$) or gold surfaces and wires (that is, literally, to $\sim Au_\infty$ compared to a given N-containing molecule). The former category of studies probed the Au–N bond with ammonia and ammonia dimer [1–4], pyridine [5–7], and DNA bases and pairs [8], whereas the latter was limited to the adsorption of ammonia on the Au(111) surface [9], the mechanical and conductance properties of 4, 4′ bipyridine attached to gold electrodes [10] or gold wires [11], and the bonding of 4-(dimethylamino) pyridine to gold nanoparticles [12]. Other works concern the short-lived (of ∼sub microseconds) dianion $Au(N_3)_4{}^{2-}$ [13a], the covalent Au–N bonding in Au–NPPh2 [13b], the gold–porphyrin cation [13c], and the interaction of $Au_n{}^{0,+1}$ clusters with nitric oxide [13d–e].

S. Lahmar et al. (eds.), Topics in the Theory of Chemical and Physical Systems, 161–191.

Very recently, it has been well established that the Au–N bond plays a key stabilization role in the hybridization of the DNA bases and Watson–Crick base pairs on gold clusters and surfaces [14–17]. Stimulated by these results, we further explore in this work the main features of the Au–N bond, and consider how the (neutral and charged) gold clusters $Au_3^Z(Z = 0, \pm 1)$ interact with the small ammonia clusters $(NH_3)_{1 \leq n \leq 3}$. Our aim is to provide a new consistent insight on the Au_3^Z–$(NH_3)_n$ bonding patterns and on their dependence on the charge states of gold clusters which have been overlooked in earlier studies, mainly focused on the cationic species.

The choice of a triangular gold cluster, Au_3^Z, a few angstroms of size, is motivated by the fact that it can be considered as the simplest reference model of gold nanoparticles that demonstrates their exceptional catalytic features [[14, 18]a–b]. We hereby partly cover the gap between the gold atom and the gold surface, rather on the former side.

The simplest complexes stabilized by the Au–N bond that have been already studied are gold–ammonia (Au–NH_3) and Au^+–NH_3, both of C_{3v} symmetry but markedly different in all other aspects. The first one is the Au–N equilibrium bond that anchors the ammonia molecule to $Au^Z(Z = 0, +1)$. In Au–NH_3, using the Douglas–Kroll method which is based on the spin-adapted single and double coupled-cluster excitation amplitudes, with perturbative estimate of the triple excitations (CCSD(T)) and the ROHF reference function [2], the Au–N bond is predicted to be equal to 2.277 Å and therefore appears to be much longer than Au^+–NH_3, with $R(Au^+–N) = 2.013$ Å (B3LYP [3]; 2.10 Å in Ref. [4]; MP2 predicts 2.028 Å). It is natural to anticipate a correlation between the Au–N bond and the binding energies of Au–NH_3 and of Au^+–NH_3: the binding energy $E_b(Au^+–NH_3) = 63.5\,kcal \cdot mol^{-1}$ with the shorter Au–N bond (B3LYP; 68.6 $kcal \cdot mol^{-1}$ for MP2; 65.3 $kcal \cdot mol^{-1}$ for CCSD(T) [3a–c]; also 54.9 $kcal \cdot mol^{-1}$ [3d]) considerably exceeds $E_b(Au–NH_3) = 9.19\,kcal \cdot mol^{-1}$ ([2]; 13–15 $kcal \cdot mol^{-1}$ in Ref. [1] using the method CCSD with 19-electron RECP), where the Au–N anchor is longer.

It appears that, in the complexes of Au–pyridine and Au^+–pyridine, the Au–N bond is longer, viz., 2.391 Å and 2.056 Å, as compared to Au–NH_3 and Au^+–NH_3. This is to be compared, on one hand, to E_b(Au–pyridine) $= 9.12\,kcal \cdot mol^{-1}$, (B3LYP/6–311 + G(d,p) (C_5H_5N)∪ LANL2DZ (Au) computational level [5]; see also Refs [6, 7]), almost equal to that of Au–NH_3; and, on the other hand, to $E_b(Au^+$–pyridine$) = 76.09\,kcal \cdot mol^{-1}$, larger than $E_b(Au^+$-$NH_3)$. In this respect it is worth noting that the electronic ground states of Au and Au^+ are [Xe]$5d^{10}6s^1$ and [Xe]$5d^{10}6s^0$, respectively. The 6*s* valence atomic orbital (AO) serves as an acceptor, in contrast to the donor valence 5*d* AOs [17]. The effective radius of the gold 6*s* orbital is approximately 1.62 Å, due to the well-known relativistic effect that causes this orbital to contract in an atom with high atomic mass, whereas the effective radius of its valence 5*d* orbitals is only ~0.86 Å. The Au–N bond of pyridine-$Au^Z(Z = 0, +1)$ complexes exhibits an efficient σ–donation bond which, in turn, leads to significant π–back-donation interaction [5]. Due to Coulomb interaction, the σ–donation obviously prevails for the cationic complex compared to the neutral one and thus explains the shortening of the Au–N bond under ionization of Au.

This paper is organized as follows. The computational framework is separately out-lined in notes [19–28]. Section 2 examines the complexes Au^Z-$(NH_3)_{1\leq n\leq 3}$($Z = 0, \pm 1$). On one hand, it partially overlaps with the recent work by Reimers and co-workers [4] on the single and double ammination of Au^+ while, on the other hand, it extends their study to the triple ammination, as well as to the similar reactions with neutral and negatevely charged gold. The primary goal of section 2 is to set the stage for the further studies of bonding patterns between gold and ammonia clusters, which are presented in section 3. It is shown that the Au–NH_3 bonding is not limited to a simple Au–N bond, but is a rather complex phenomenon which involves also the nonconventional hydrogen bond of N–H$\cdots$Au type where gold acts a nonconventional proton acceptor. This type of interaction controls the bonding patterns in the systems studied, especially for the negative charge state of the gold cluster and for the double and higher amminations of the neutral cluster. Section 4 illustrates how the computed potential energy surfaces (PESs) of $Au_{1,3}{}^Z$-$(NH_3)_{1\leq n\leq 3}$($Z = 0, \pm 1$) can be used to implement molecular qubit logic gate operations, and enables quantum information processing. In section 5 we summarize the key results of this work and outline perspectives particularly related to further design of molecular logic gates.

2. COMPLEXES Au^Z-$(NH_3)_{1\leq n\leq 3}$($Z=0, \pm 1$)

The equilibrium geometry of the complex Au–NH_3 is displayed in Figure 1. Its formation results from a so-called "anchoring" Au–N bond, with R(Au-N) = 2.354 Å. The binding energy E_b of Au–NH_3 amounts to 9.55 kcal · mol^{-1} without the ZPVE correction, which agrees with the previously reported values [1,2], and to 7.83 kcal · mol^{-1} after including ZPVE (notice that the E_b's reported hereafter are ZPVE-corrected). A charge transfer from the nitrogen to the gold atom which in terms of Mulliken charge amounts to $\Delta q_M(N) = 0.024\,e$ is one of the major effects that leads to the Au–N anchoring. The formation of the latter obviously weakens the N–H bonds, which undergo a small contraction and whose N–H stretching frequencies are lowered. Due to Coulomb repulsion, such anchoring is no longer possible in the Au^-–NH_3 complex – this is in contrast to a variety of anchoring patterns arising between the gold cation and the ammonia dimer or trimer as illustrated in Table 1. These patterns are closely related to the phenomenon of successive ammination of coinage metals [29] (also Ref. [4] and references therein).

The first ammination reaction of the cation Au^+ yields the complex Au^+–NH_3 with a shorter anchoring Au–N bond, R(Au^+–N) = 2.100 Å, and larger binding energy, $E_b(Au^+$–$NH_3) \approx 63.8$ kcal·mol^{-1}, than in Au–NH_3 (Table 1). Under the second ammination, Au^+ forms the more stable complex Au^+–$(NH_3)_2{}^I$, D_{3d} symmetry (see Table 1 for geometry), whose Au–N bond is further shortened by 0.014 Å with respect to Au^+–NH_3. Such a shortening could be interpreted as a sign favoring a second ammination. However, this is not the case in thermodynamical terms because, relative to the first ammination, the second ammination, at room temperature, is characterized by the positive difference in enthalpy $\Delta\Delta H_{298}{}^\circ \approx 0.9$ kcal · mol^{-1} and the negative difference in entropy $\Delta\Delta S^\circ = -49.4$ cal · K^{-1} · mol^{-1} which altogether

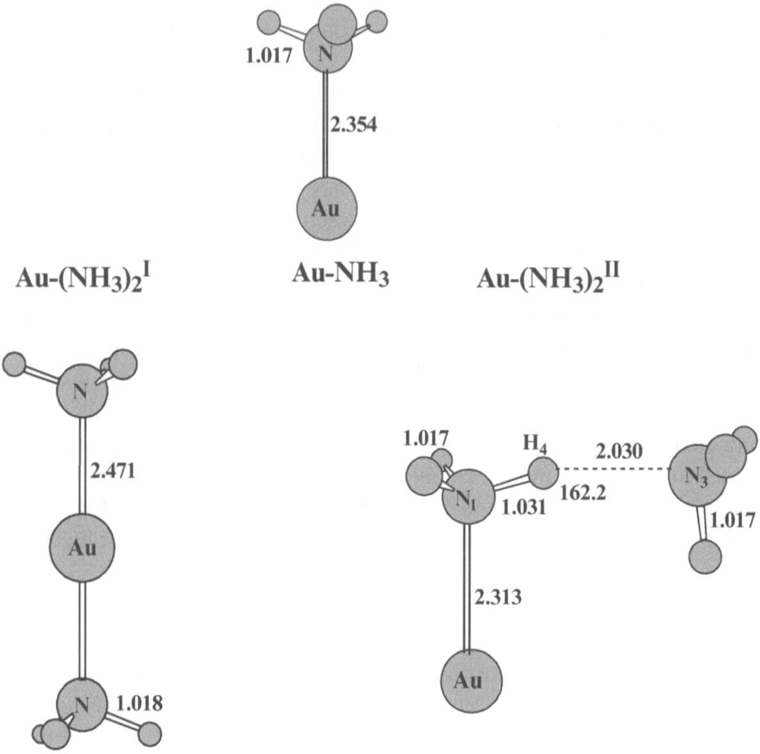

Figure 1. Complexes Au–$(NH_3)_{1\leq n\leq 2}$. The enthalpy of formation $\Delta H_{298}°$(Au–NH_3) = −7.21 kcal · mol^{-1}, $\Delta S°$(Au–NH_3) = 22.33 cal · K^{-1} · mol^{-1}, and the Gibbs free energy $\Delta G_{298}°$(Au–NH_3) = −13.86 kcal · mol^{-1}. Relative to asymptote Au–NH_3 + NH_3, the binding energy of Au–$(NH_3)_2{}^{II}$ amounts to 4.84 kcal · mol^{-1} (also $\Delta H_{298}°$(Au–$(NH_3)_2{}^{II}$) = −5.09 kcal·mol^{-1}, $\Delta S°$(Au–$(NH_3)_2{}^{II}$) = −23.96 cal · K^{-1} · mol^{-1}, and $\Delta G_{298}°$(Au–$(NH_3)_2{}^{II}$) = +2.05 kcal · mol^{-1}), whereas relative to Au + $(NH_3)_2$ it is equal to 10.74 kcal · mol^{-1}. The energy difference ΔE_{I-II} between Au–$(NH_3)_2{}^{I}$ and Au–$(NH_3)_2{}^{II}$ is 4.31 kcal · mol^{-1}. Comparing with the ammonia dimer, $\Delta\nu(N_1–H_4)$ of Au–$(NH_3)_2{}^{II}$ is equal to −131 cm^{-1} and IR activity of the $\nu(N_1–H_4)$ stretch is enhanced by a factor of ∼ 3.2. The H-bridge stretching vibrational mode $\nu_\sigma(H_4 \cdots N_3)$ = 198 cm^{-1}. The Au-N stretching mode ν(Au–N) = 314 cm^{-1} in Au–$(NH_3)_2{}^{II}$ and splits into ν^{sym}(Au–N) = 155 cm^{-1} and ν^{asym}(Au–N) = 175 cm^{-1} in the complex Au–$(NH_3)_2{}^{I}$. Bond lengths are given in Å and bond angles in degrees.

result in the Gibbs free energy difference $\Delta\Delta G_{298}° \approx 15.6$ kcal · mol^{-1}. Therefore, the second ammination of Au^+ can be regarded as primarily unfavorable compared to the first one [4] (see [30]).

Note that in a solution, where association of ammonia molecules can occur, there may exist another pathway for the second ammination reaction, where Au^+ anchors the H-bonded ammonia dimer and forms the complex Au^+–$(NH_3)_2{}^{II}$ (Table 1 for its geometry). This latter, however, is less stable, by ca. 43.6 kcal · mol^{-1}, than Au^+–$(NH_3)_2{}^{I}$. Such a reaction pathway, thermodynamically characterized by $\Delta\Delta H_{298}°$ = −16.7 kcal · mol^{-1}, $\Delta\Delta S°$ = −10.2 cal · K^{-1} · mol^{-1}, and

Table 1. Multiple amminations of Au^+ (compare with the neutral analogs in Figure 1). The corresponding reactant asymptotes are indicated in bold.

Complex	Selected features	$\Delta G_{298}°$(kcal· mol^{-1})	$\Delta H_{298}°$(kcal· mol^{-1})$\Delta S°$(cal· K^{-1} · mol^{-1})
$Au^+–NH_3$ 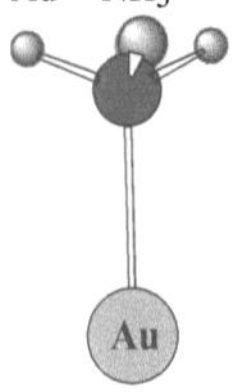	$R(Au^+–N) = 2.100$ Å $R(N–H)^a = 1.022$ Å $E_b(Au^+ + NH_3) =$ 63.76 kcal · mol^{-1} $IE(Au–NH_3)^b = 6.94$ eV	**$Au^+ + NH_3$**: −69.01 −56.8 (B3LYP [8][c]) −65.4 (PW91 [8][c])	**$Au^+ + NH_3$**: −63.49 −63.7 (B3LYP [8][c]) −72.4 (PW91 [8][c]) 18.53 −23.2 (B3LYP [8][c]) −23.5 (PW91 [8][c])
$Au^+–(NH_3)_2{}^I(D_{3d})$ 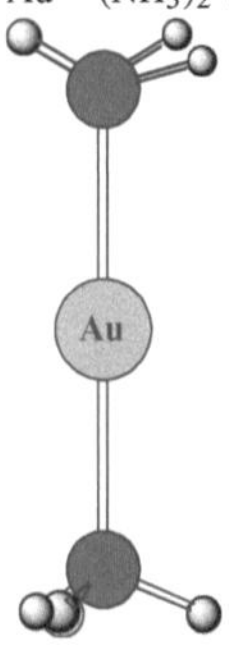	$R(Au^+–N) = 2.086$ Å $R(N–H) = 1.021$ Å $E_b(Au^+ + 2NH_3) =$ 125.58 kcal · mol^{-1} $E_b(Au^+–NH_3 + NH_3) =$ 61.82 kcal · mol^{-1}	**$Au^+ + 2NH_3$**: −122.42 −107.9 (B3LYP [8][c]) −120.8 (PW91 [8][c]) **$Au^+–NH_3 + NH_3$**: −53.41 −51.2 (B3LYP [8][c]) −55.4 (PW91 [8][c])	**$Au^+ + 2NH_3$**: −126.10 −124.7 (B3LYP [8][c]) −139.5 (PW91 [8][c]) −12.33 −56.2 (B3LYP [8][c]) −62.8 (PW91 [8][c]) **$Au^+–NH_3 + NH_3$**: −62.61 −61.0 (B3LYP [8][c]) −67.1 (PW91 [8][c]) −30.87 −33.0 (B3LYP [8][c]) −39.4 (PW91 [8][c])
$Au^+–(NH_3)_2{}^{II}$ 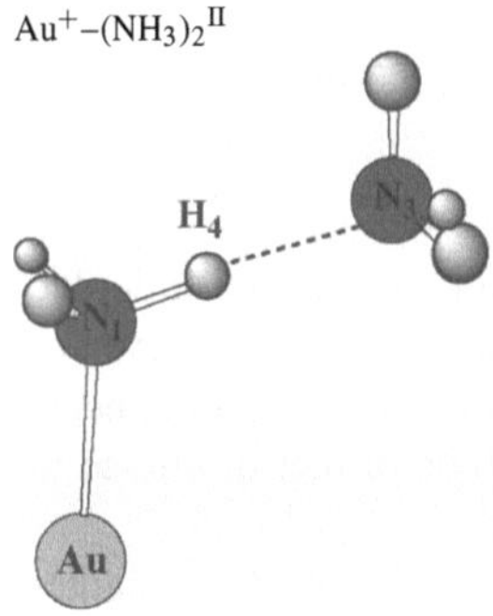	$R(Au^+–N) = 2.072$ Å $R(N_1–H_4)^d = 1.077$ Å $r(H_4 \cdots N_3)^d = 1.727$ Å $\angle N_1H_4N_3{}^d = 176.6°$ $\Delta\nu(N_1–H_4)^d = -874$ cm^{-1} $A_{IR}(Au^+–(NH_3)_2{}^{II})/A_{IR}((NH_3)_2)^d = 17.8$ $E_b(Au^+ + (NH_3)_2) =$ 80.07 kcal · mol^{-1} $\Delta E_{I–II} = -43.59$ kcal·mol^{-1}	**$Au^+ + (NH_3)_2$**: −82.72	**$Au^+ + (NH_3)_2$**: −80.23 8.36

Table 1. continued

$Au^+–(NH_3)_3{}^I$ 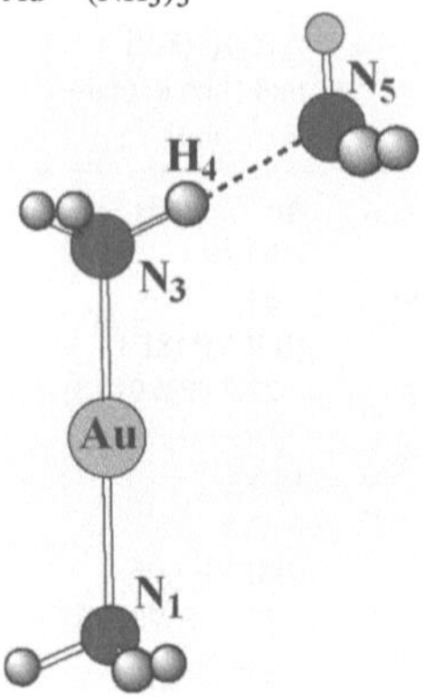	$R(Au^+–N_1) = 2.094$ Å $R(Au^+–N_3) = 2.070$ Å $R(N_3–H_4) = 1.056$ Å $r(H_4 \cdots N_5) = 1.844$ Å $\angle N_3H_4N_5 = 176.9°$ $\Delta\nu(N_3–H_4)^d = -533\,cm^{-1}$ $A_{IR}(Au^+–(NH_3)_3{}^I)/A_{IR}((NH_3)_2)^d = 11.4$ $E_b(Au^+–(NH_3)_2{}^I + NH_3) = 13.57\,kcal \cdot mol^{-1}$	**$Au^+–(NH_3)_2{}^I + NH_3$:** −6.40	**$Au^+–(NH_3)_2{}^I + NH_3$:** −13.96 −25.36
$Au^+–(NH_3)_3{}^{II}$ 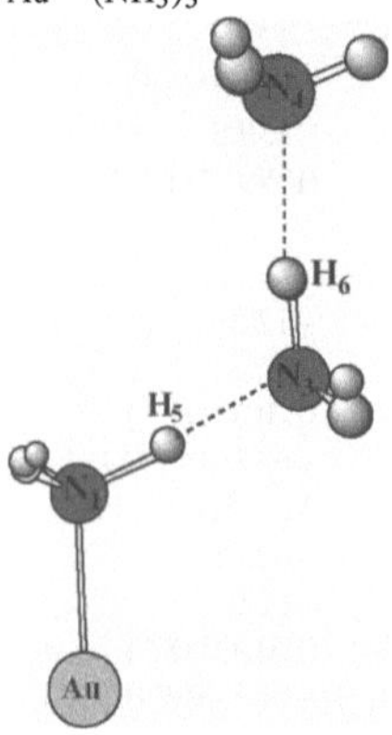	$R(Au^+–N_1) = 2.064$ Å $R(N_1–H_5) = 1.118$ Å $R(N_3–H_6) = 1.039$ Å $r(H_5 \cdots N_3) = 1.586$ Å $r(H_6 \cdots N_4) = 2.005$ Å $\angle N_1H_5N_3 = 176.8°$ $\angle N_3H_6N_4 = 177.4°$ $\Delta\nu(N_1–H_5)^d = -1481\,cm^{-1}$ $\Delta\nu(N_3–H_6)^d = -230\,cm^{-1}$ $E_b(Au^+–(NH_3)_2{}^{II} + NH_3) = 9.43\,kcal \cdot mol^{-1}$ $\Delta E_{I–II} = -47.72\,kcal \cdot mol^{-1}$	**$Au^+–(NH_3)_2{}^{II} + NH_3$:** −2.45	**$Au^+–(NH_3)_2{}^{II} + NH_3$:** −9.75 −24.48

[a] $NH_3 : R(N–H) = 1.016$ Å.
[b] The first ionization energy IE(Au) = 9.42 eV.
[c] Basis set [8]: aug-cc-pVDZ (NH_3) ∪ SDD (Au).
[d] Relative to $(NH_3)_2$: $R(N_1–H_4) = 1.022$ Å, $r(H_4 \cdots N_3) = 2.229$ Å, $\angle N_1H_4N_3 = 167.6°$, $\Delta\nu(N_1–H_4) = -63\,cm^{-1}$, the ratio of IR activities $A_{IR}((NH_3)_2)/A_{IR}(NH_3) = 43.7$, $E_b((NH_3)_2 \rightarrow 2NH_3) = 1.92\,kcal \cdot mol^{-1}$.

$\Delta\Delta G_{298}° = -13.7\,kcal \cdot mol^{-1}$, reveals some preference over the successive double ammination of Au^+. It is worth noting also that the neutral charge state of Au reverses the order of stability of $Au–(NH_3)_2{}^I$ and $Au–(NH_3)_2{}^{II}$ (Figure 1): the latter appears to be more stable by $4.3\,kcal \cdot mol^{-1}$. This phenomenon will be further discussed in section 4.

The third ammination pathway illustrated in Table 1 shows, first, that Au^+ enables to directly accommodate at most two anchoring Au–N bonds and, second, that the scenario of all higher ($n \geq 3$) ammination reactions proceeds through H-associated ammonia clusters. The weakness of the intramolecular H-bonding in ammonia clusters predetermines the weaker energetics of all higher ammination reactions.

The strength of the anchoring Au–N bond is therefore largely predetermined by the charge which transfers from the nitrogen moiety to the gold atom while this bond is formed – obviously, the charge transfer is larger for Au^+ than for Au since $EA(Au^+) \equiv IE(Au) = 9.42\,eV > EA(Au) = 2.13\,eV$. This explains the stronger anchoring of ammonia clusters with Au^+ and the considerable weakening of N–H bonds adjacent to gold. If one of the N–H bonds is weakly H-bonded to a neighboring ammonia molecule, the anchoring bond increases the antibonding character of the N–H bond and causes the corresponding proton to transfer towards the proton acceptor, strengthening this hydrogen bond and converting it to the class of the moderate ones. The complex $Au^+–(NH_3)_2{}^{II}$ is likely the most impressive example of modulation of the H-bonding network of ammonia clusters by its anchoring to gold. Herein, the $N_1–H_4$ bond is elongated by 0.061 Å, the hydrogen bond $H_4 \cdots N_3$ is shortened by 0.502 Å and the $\nu(N_1–H_4)$ stretch downshifts by $874\,cm^{-1}$, compared to the ammonia dimer.

In contrast to the neutral and positive charge states of Au, the gold anion Au^- interacts with the ammonia molecule in a different fashion – as seen in Figure 2, the anchoring Au–N bond is no longer formed. Au^-, however, is bonded to NH_3 via the contact bridge N–H...Au^-, which involves the hydrogen atom and resembles, by all features, a conventional hydrogen bond [31]. To be more specific (compare with the necessary and sufficient conditions listed in Refs. [14,26]):

(i) the $N_1–H_3$ bond is elongated by 0.015 Å; (ii) the $H_3 \cdots Au_2$ separation is typical of the strength of H-bond. It is equal to 2.690 Å and is smaller than the sum of the van der Waals radii of H and Au (the concomitant H-bridge vibrational mode $\nu_\sigma = 92\,cm^{-1}$) and (iii) the $\nu_1(N_1–H_3)$ stretch is red-shifted by $160\,cm^{-1}$. Therefore, the interaction between Au^- and NH_3 is of a hydrogen bonding type where the gold anion acts as a nonconventional proton acceptor to form a nonconventional N–H $\cdots$ Au hydrogen bond with typical conventional proton donor group N–H.

The molecular orbital picture of the formation of the nonconventional $N_1–H_3 \cdots Au_2$ hydrogen bond in $Au^-–NH_3$ provides an insight on the propensity of Au^- to behave as a nonconventional proton acceptor. The outermost occupied AOs of open-shell ground-state electronic configuration $[Xe]5d^{10}6s^1$ of the gold atom are the 5-degenerate $d_{\pm2,\pm1,0} - AO_\alpha{}'$ s 5–9, which consist of the spin-up $HOAO - 1_\alpha$ with the AO energy $\varepsilon_{5–9}{}^\alpha$ of $-8.30\,eV$, the $s - HOAO_\alpha$ with $\varepsilon_{10}{}^\alpha = -6.62\,eV$, and the similar 5-degenerate $d_{\pm2,\pm1,0} - AO_\beta{}'$s 5–9, the spin-down $HOAO_\beta$ with $\varepsilon_{5–9}{}^\beta = -8.06\,eV$ (see Figure 3). The lowest unoccupied $p_y - LUAO_\alpha$ and s–$LUAO_\beta$ of Au are correspondingly characterized by the orbital energies $\varepsilon_{11}{}^\alpha = +1.50\,eV$ and $\varepsilon_{10}{}^\beta = -4.59\,eV$. The two outermost occupied AOs, the d-HOAO-1 being split into the $d_{\pm2,\pm1,0}$–AOs5–9 and the s–HOAO determine the ground-state electronic configuration $[Xe]5d^{10}6s^2$ of Au^- (Figure 4). Their AO energies amount to $-0.87(-0.0319\,au)$ and $+0.25\,eV(+0.0092\,au)$, respectively.

The molecular orbital (MO) picture of the formation of the nonconventional hydrogen bond $N_1–H_3 \cdots Au_2$ between the ammonia molecule and Au^- is illustrated

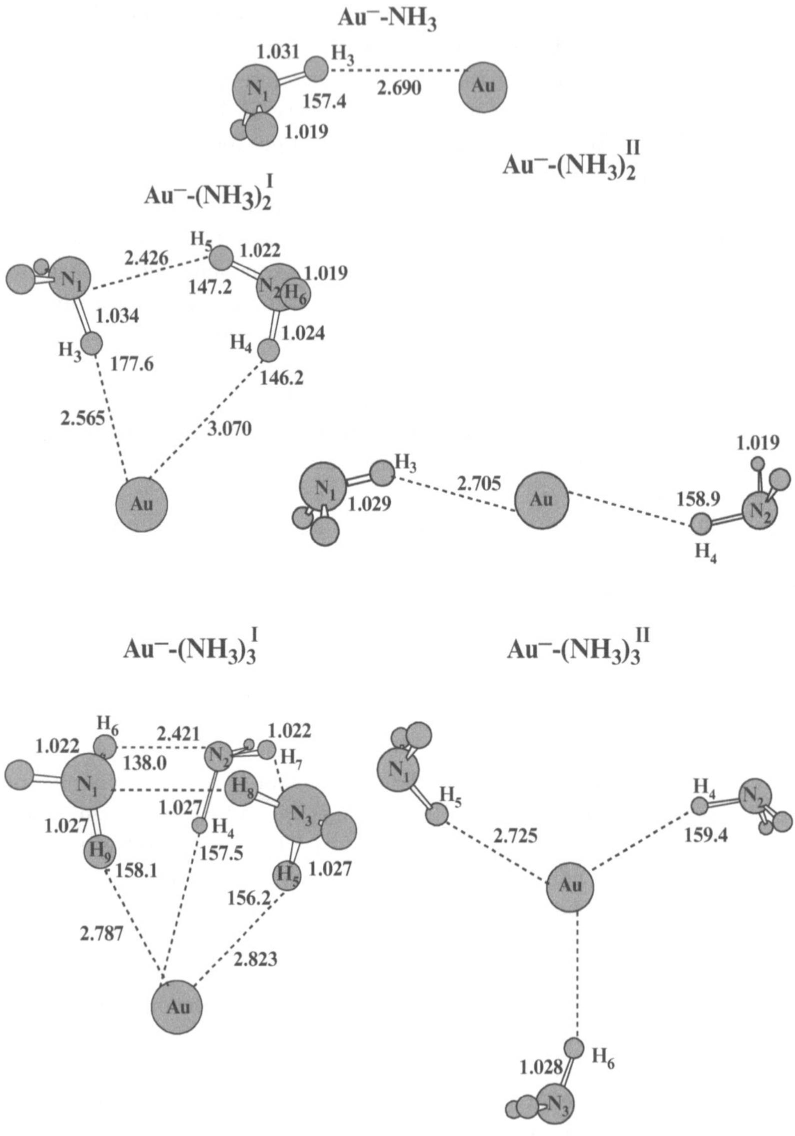

Figure 2. The complexes Au^{-}–$(NH_3)_{1\leq n\leq 3}$. Their selected vibrational modes: (i) Au^{-}–$(NH_3)_2{}^{I}$: $\nu(N_1–H_3) = 3253\,cm^{-1}$ $(467\,km{\cdot}mol^{-1})$; $\nu(N_2–H_{5–7}) = 3415\,cm^{-1}$ $(65\,km{\cdot}mol^{-1})$, $3504\,cm^{-1}$ $(4\,km{\cdot}mol^{-1})$, and $3508\,cm^{-1}$ $(44\,km\cdot mol^{-1})$; $\nu_\sigma(H_7\cdots N_1) = 118\,cm^{-1}$; $\nu_\sigma(H_3\cdots Au) = 100\,cm^{-1}$; (ii) Au^{-}–$(NH_3)_2{}^{II}$: $\nu(N_1–H_3,\ N_2–H_5) = 3345\,cm^{-1}$ $(558\,km\cdot mol^{-1})$, $3351\,cm^{-1}$ $(0\,km\cdot mol^{-1})$; $\nu_\sigma(H_{3,5}\cdots Au) = 84, 89\,cm^{-1}$; (iii) Au^{-}–$(NH_3)_3{}^{I}$: $\nu(N_2–H_9) = 3360\,cm^{-1}$ $(157\,km\cdot mol^{-1})$; $\nu(N_1–H_{10}) = 3366\,cm^{-1}$ $(141\,km\cdot mol^{-1})$; $\nu(N_3$- $H_8) = 3370\,cm^{-1}$ $(226\,km\cdot mol^{-1})$; $\nu(N_3–H_6) = 3492\,cm^{-1}$ $(50\,km{\cdot}mol^{-1})$; $\nu(N_2–H_{12}) = 3494\,cm^{-1}$ $(41\,km{\cdot}mol^{-1})$; $\nu(N_1–H_5) = 3494\,cm^{-1}$ $(59\,km{\cdot}mol^{-1})$; (iv) Au^{-}–$(NH_3)_3{}^{II}$: $\nu(N_1–H_5,\ N_3–H_7,\ N_{10}–H_{11}) = 3359\,cm^{-1}$ $(350\,km\cdot mol^{-1})$, $3360\,cm^{-1}$ $(342\,km\cdot mol^{-1})$, $3363\,cm^{-1}$ $(5\,km\cdot mol^{-1})$. Bond lengths are given in Å and bond angles in deg.

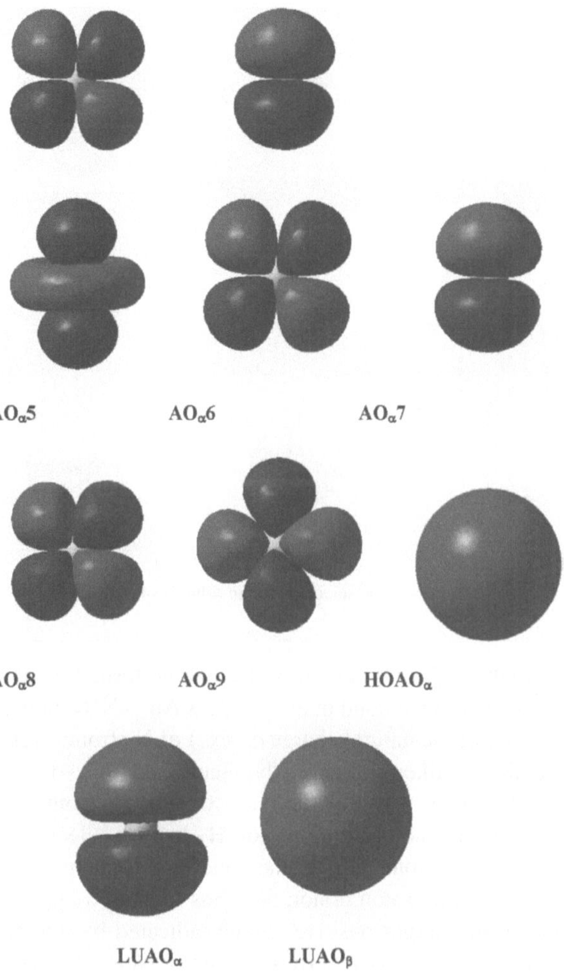

Figure 3. The selected atomic orbitals of Au.

in Figure 5 in terms of the two bonding MO8 ($\varepsilon_8 = -8.27\,\text{eV}$) and MO9 ($\varepsilon_9 = -3.54\,\text{eV}$) and the two antibonding MO10 ($\varepsilon_{10} = -1.41\,\text{eV}$) and MO12 ($\varepsilon_{12} = -1.39\,\text{eV}$). On Au^-, the bonding MO8 and MO9 are correspondingly represented by the hybridized $s(\sim 84\%)$ and $d_{+2}(\sim 15\%)$, and the $s(\sim 70\%)$, $d_0(\sim 19\%)$, and $d_{0+2}(\sim 21\%)$ AOs, and are therefore characterized by the dominant long-range s character. The antibonding MO10 and MO12 are decomposed into the $s(\sim 3\%)$, $d_0(\sim 12\%)$, $d_{+1}(\sim 30\%)$, $d_{+2}(\sim 55\%)$, and $d_0(\sim 20\%)$, $d_{+1}(\sim 60\%)$, and $d_{+2}(\sim 20\%)$ AOs, respectively, and in contrast reveal the leading short-range d character. The HOMO of the Au^-–NH_3 complex with $\varepsilon_{15} = -0.20\,\text{eV}$ has a dominant s and p character.

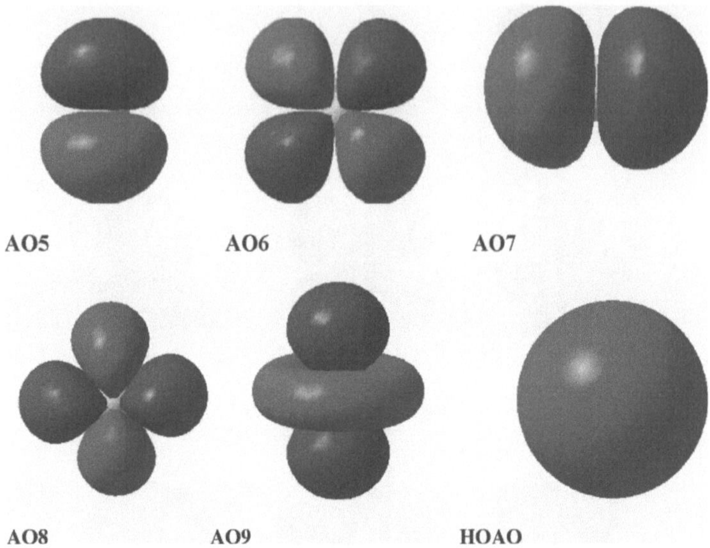

Figure 4. The selected atomic orbitals of Au^-.

Natural Bond Analysis (NBO) demonstrates that the formation of the nonconventional N_1–$H_3 \cdots Au_2$ hydrogen bond in the complex Au^-–NH_3 induces a charge flow resulting in a decrease of the natural charge $q_N(N_1)$ of N_1 from $-1.138e$ to $-1.171e$ (note that, instead, the Mulliken charge $q_M(N_1)$ increases from $-0.848e$ to $-0.766e$), and an increase of $q_N(H_3)$ by $\Delta q_N(H_3) = 0.034e$ (in contrast, $\Delta q_M(H_3) = -0.083e$). In other words, the bridging proton H_3 loses its electron density and the electron density flows from lone pairs of the nonconventional proton acceptor to the σ^*–antibonding MO of the proton donor, and thus induces a larger negative charge on N_1. Simultaneously, a larger positive charge indicated by $\Delta q_N(Au_2) = 0.036e$ is induced on the nonconventional proton acceptor Au_2. The occupancy of the $\sigma^*(N_1$–$H_3)$ MO increases by $+2.4$ me which causes the N_1–H_3 bond weakening and the concomitant red-shifting of its stretching vibrational frequency. The *s* character of the $\sigma^*(N_1$–$H_3)$ MO amounts to 27.7 %.

The binding energy $E_b(Au^-$–$NH_3)$, which fully arises from the nonconventional N–H…Au–hydrogen bonding, amounts to 6.01 kcal · mol^{-1}, three times larger than the conventional N–H $\cdots$ N H–bond of the ammonia dimer. The related enthalpy of formation $\Delta H_{298}{}^\circ(Au^-$–$NH_3) = -4.90$ kcal · mol^{-1}. A positive character of the entropy of formation $\Delta S^\circ(Au^-$–$NH_3)$ and its large magnitude, 26.41 cal · K^{-1} · mol^{-1}, leads to Gibbs free energy of the first ammination of Au^-, equal to $\Delta G_{298}{}^\circ(Au^-$–$NH_3) = -12.77$ kcal · mol^{-1}. This value is considerably larger than $\Delta H_{298}{}^\circ$. Therefore, due to a substantial entropy effect, the formation of the nonconventional hydrogen bond in Au^-–NH_3 is highly favored.

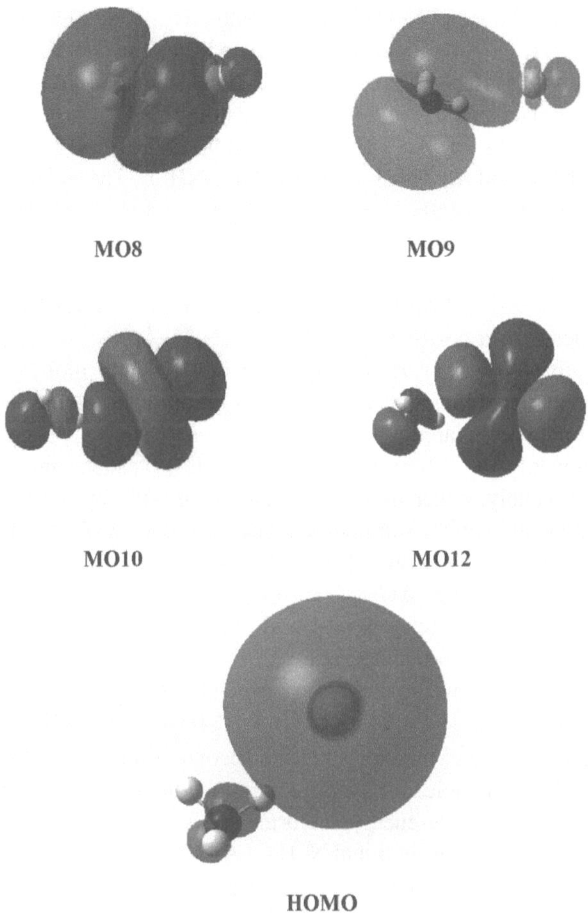

Figure 5. The selected molecular orbital patterns of Au^{-}–NH_3.

The nonconventional N–H $\cdots$ Au^{-} hydrogen bond coexists with the conventional one in the ammonia dimer when the latter forms with Au^{-} the complex Au^{-}–$(NH_3)_2{}^{I}$ on the PES of Au^{-}–$(NH_3)_2$ (see Figure 2). This latter is quasi isoenergetic to the chain complex Au^{-}–$(NH_3)_2{}^{II}$; the small energy difference, $\Delta E_{I\text{-}II}$, between Au^{-}–$(NH_3)_2{}^{I}$ and Au^{-}–$(NH_3)_2{}^{II}$ amounts to only $-0.34\,\text{kcal} \cdot \text{mol}^{-1}$. Since $\Delta S_{I\text{-}II}{}^{\circ} = -11.19\,\text{cal} \cdot \text{K}^{-1} \cdot \text{mol}^{-1}$, the order of stability of these complexes is reversed at $T \geq 30\,\text{K}$.

Adding the second NH_3 to the dimer Au^{-}–NH_3 leads to the formation of the complex Au^{-}–$(NH_3)_2{}^{I}$, and further strengthens the dimeric N_1–$H_3 \cdots Au^{-}$ H–bond, as indicated by a small elongation of $R(N_1$–$H_3)$, 0.003 Å, a shortening of $r(H_3 \cdots Au)$, by 0.135 Å, an increase of the H-bond angle $\angle N_1H_3Au$ from 157.4° to 177.6°, and a further red shift of the $\nu(N_1$–$H_3)$ stretching vibrational mode, by $72\,\text{cm}^{-1}$. Even though the additional contact N_2–$H_5 \cdots Au^-$ with r($H_5 \cdots Au^-$) = 3.070 Å is much weaker than N_1–$H_3 \cdots Au^-$ it can also be treated, in some sense, as an extremely weak nonconventional H-bond, since it largely obeys the conditions imposed on the conventional H-bond as well (see Ref. [31f] and note [21]).

The binding energy of Au^-–$(NH_3)_2{}^I$ is 4.90 kcal · mol^{-1} with respect to Au^-–NH_3 + NH_3 and is lower than $E_b(Au^-$–$NH_3)$. The related enthalpy of formation, $\Delta H_{298}{}^\circ(Au^-$–$(NH_3)_2{}^I)$ = −5.01 kcal · mol^{-1}, is slightly larger than $\Delta H_{298}{}^\circ(Au^-$–$NH_3)$ in absolute value (note that $\Delta H_{298}{}^\circ(Au^-$–$(NH_3)_2{}^I)$ − $\Delta H_{298}{}^\circ(Au^-$–$(NH_3)_2{}^{II})$ = −0.25 kcal · mol^{-1}). A negative entropy effect, ΔS° $(Au^-$–$(NH_3)_2{}^I)$ = −23.81 cal·K^{-1}·mol^{-1} and $\Delta S^\circ(Au^-$–$(NH_3)_2{}^{II})$ = −12.62 cal· K^{-1} · mol^{-1}, leads to a positive $\Delta G_{298}{}^\circ(Au^-$–$(NH_3)_2{}^I)$ = 2.09 kcal · mol^{-1} and a slightly negative $\Delta G_{298}{}^\circ(Au^-$–$(NH_3)_2{}^{II})$ = −1.10 kcal · mol^{-1}. These results demonstrate a noticeable resilience of the gold anion to its second ammination at room temperature as compared to the first one.

The formation of Au^-–$(NH_3)_2{}^I$ and Au^-–$(NH_3)_2{}^{II}$ can be interpreted in two different ways, namely, either as the ammination of Au^- by the ammonia dimer or as its simultaneous double ammination. The energetics of the former reaction is characterized by E_b = 8.99 kcal · mol^{-1}, $\Delta H_{298}{}^\circ$ = −7.84 kcal · mol^{-1}, ΔS° = 19.56 cal · K^{-1} · mol^{-1}, and $\Delta G_{298}{}^\circ$ = −13.67 kcal · mol^{-1}. The latter value of $\Delta G_{298}{}^\circ$ marks a little preference (≈−1 kcal · mol^{-1}) of the dimer ammination of Au^- over the first one at room temperature.

Interestingly, the difference in enthalpy between the second and first amminations, $\Delta_{2\text{–}1}\Delta H_{298}{}^\circ(Au^-) \equiv \Delta H_{298}{}^\circ(Au^-$–$(NH_3)_2^I) - \Delta H_{298}{}^\circ(Au^-$–$NH_3)$ = −0.11 kcal· mol^{-1}, is slightly negative, meaning that the second ammination at 0 K is slightly preferential. Such a trend holds for the third ammination of Au^- whose PES shown in Figure 2 consists of two lower-energy conformers, Au^-–$(NH_3)_3{}^I$ and Au^-–$(NH_3)_3{}^{II}$. They both have three non conventional N–H $\cdots Au^-$ hydrogen bonds where the gold anion plays a role of a triple nonconventional proton acceptor. The former can be geometrically viewed as Au^- triply H-bonded to the ammonia trimer whereas the latter is merely the result of the third ammination of Au^-–$(NH_3)_2{}^{II}$.

With respect to Au^-–$(NH_3)_2{}^I$ + NH_3, $E_b(Au^-$–$(NH_3)_3{}^I)$ amounts to 4.56 kcal · mol^{-1}. The corresponding enthalpy of formation $\Delta H^\circ_{298}(Au^-$–$(NH_3)_3{}^I)$ = −5.12 kcal · mol^{-1} and hence, $\Delta_{3\text{–}2}\Delta H_{298}{}^\circ(Au^-) \equiv \Delta H_{298}{}^\circ(Au^-$–$(NH_3)_3{}^I)$ − $\Delta H^\circ_{298}(Au^-$–$(NH_3)_2{}^I)$ = −0.11 kcal · mol^{-1}. A large entropy factor ΔS° $(Au^-$–$(NH_3)_3{}^I)$ = −32.99 cal · K^{-1} · mol^{-1} diminishes however the enthalpy contribution and leads to a positive $\Delta G_{298}{}^\circ(Au^-$–$(NH_3)_3{}^I)$ = 4.71 kcal · mol^{-1}. The energy difference $\Delta E_{\text{I-II}}$ between Au^-–$(NH_3)_3{}^I$ and Au^-–$(NH_3)_3{}^{II}$ is only −0.25 kcal · mol^{-1}; hence, we can treat these two complexes as nearly isoenergetic, at least at 0 K.

The complex Au^-–$(NH_3)_3{}^{II}$ is less compact than Au^-–$(NH_3)_3{}^I$ and is therefore characterized by a larger entropy, which determines a negative difference, $\Delta S_{\text{I-II}}{}^\circ$ = −35.64 cal · K^{-1} · mol^{-1}, and reverses the order of stability of these complexes, already at T ≥ 7 K. A large entropy of Au^-–$(NH_3)_3{}^{II}$ has a remarkable effect on the third ammination of the gold anion. As $\Delta H_{298}°(Au^{-}\text{–}(NH_3)_3{}^{II}) = -3.98\,kcal \cdot mol^{-1}$ relative to $Au^{-}\text{-}(NH_3)_2{}^{II} + NH_3$, the small magnitude of $\Delta S°(Au^{-}\text{–}(NH_3)_3{}^{II}) = -8.54\,cal \cdot K^{-1} \cdot mol^{-1}$ has a smaller effect on $\Delta G_{298}°(Au^{-}\text{–}(NH_3)_3{}^{II})$, amounting to $-1.44\,kcal \cdot mol^{-1}$. As a consequence, $\Delta_{3-2}\Delta G_{298}°(Au^{-}) \equiv \Delta G_{298}°(Au^{-}\text{–}(NH_3)_3{}^{II}) - \Delta G_{298}°(Au^{-}\text{–}(NH_3)_2{}^{II})$, equal to $-0.34\,kcal \cdot mol^{-1}$, reveals a small anomaly of the third ammination of Au^{-}, by analogy with the anomalous second ammination of the silver cation [29].

3. COMPLEXES $Au_3{}^{Z}\text{–}(NH_3)_{1\leq n\leq 3}(Z = 0, \pm 1)$

The bonding patterns formed between the triangular gold cluster and the ammonia molecule, illustrated in Table 2, differ from those described for $Au^{Z}\text{–}NH_3$, especially for Au_3. This can be anticipated from the more smeared redistribution of the electron charge over the entire cluster and the different shapes of the high occupied and low unoccupied frontier molecular orbitals, which determine its reactivity (see Ref. [26]c).

We observe two opposite trends in the bonding patterns when increasing the cluster size from Au^{Z} to $Au_3{}^{Z}$. On one hand, the neutral Au_3 anchors NH_3 more strongly than Au: $E_b(Au_3\text{–}NH_3) = 24.42\,kcal \cdot mol^{-1} > E_b(Au\text{–}NH_3) = 7.83\,kcal \cdot mol^{-1}$, and the anchor bond is contracted by 0.16 Å with respect to that in $Au\text{–}NH_3$. On the other hand, both charged species, $Au_3{}^{+}$ and $Au_3{}^{-}$, bind NH_3 in a weaker way (Table 2): (i) the anchoring Au-N bond of $Au_3{}^{+}\text{–}NH_3$ is longer by 0.06 Å than $Au^{+}\text{–}NH_3$; (ii) the nonconventional $N\text{–}H \cdots Au$ hydrogen bond that

Table 2. Multiple amminations of Au_3^{Z} in terms of the most stable complexes (see Figures 6–11).

n	$Au_3\text{–}(NH_3)_{1\leq n\leq 3}$	$Au_3{}^{+}\text{–}(NH_3)_{1\leq n\leq 3}$	$Au_3{}^{-}\text{–}(NH_3)_{1\leq n\leq 3}$
1	**$Au_3 + NH_3$**: R(Au–N) = 2.194 Å R(Au–Au) = 2.711, 2.670, 2.781 Å $E_b = 24.42\,kcal \cdot mol^{-1}$ $\Delta H_{298}° = -24.84\,kcal \cdot mol^{-1}$ $\Delta S° = -27.37\,cal \cdot K^{-1} \cdot mol^{-1}$ $\Delta G_{298}° = -16.68\,kcal \cdot mol^{-1}$	**$Au_3{}^{+} + NH_3$**: R(Au–N) = 2.161 Å R(Au–Au) = 2.722, 2.722, 2.623 Å $E_b = 43.81\,kcal \cdot mol^{-1}$ $\Delta H_{298}° = -44.30\,kcal \cdot mol^{-1}$ $\Delta S° = -29.82\,cal \cdot K^{-1} \cdot mol^{-1}$ $\Delta G_{298}° = -35.42\,kcal \cdot mol^{-1}$	**$Au_3{}^{-} + NH_3$**: R(N–H) = 1.022 Å R(H...Au) = 2.993 Å ∠NHAu = 159.7° ν(N-H) = 3445 cm^{-1} R(Au–Au) = 2.634, 2.630 Å $E_b = 3.21\,kcal \cdot mol^{-1}$ $\Delta H_{298}° = -2.65\,kcal \cdot mol^{-1}$ $\Delta S° = -9.97\,cal \cdot K^{-1} \cdot mol^{-1}$ $\Delta G_{298}° = +0.32\,kcal \cdot mol^{-1}$
2	**$Au_3\text{–}NH_3 + NH_3$**: $E_b = 14.85\,kcal \cdot mol^{-1}$ $\Delta H_{298}° = -15.17\,kcal \cdot mol^{-1}$ $\Delta S° = -29.17\,cal \cdot K^{-1} \cdot mol^{-1}$ $\Delta G_{298}° = -6.48\,kcal \cdot mol^{-1}$	**$Au_3{}^{+}\text{–}NH_3 + NH_3$**: $E_b = 38.74\,kcal \cdot mol^{-1}$ $\Delta H_{298}° = -39.20\,kcal \cdot mol^{-1}$ $\Delta S° = -27.67\,cal \cdot K^{-1} \cdot mol^{-1}$ $\Delta G_{298}° = -30.95\,kcal \cdot mol^{-1}$	**$Au_3{}^{-}\text{–}NH_3 + NH_3$**: $E_b = 3.71\,kcal \cdot mol^{-1}$ $\Delta H_{298}° = -3.88\,kcal \cdot mol^{-1}$ $\Delta S° = -25.11\,cal \cdot K^{-1} \cdot mol^{-1}$ $\Delta G_{298}° = +3.60\,kcal \cdot mol^{-1}$
3	**$Au_3\text{–}(NH_3)_2{}^{I} + NH_3$**: $E_b = 7.44\,kcal \cdot mol^{-1}$ $\Delta H_{298}° = -7.91\,kcal \cdot mol^{-1}$ $\Delta S° = -31.12\,cal \cdot K^{-1} \cdot mol^{-1}$ $\Delta G_{298}° = +1.36\,kcal \cdot mol^{-1}$	**$Au_3{}^{+}\text{–}(NH_3)_2{}^{I} + NH_3$**: $E_b = 31.63\,kcal \cdot mol^{-1}$ $\Delta H_{298}° = -32.07\,kcal \cdot mol^{-1}$ $\Delta S° = -28.79\,cal \cdot K^{-1} \cdot mol^{-1}$ $\Delta G_{298}° = -23.49\,kcal \cdot mol^{-1}$	**$Au_3{}^{-}\text{–}(NH_3)_2{}^{II} + NH_3$**: $E_b = 2.44\,kcal \cdot mol^{-1}$ $\Delta H_{298}° = -3.03\,kcal \cdot mol^{-1}$ $\Delta S° = -35.61\,cal \cdot K^{-1} \cdot mol^{-1}$ $\Delta G_{298}° = +7.58\,kcal \cdot mol^{-1}$

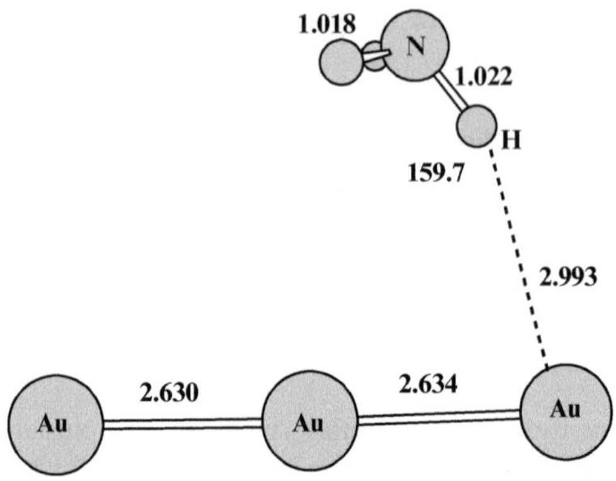

Figure 6. The complex Au_3^-–NH_3. Notice that no bonding of NH_3 to the central atom of the Au_3^- cluster is found. Bond lengths are given in Å and bond angles in degrees.

solely stabilizes the complex Au_3^-–NH_3 (see Figure 6) is weaker. It is characterized by a smaller elongation of the involved N–H bond (ΔR(N–H) $= 0.006$ Å in Au_3^-–NH_3 vs. ΔR(N–H) $= 0.015$ Å in Au^-–NH_3), by a much longer H-bond (r(H···Au) $= 2.993$ Å vs. r(H···Au) $= 2.690$ Å), and finally by a less noticeable red shift ($\Delta\nu$(N–H) $= -40\,cm^{-1}$ vs. $\Delta\nu$(N–H) $= -160\,cm^{-1}$). Table 2 presents the thermodynamics of the first ammination of Au_3^Z and particularly indicates a preference of neutral Au_3($\Delta G_{298}{}^\circ = -16.68\,kcal \cdot mol^{-1}$) over Au($\Delta G_{298}{}^\circ = -13.86\,kcal \cdot mol^{-1}$) to the first ammination, despite a disfavoring entropy effect ($\Delta S^\circ = -27.37\,cal \cdot K^{-1} \cdot mol^{-1}$ vs. $\Delta S^\circ = 22.33\,cal \cdot K^{-1} \cdot mol^{-1}$).

Three different low-energy pathways, displayed in Figures 7 and 8, govern the second ammination of the neutral and charged $Au_3{}^Z$. On the neutral and cationic sheets of the total PES of $Au_3{}^Z$–$(NH_3)_2$, the most favorable pathways proceed to the structures $Au_3{}^{0,+1}$–$(NH_3)_2{}^I$ – they lie above the asymptote $Au_3{}^{0,+1}$–$NH_3 + NH_3$ by $E_b(Au_3{}^0–(NH_3)_2{}^I) = 14.85\,kcal \cdot mol^{-1}$ and $E_b(Au_3{}^+–(NH_3)_2{}^I) = 38.74\,kcal \cdot mol^{-1}$, respectively. The thermo-dynamics of these reaction pathways presented in Table 2 demonstrates that, by analogy with $Au^{0,+1}$, the second ammination of $Au_3{}^{0,+1}$ is less favorable too. Interestingly, the second ammination of Au_3 reduces its ionization energy IE(Au_3), defined as the energy difference between Au_3–$(NH_3)_2{}^I$ and $Au_3{}^+$–$(NH_3)_2{}^I$, to 5.15 eV, i.e. by 1.88 eV.

In both structures $Au_3{}^{0,+1}$–$(NH_3)_2{}^I$, each of two gold atoms anchors a single ammonia molecule. Their formation markedly changes the gold clusters: in the neutral Au_3–$(NH_3)_2{}^I$, the gold–gold bond that couples the anchoring bonds is considerably strengthened whereas the other two are slightly weakened, and the opposite effect is predicted for the cationic $Au_3{}^+$–$(NH_3)_2{}^I$ complex.

Less favorable are the pathways leading to either $Au_3{}^{0,+1}$–$(NH_3)_2{}^{II}$($\Delta E_{I–II}{}^{0,+1} = -6.61$ and $-26.00\,kcal \cdot mol^{-1}$) or $Au_3{}^{0,+1}$–$(NH_3)_2{}^{III}$($\Delta E_{II–III}{}^{0,+1} = -8.12$ and $+1.64\,kcal \cdot mol^{-1}$). A partial structural resemblance of $Au_3{}^{0,+1}$–$(NH_3)_2{}^{II}$ involving the H-bonded ammonia dimer with $Au^{0,+1}$–$(NH_3)_2{}^{II}$ can readily be noticed for

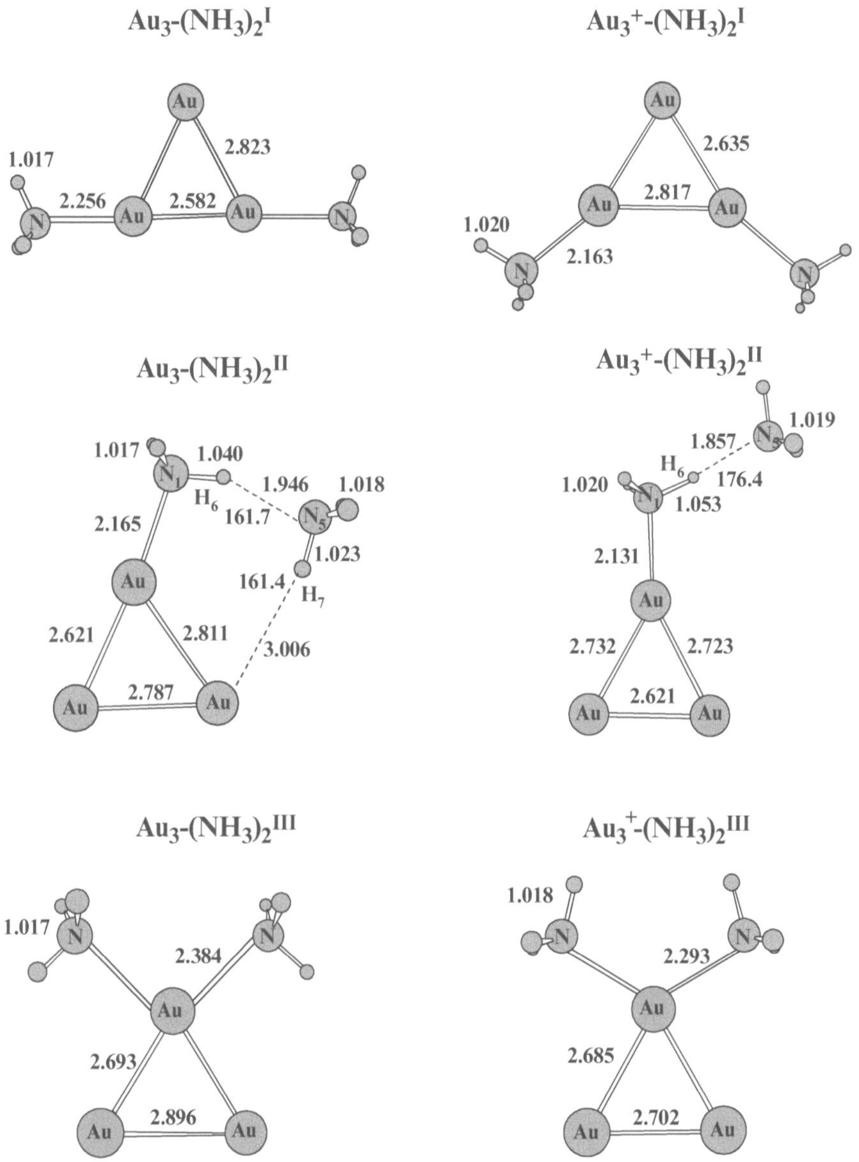

Figure 7. The low-energy complexes $Au_3{}^{0,+1}$–$(NH_3)_2$. The selected vibrational modes of $Au_3{}^{0,+1}$–$(NH_3)_2{}^{II}$: (i) $Z = 0$: $\nu(N_1–H_6) = 3145\,cm^{-1}$ ($504\,km \cdot mol^{-1}$); $\nu(N_5–H_7) = 3429\,cm^{-1}$ ($89\,km \cdot mol^{-1}$); (ii) $Z = +1$: $\nu(N_1–H_6) = 2933\,cm^{-1}$ ($1667\,km \cdot mol^{-1}$). Bond lengths are given in Å and bond angles in degrees.

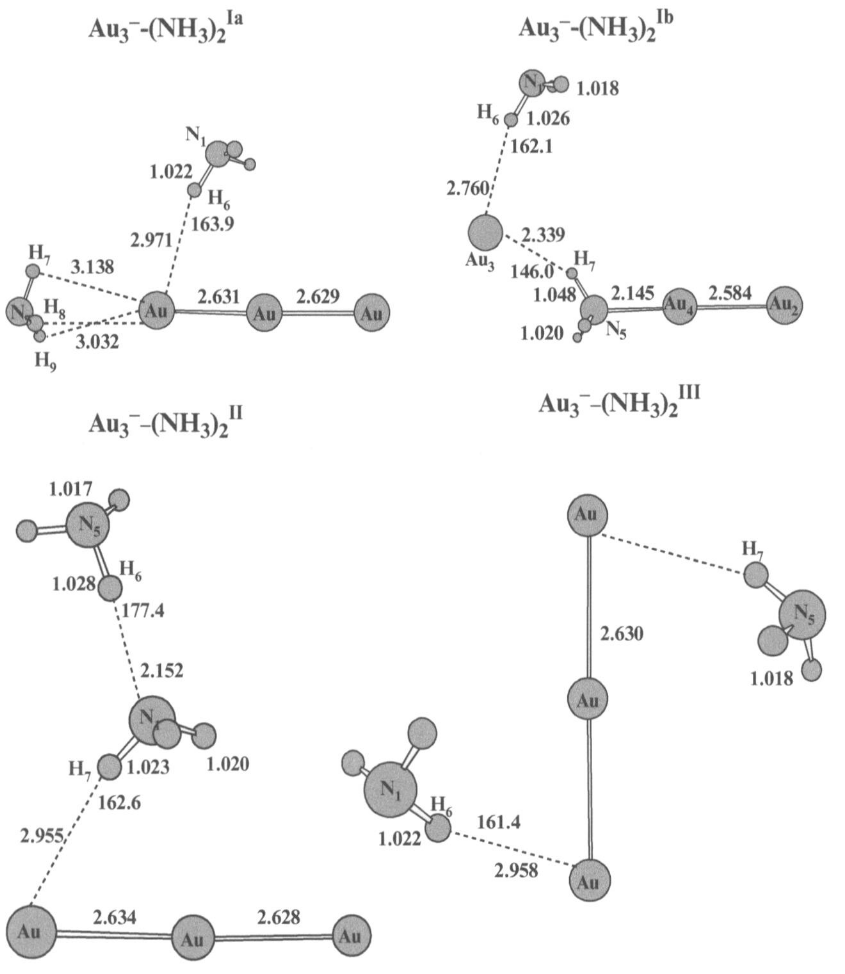

Figure 8. The low-energy complexes Au_3^{-}–$(NH_3)_2$. Their selected vibrational modes: (i) Au_3^{-}–$(NH_3)_2^{II}$: $\nu(N_5–H_6) = 3348\,cm^{-1}(385\,km \cdot mol^{-1})$; $\nu(N_1–H_7) = 3435\,cm^{-1}(26\,km \cdot mol^{-1})$; (ii) Au_3^{-}–$(NH_3)_2^{Ia}$: $\nu(N_1–H_6) = 3443\,cm^{-1}(12\,km \cdot mol^{-1})$; $\nu(N_5–H_{7–9}) = 3464\,cm^{-1}(41\,km \cdot mol^{-1})$; (iii) Au_3^{-}–$(NH_3)_2^{Ib}$: $\nu(N_5–H_7) = 3031\,cm^{-1}(1334\,km \cdot mol^{-1})$; $\nu(N_1–H_6) = 3389\,cm^{-1}(144\,km \cdot mol^{-1})$; (iv) Au_3^{-}–$(NH_3)_2^{III}$: $\nu(N_5–H_7;\ N_1–H_6) = 3443, 3446\,cm^{-1}(13, 9\,km \cdot mol^{-1})$. Units as in Figure 7.

each value of Z. Their differences are easily seen as well, mostly in their H-bonding part, which bonding patterns are governed by the total charge of the complex. In the neutral case, the anchoring bond of Au_3–$(NH_3)_2^{II}$ is much shorter and strongly activates the adjacent $N_1–H_6$ bond, as is also the case for the anchoring bond in Au–$(NH_3)_2^{II}$, which strengthens $N_1–H_6 \cdots N_5$ hydrogen bond, causing the $N_1–H_6$ stretch to be red-shifted by $177\,cm^{-1}$ with respect to the ammonia dimer. The cationic state exhibits the opposite effect. The difference between the neutral and cationic states of Au_3–$(NH_3)_2{}^{II}$ is also manifested by the orientation of the second ammonia molecule relative to the gold cluster: when $Z = 0$ this molecule forms a weak nonconventional N_5–H_7...Au_3 hydrogen bond, for $Z = 1$ the latter breaks and the second ammonia molecule rotates along the intramolecular N_1–$H_6 \cdots N_5$ bond from the gold. These are the key features that allow to distinguish the neutral and cationic states on the PES of $Au_3{}^Z$–$(NH_3)_2$. We discuss in section 4 how these features can be used to encode qubits.

Switching the neutral charge state on the PES of the second ammination to the negative one drastically changes the reaction and bonding patterns (see Figure 8). Instead of $Au_3{}^0$–$(NH_3)_2{}^I$, which energetically prevails in the neutral charge state, the lowest-energy complex in the anionic state turns out to be the $Au_3{}^-$–$(NH_3)_2{}^{II}$ complex, which originates from the less favorable neutral parent $Au_3{}^0$–$(NH_3)_2{}^{II}$ as a result of alternating the anchoring Au–N bond by the nonconventional N–H$\cdots$Au hydrogen bond.

The enthalpy of such an alternation is estimated as equal to 2.46 eV, that is, about 1.18 eV smaller than EA(Au_3). In the other words, the ammonium microsolvation leads to considerable decrease of the electron affinity of a given gold cluster.

In addition, the neutral reaction pathway linked to $Au_3{}^0$–$(NH_3)_2{}^I$ bifurcates on the negatively charged sheet, giving a rise to $Au_3{}^-$–$(NH_3)_2{}^{Ia}$ and $Au_3{}^-$–$(NH_3)_2{}^{Ib}$. The former complex is slightly less stable, by 1.40 kcal $\cdot$ mol^{-1}, than $Au_3{}^-$–$(NH_3)_2{}^{II}$ and more stable by 7.31 kcal $\cdot$ mol^{-1} compared to $Au_3{}^-$–$(NH_3)_2{}^{I\,b}$. Despite lower stability of $Au_3{}^-$–$(NH_3)_2{}^{I\,b}$, its formation is quite peculiar since it involves breaking two Au–Au bonds of the triangular neutral gold cluster in order to form a new Au_4–N_5 bond, which bridges the fragmented Au_3 atom via the two strong nonconventional hydrogen bonds N_5–$H_7 \cdots Au_3$ and N_1–$H_6 \cdots Au_3$. It is also worth mentioning the complexes $Au_3{}^-$–$(NH_3)_2{}^{Ia}$ and $Au_3{}^-$–$(NH_3)_2{}^{III}$ are almost degenerate (within 0.67 kcal $\cdot$ mol^{-1}). In the latter one, two separate ammonia molecules form nonconventional N- H$\cdots$Au H-bonds with different terminal atoms of the $Au_3{}^-$ chain (see Figure 8 and corresponding legend).

The energetics of the 3rd ammination of the $Au_3{}^Z$ cluster is determined by the low-energy portion of the PES displayed in Figures 9–11. The neutral charge state of this PES exhibits two branches of conformers, both originating from the complexes Au_3–$(NH_3)_2{}^I$ and Au_3–$(NH_3)_2{}^{II}$ and being apart by approximately 6 kcal$\cdot$ mol^{-1}. The higher-energy branch consists of the cyclic complex Au_3–$(NH_3)_3{}^I$, whose ammonia trimer is Au–N anchored to the triangular gold cluster on the one side and forms the nonconventional N–H$\cdots$Au bond on the other, and the complex Au_3–$(NH_3)_3{}^{II}$ which involves three anchoring Au–N bonds. The lower-energy branch is represented by the complexes Au_3–$(NH_3)_3{}^{III}$ and Au_3–$(NH_3)_3{}^{IV}$ whose binding energies are respectively equal to 7.44 and 5.36 kcal $\cdot$ mol^{-1} relative to the asymptote Au_3–$(NH_3)_2{}^I + NH_3$ (see also Table 2 for the related thermodynamics). Structurally, these two complexes can be viewed as formed from Au_3–$(NH_3)_2{}^I$ and the third ammonia molecule via the N_1–$H_5 \cdots N_6$ hydrogen bond. The latter is significantly stronger than the one that dimerizes two ammonia molecules.

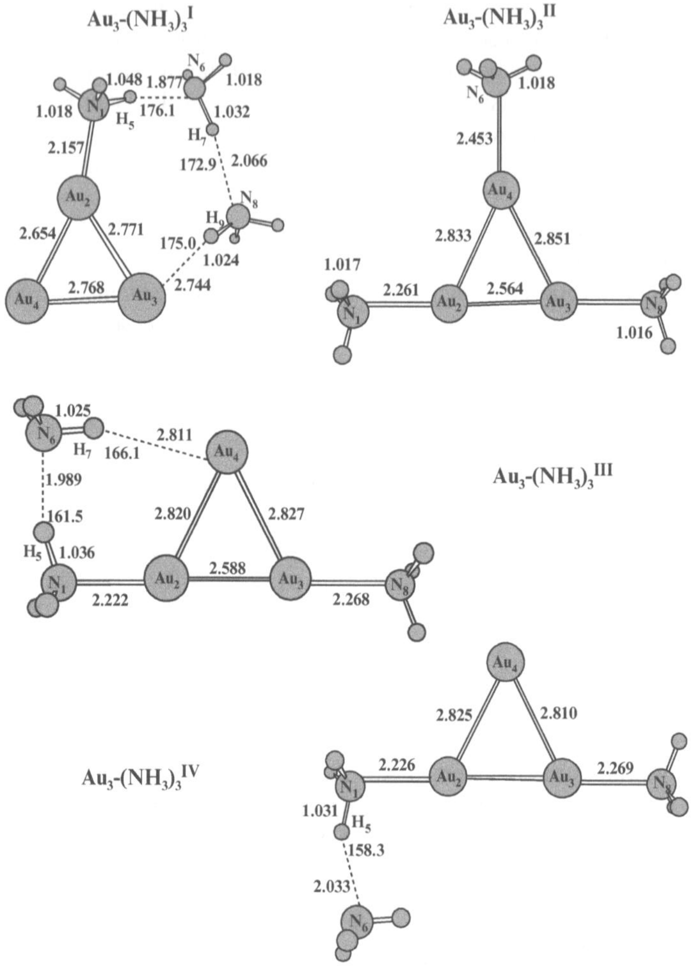

Figure 9. The low-energy complexes Au_3–$(NH_3)_3$. Their selected vibrational modes: (i) Au_3–$(NH_3)_3^{I}$: $\nu(N_1–H_5) = 3013\,cm^{-1}(1077\,km \cdot mol^{-1})$; $\nu(N_6–H_7) = 3289\,cm^{-1}(375\,km \cdot mol^{-1})$; $\nu(N_8–H_9) = 3409\,cm^{-1}(108\,km \cdot mol^{-1})$; (ii) Au_3–$(NH_3)_3{}^{III}$: $\nu(N_1–H_5) = 3214\,cm^{-1}(362\,km \cdot mol^{-1})$; $\nu(N_6–H_7) = 3401\,cm^{-1}(159\,km \cdot mol^{-1})$; (iii) Au_3–$(NH_3)_3{}^{IV}$: $\nu(N_1–H_5) = 3294\,cm^{-1}(446\,km \cdot mol^{-1})$. Bond lengths are given in Å and bond angles in degrees.

For example, in Au_3–$(NH_3)_3{}^{III}$ the N_1–$H_5 \cdots N_6$ H-bond is characterized, with respect to the ammonia dimer, by $\Delta R(N_1–H_5) = 0.014\,Å$, $\Delta r(H_5 \cdots N_6) = -0.240\,Å$, $\Delta\nu(N_1–H_5) = -208\,cm^{-1}$. By simply juxtaposing the geometries of Au_3–$(NH_3)_3{}^{III}$ and Au_3–$(NH_3)_3{}^{IV}$, we can get a rough estimate of the energy of formation of the nonconventional N_6–$H_7 \cdots Au_4$ hydrogen bond, which amount to the difference in their binding energies and is equal to $\sim 2\,kcal \cdot mol^{-1}$. It is, on one hand, weaker than in Au^-–NH_3 and, on the other hand, stronger than in $Au_3{}^-$–NH_3. Its

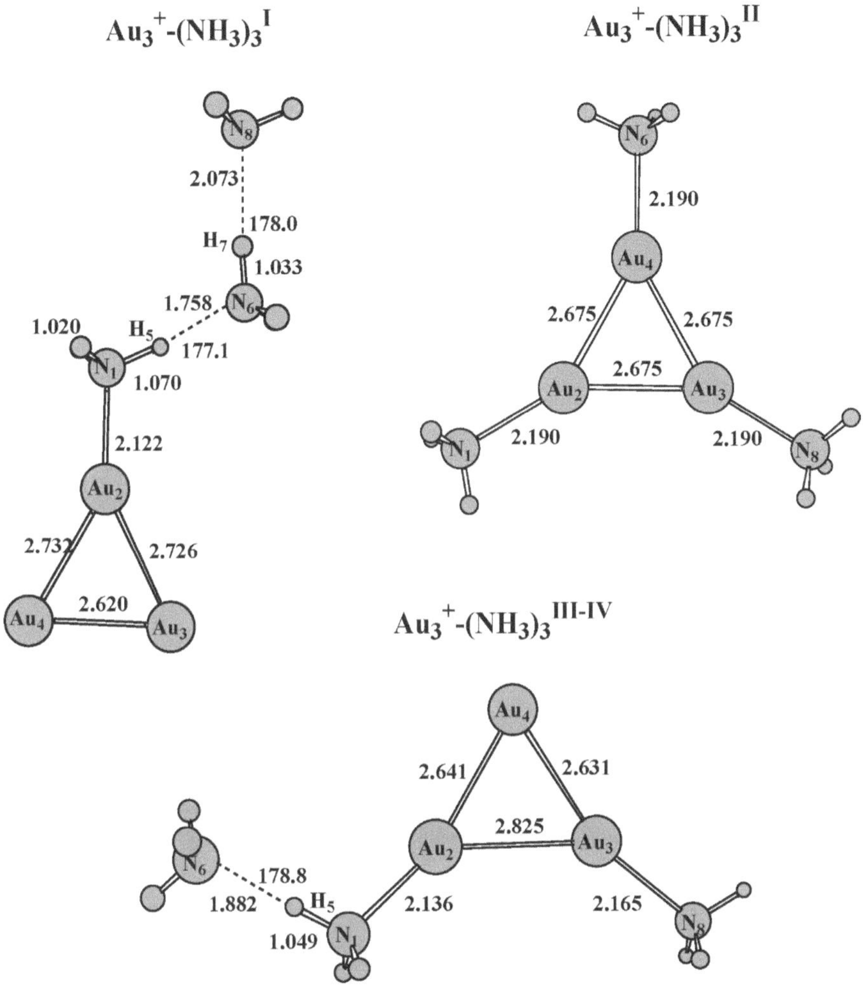

Figure 10. The low-energy complexes $Au_3^+ - (NH_3)_3$. Their selected vibrational modes: (i) $Au_3^+ - (NH_3)_3^I$: $\nu(N_1–H_5) = 2656\,cm^{-1}(2588\,km \cdot mol^{-1})$; $\nu(N_6–H_7) = 3281\,cm^{-1}(514\,km \cdot mol^{-1})$; (ii) $Au_3^+–(NH_3)_3^{III–IV}$: $\nu(N_1–H_5) = 2999\,cm^{-1}(1501\,km \cdot mol^{-1})$. Units as in Figure 9.

strength can be estimated with respect to the ammonia dimer: $\Delta R(N_6–H_7) = 0.008$ Å and $\Delta\nu(N_6–H_7) = -215\,cm^{-1}$. We may therefore infer that two bonding factors determine the larger stability of the structure $Au_3–(NH_3)_3^{III}$, on the neutral sheet of the PES which governs the third ammination of triangular gold cluster. A leading role is attributed to the anchoring Au–N bond and a minor one to the nonconventional N–H $\cdots$ Au H–bond. The former causes the redistribution of the electron charge over the entire gold cluster, which directs the unanchored atom to act as a nonconventional proton acceptor and also activates the adjacent N–H bond. Both effects occurring in

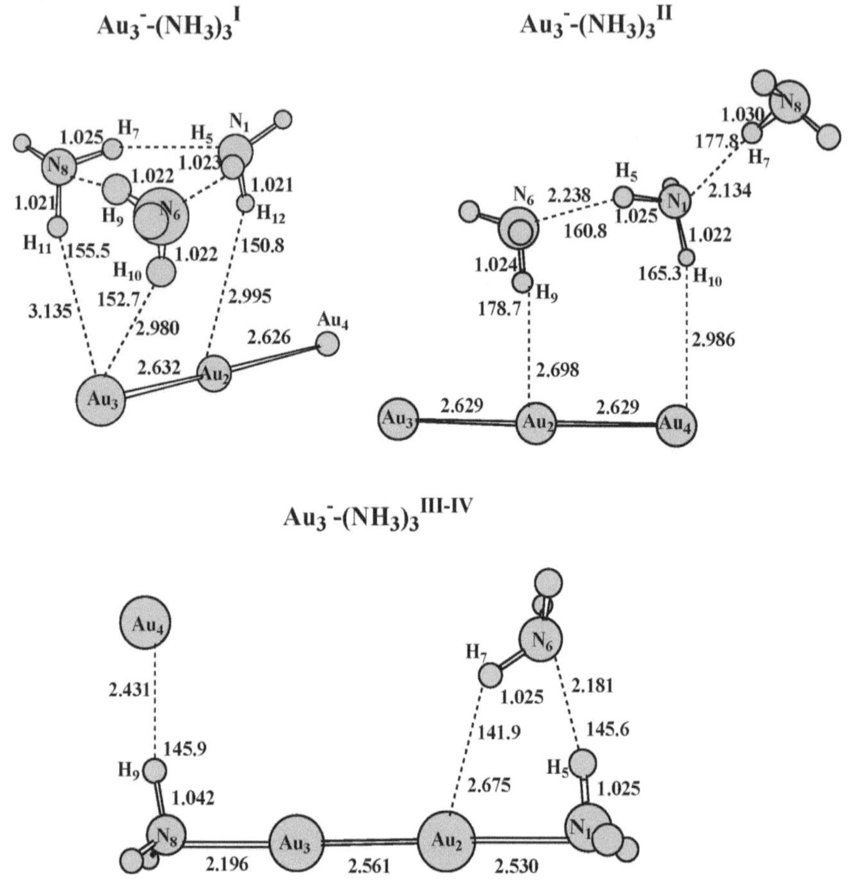

Figure 11. The low-energy complexes $Au_3{}^{-}$–$(NH_3)_3$. Their selected vibrational modes: (i) $Au_3{}^{-}$–$(NH_3)_3{}^{I}$: $\nu(N_8–H_7) = 3391\,cm^{-1}$ $(89\,km \cdot mol^{-1})$; $\nu(N_1–H_5) = 3414\,cm^{-1}$ $(53\,km \cdot mol^{-1})$; $\nu(N_6–H_{9,11}) = 3420, 3526\,cm^{-1}$ $(77, 61\,km \cdot mol^{-1})$; $\nu(N_1–H_{5,10}) = 3530\,cm^{-1}$ $(71\,km \cdot mol^{-1})$; (ii) $Au_3{}^{-}$–$(NH_3)_3{}^{II}$: $\nu(N_8–H_7) = 3326\,cm^{-1}$ $(473\,km \cdot mol^{-1})$; $\nu(N_1–H_5) = 3396\,cm^{-1}$ $(90\,km \cdot mol^{-1})$; $\nu(N_6–H_9) = 3412\,cm^{-1}$ $(76\,km \cdot mol^{-1})$; $\nu(N_1–H_{10}) = 3497\,cm^{-1}$ $(45\,km \cdot mol^{-1})$; (iii) $Au_3{}^{-}$–$(NH_3)_3{}^{III\text{-}IV}$: $\nu(N_8–H_9) = 3134\,cm^{-1}$ $(1120\,km \cdot mol^{-1})$; $\nu(N_1–H_5) = 3378\,cm^{-1}$ $(42\,km \cdot mol^{-1})$; $\nu(N_6–H_7) = 3393\,cm^{-1}$ $(270\,km \cdot mol^{-1})$. Units as in Figure 9.

the ammonia dimer, the substantial activation of the N_1–H_6 bond and the formation of a nonconventional H-bond, significantly reinforce the anchoring bond. This can be seen from a comparison of the two anchoring bonds, Au_2–N_1 and Au_3–N_8 (see Figure 9), which differ by 0.046 Å.

The landscape of the PES of cationic $Au_3{}^{+}$–$(NH_3)_3$ is markedly different from that of its neutral cousin (see Figure 10).

First, the most stable complex turns out to be $Au_3{}^{+}$–$(NH_3)_3{}^{II}$, whose neutral parent resides on the higher-energy branch. Relative to the reaction asymptote of

$Au_3^+–(NH_3)_2^I + NH_3$, $Eb(Au_3^+–(NH_3)_3^{II}) = 31.63\,kcal \cdot mol^{-1}$, which, compared to $E_b(Au_3^+–(NH_3)_2^I) = 38.74\,kcal \cdot mol^{-1}$, demonstrates a small decrease of the third ammination for Au_3^+. The thermodynamics of the formation of $Au_3^+–(NH_3)_3^{II}$, quantitatively described by $\Delta H^\circ_{298}(Au_3^+–(NH_3)_3^{II}) = -32.07\,kcal \cdot mol^{-1}$, $\Delta S^\circ(Au_3^+–(NH_3)_3^{II}) = -28.79\,cal \cdot K^{-1} \cdot mol^{-1}$, $\Delta G^\circ_{298}(Au_3^+–(NH_3)_3^{II}) = -23.49\,kcal \cdot mol^{-1}$ (see Table 2), shows the same trend. However, comparing with Au^+ (Table 1), the third ammination of Au_3^+ is more likely since more gold atoms are available for Au–N anchoring.

Secondly, the ionization of $Au_3–(NH_3)_3^{III}$ and $Au_3–(NH_3)_3^{IV}$ converges to the same cation $Au_3^+–(NH_3)_3^{III-IV}$, which is $20.21\,kcal \cdot mol^{-1}$ higher than $Au_3^+–(NH_3)_3^{II}$. The intramolecular $N_1–H_5 \cdots N_6$ hydrogen bond which exists in $Au_3–(NH_3)_3^{III}$ and $Au_3–(NH_3)_3^{IV}$ is significantly strengthened under ionization (for example, $\nu(N_1–H_5)$ of $Au_3–(NH_3)_3^{III}$ is red-shifted by $212\,cm^{-1}$).

Third, the ionization substantially deepens, to $50.65\,kcal \cdot mol^{-1}$, the energy gap between $Au_3–(NH_3)_3^{II}$ and $Au_3–(NH_3)_3^{I}$.

The order of stability established for the neutral charge state of the PES of $Au_3^Z–(NH_3)_3$ is completely reversed for the negative state (Figure 11). Moreover, by analogy with $Au_3^-–(NH_3)_2$, the lowest-energy state is almost doubly degenerate (at 0 K) and occupied by the complexes $Au_3^-–(NH_3)_3^I$ and $Au_3^-–(NH_3)_3^{II}$, which are on the high-energy branch of the neutral charge state. These complexes are quite different in that, while the former is actually the nonconventionally H-bonded complex between Au_3^- and the cyclic $(NH_3)_3$, the latter is that between Au_3^- and the open one. The binding energy of $Au_3^-–(NH_3)_3^I$ with respect to $Au_3^-–(NH_3)_2^{II} + NH_3$ is $2.44\,kcal \cdot mol^{-1}$. This value almost coincides with its binding energy ($2.21\,kcal \cdot mol^{-1}$) taken relative to the asymptote corresponding to the infinitely separated Au_3^- and the cyclic ammonia trimer. The energy difference ΔE_{I-II} between $Au_3^-–(NH_3)_3^I$ and $Au_3^-–(NH_3)_3^{II}$ is mostly due to the ZPVE effect, which amounts only to $0.89\,kcal \cdot mol^{-1}$, and hence we can treat these two complexes as nearly isoenergetic, at least at 0 K. A lesser compactness of $Au_3^-–(NH_3)_3^{II}$ with respect to $Au_3^-–(NH_3)_3^I$ results in a negative difference $\Delta S_{I-II}^\circ = -10.76\,cal \cdot K^{-1} \cdot mol^{-1}$, whose contribution to the Gibbs free energy difference amounts to $3.21\,kcal \cdot mol^{-1}$ at room temperature (for thermodynamic characteristics of $Au_3^-–(NH_3)_3^I$, see Table 2).

Both lowest-energy neutral conformers $Au_3–(NH_3)_3^{III}$ and $Au_3–(NH_3)_3^{IV}$ are transformed into the negatively charged complex $Au_3^-–(NH_3)_3^{III-IV}$, structurally similar to $Au_3^-–(NH_3)_2^{Ib}$. The difference in energy between $Au_3^-–(NH_3)_3^I$ and $Au_3^-–(NH_3)_3^{III-IV}$ amounts to $5.79\,kcal \cdot mol^{-1}$.

The structural aspects of negatively charged complexes $Au_3^-–(NH_3)_3^I$, $Au_3^-–(NH_3)_3^{II}$ and $Au_3^-–(NH_3)_3^{III-IV}$ can be readily seen on Figure 11. The selected frequencies listed in the legend allow distinguishing the former two and both from the latter. The lowest stretching vibrational modes of $Au_3^-–(NH_3)_3^{II}$, centered at 3326 and $3396\,cm^{-1}$, are referred to the stretches of the intramolecular H-bonds, $N_8–H_7 \cdots N_1$ and $N_1–H_5 \cdots N_6$, of the open ammonia trimer. The next stretching mode, peaked at $3412\,cm^{-1}$, is associated with the nonconventional H-bond N_6–$H_9 \cdots Au_2$, where the central atom of the Au_3^- gold chain acts as a nonconventional proton acceptor and which, according to its parameters $R(N_6–H_9) = 1.024$ Å and $r(H_9 \cdots Au_2) = 2.698$ Å, is stronger than in $Au_3{}^-–(NH_3)_2{}^{II}$. Three nonconventional H-bonds of $Au_3{}^-–(NH_3)_3{}^{I}$, which bind the cyclic ammonia trimer to $Au_3{}^-$, are generally weaker than in $Au_3{}^-–(NH_3)_2{}^{II}$. Also, the nonconventional H-bond N_8–$H_9 \cdots Au_4$ of the less stable complex $Au_3{}^-–(NH_3)_3{}^{III-IV}$ is weaker than the analogous one in $Au_3{}^-–(NH_3)_2{}^{Ib}$, but another one, N_6–$H_7 \cdots Au_2$, in this complex is stronger and comparable with the strongest intramolecular bond N_8–$H_7 \cdots N_1$ of $Au_3{}^-–(NH_3)_3{}^{I}$.

4. APPLICATION TO QUANTUM LOGIC: MOLECULAR SWITCHES

The aim of this section is to illustrate how the calculated landscape of the PESs of $Au_{k=1,3}{}^{Z}–(NH_3)_{1\leq n\leq 3}(Z = 0, \pm 1)$ can be used to implement molecular qubit logic gate operations. Quantum information processing [32] has recently been under extensive studies in both areas of physics and chemistry [33]. A novelty of the present approach is based on our findings reported in sections 2 and 3 that the different charged states of the Au atom and the triangular Au_3 gold cluster admit specific and fundamentally different bonding patterns with ammonia molecule and its oligomers, switching from the anchoring Au–N bond to the nonconventional N–H $\cdots$ Au hydrogen bond. The charge of the species $Au_{1,3}$ can, for instance, be varied using different metallic supporters and /or applied voltages and the NeNePo ("negative ion – to neutral – to positive ion") experimental technique [34]. The description of the NeNePO experimental setup is given in Refs. [34, c–d]. A switching from the neutral state of a given complex $Au_k{}^0–(NH_3)_{1\leq n\leq 3}(k = 1, 3)$ to its cationic state $Au_k{}^+–(NH_3)_{1\leq n\leq 3}$ can also be initiated by resonant photoionization [35].

Two basic and essentially different ingredients of the bonding patterns discussed in sections 2 and 3 are used to design logic gates: (i) the anchoring Au–N bond, and (ii) the nonconventional N–H $\cdots$ Au hydrogen bond. The bonding of the ammonia oligomers $(NH_3)_2$ and $(NH_3)_3$ to Au_1 or Au_3 can be switched to one or the other by alternating the neutral or positively charged sheet of the PES to the negatively charged one. Herein, we give insight on how these ingredients could be tailored experimentally by defining a set of control parameters of the model, in order to realize a gate set. Additionally, as far as the number of ammonia molecules increases, different dispositions of a so-called "armed" ammonia chains and the inter-ammonia hydrogen bonding patterns can be also used for encoding.

We start by discussing the simplest situation of the complex $Au^Z–NH_3$, that contains a single atom of gold. The dependence of its structural and energetic properties on the charge state is discussed in detail in section 2. We model the complex $Au^Z–NH_3$ by a two-level system with total Hamiltonian:

$$(1) \qquad \hat{H}^Z_{11} = \hat{H}^Z_{Au} + \hat{H}_{NH3} + \lambda V_{int,11}{}^Z$$

The subscript 11 stands for one Au atom and one NH_3 molecule (precisely, k = 1 and n = 1). $\hat{H}_{Au}{}^{Z}$ and $\hat{H}_{NH3}$ are the sub-Hamiltonians of the Au^Z and NH_3 subsystems, respectively, and $\lambda V_{int,11}{}^{Z}$ is the interaction energy, λ being the coupling constant. The following two scenarios are possible, depending on the charge state of the Au atom:

(i) When Z = 0 or +1, there can be an Au–N anchoring in $Au^{0,+1}$–NH_3. In that case $\hat{H}_{11}{}^{Z}$ acts on the "anchoring" two-dimensional Hilbert space A (stands for "anchoring"), spanned by the states $|0_A\rangle$ and $|1_A\rangle$ over the real field $\Re^1$. Varying λ results in a mixed "anchor" qubit $\alpha|0_A\rangle + \beta|1_A\rangle(\alpha^2 + \beta^2 = 1; \alpha, \beta \in \Re^1)$. In other words, λ controls the probability of anchoring: if λ is "off" ($\lambda = 0$), we have a "no-anchor" qubit $|0_A\rangle$; the opposite case, when λ is "on" ($\lambda = 1$), yields a pure "anchor" qubit $|1_A\rangle$.

(ii) When Z = −1, we are located on the negative sheet of the PES of Au–NH_3, and $\hat{H}_{11}{}^{-1}$ acts on the "H-bonding" two-dimensional Hilbert space H spanned by the states $|0_H\rangle$ and $|1_H\rangle$ over $\Re^1$. $|0_H\rangle$ is the state with "no nonconventional hydrogen bonding", whereas $|1_H\rangle$ stands for the case where a hydrogen bond is present.

Let us now consider the electronic ground-state PES of the system Au^Z–$(NH_3)_2$ consisting of a single Au atom and an ammonia dimer $(NH_3)_2$. Its total Hamiltonian $\hat{H}_{12}{}^{Z}$ is written as

$$(2) \qquad \hat{H}_{12}{}^{Z} = \hat{H}_{Au}{}^{Z} + \hat{H}_{(NH3)2} + \lambda V_{int,12}{}^{Z}.$$

It was shown in section 2 that the low-energy portion of the PES of Au^Z–$(NH_3)_2$ admits either two different anchoring Au–N bonding patterns in the neutral or positively charged complexes $Au^{0,+}$–$(NH_3)_2{}^{I}$ and $Au^{0,+}$–$(NH_3)_2{}^{II}$ or two different nonconventional N-H $\cdots$ Au H-bonds as it takes place in the complexes Au^-–$(NH_3)_2{}^{I}$ and Au^-–$(NH_3)_2{}^{II}$.

As can be seen on Figure 1, the higher-energy conformer Au–$(NH_3)^{I}$ exhibits two anchor bonds while the most stable complex Au–$(NH_3)^{II}$ has only one. The order of stability of the two conformers is inverted on the positively charged PES: Au^+–$(NH_3)^{I}$ that retains two anchoring bonds becomes the most stable complex, whereas the less stable one, Au^+–$(NH_3)^{II}$, has only one as its neutral parent. For the negatively charged complexes of Au^-–(NH_3) shown in Figure 2, Au^-–$(NH_3)^{I}$ exhibits two nonconventional H-bonds with the dimer $(NH_3)_2$, while Au^-–$(NH_3)^{II}$ has two such H-bonds with the non-dimerized NH_3 molecules. Therefore, for a given Z, $\hat{H}_{12}{}^{Z}$ acts upon the bipartite Hilbert spaces: (i) if Z = 0, +1, it is $A_1 \otimes A_2$ spanned over the basis composed of the direct product states $|i_{A1}\rangle \otimes |j_{A2}\rangle \equiv |i_{A1}j_{A2}\rangle$, and (ii) if Z = −1, it is $H_1 \otimes H_2$ spanned over the basis $|i_{H1}\rangle \otimes |j_{H2}\rangle \equiv |i_{H1}j_{H2}\rangle(i, j = 0, 1)$. An additional substructure of the basis originates from the sub-Hamiltonian $\hat{H}_{(NH3)2} = 2\hat{H}_{NH3} + \lambda' V_{int}{}'$ which can also be treated as acting on the states $|0_N\rangle$ and $|1_N\rangle$ belonging to N: the former describes two infinitely separated NH_3 molecules ("the intramolecular H-bond is off"; $\lambda' = 0$) and the latter describes the H-bonded ammonia dimer $(NH_3)_2$ ("the intramolecular H-bond is on"; $\lambda' = 1$).

Alternation of the neutral and charge states on the general PES of Au^Z–$(NH_3)_2$ is associated with the Hamiltonian

(3) $$\hat{H}_{12}{}^{switch} = P_0 \hat{H}_{12}{}^{0} P_0 + P_{+1} \hat{H}_{12}{}^{+1} P_{+1} + P_{-1} \hat{H}_{12}{}^{-1} P_{-1}$$

where P_Z is the projector on the Zth charge state. In the NeNePo process the interaction of $\hat{H}_{12}{}^{switch}$ with the femtosecond laser pulse specifically selects one of the projectors: the inversion of the charge state occurs sequentially from the negative (P_{-1}) via neutral (P_0) to the positive one (P_{+1}). Note that the cationic state can also be prepared from the neutral state by resonant photoionization [35].

We assume that, generally speaking, $\hat{H}_{12}{}^{switch}$ acts upon the 5-qubit product Hilbert space $H = A_1 \otimes A_2 \otimes H_1 \otimes H_2 \otimes N$, encodes the qubits and realizes the following two qubit gates, which could be controlled by an applied current and IR spectroscopy – the latter probes the charge states in the NeNePo process [34]. It is worth mentioning that the 5-qubit product Hilbert space allows a better experimental determination of the logic state of the system and can be projected onto a lower-dimensional one, for example onto $A_1 \otimes A_2 \otimes H_1$, to obtain a cruder description.

To build a NOT gate, we encode the target qubit, which can be either $|0\rangle = \begin{pmatrix} 1 \\ 0 \end{pmatrix}$ or $|1\rangle = \begin{pmatrix} 0 \\ 1 \end{pmatrix}$, into the two lowest-energy conformers, $Au–(NH_3)_2{}^{II}$ and $Au–(NH_3)_2{}^{I}$, on the neutral sheet of the PES of the $Au^Z–(NH_3)_2$ cluster. As noticed section 2, $Au–(NH_3)_2{}^{II} \equiv |0\rangle^0$ lies lower than $Au–(NH_3)_2{}^{I} \equiv |1\rangle^0$ by 4.3 kcal · mol^{-1}. In the 5-qubit encoding they are represented as $|0\rangle^0 \equiv |1_{A1}0_{A2}0_{H1}0_{H2}1_N\rangle$ and $|1\rangle^0 \equiv |1_{A1}1_{A2}0_{H1}0_{H2}0_N\rangle$. The quantum NOT gate [32] is a one-qubit gate associated with the matrix

(4) $$\mathbf{X} = \begin{bmatrix} 0 & 1 \\ 1 & 0 \end{bmatrix}$$

So that if the initial value of the qubit is $|0\rangle = \begin{pmatrix} 1 \\ 0 \end{pmatrix}$ we get the NOT logic operation, as $\mathbf{X}\begin{pmatrix} 1 \\ 0 \end{pmatrix} = \begin{pmatrix} 0 \\ 1 \end{pmatrix} \equiv |1\rangle$, and vice versa if the initial value of the quibit is $|1\rangle$. In the 5-qubit encoding, the action of **X** is to simultaneously flip the pair of the qubits $|0_{A2}\rangle$ and $|1_N\rangle(\in |0\rangle^0)$ into $|1_{A2}\rangle$ and $|0_N\rangle(\in |1\rangle^0)$. The action of the NOT gate can therefore be implemented on the neutral PES by using an IR laser pulse.

On the positively charged PES, the conformer $Au^+–(NH_3)_2{}^{I}$ is the global minimum, that is, $Au^+–(NH_3)_2{}^{I} \equiv |0\rangle^{+1} \equiv |1_{A1}1_{A2}0_{H1}0_{H2}0_N\rangle$, whereas a higher-energy conformer is $Au^+–(NH_3)_2{}^{II} \equiv |1\rangle^{+1} \equiv |1_{A1}0_{A2}0_{H1}0_{H2}1_N\rangle$. Their energy difference amounts to 43.6 kcal · mol^{-1}. Clearly, a simultaneous flip of the qubits $|1_{A2}\rangle$ and $|0_N\rangle(\in |0\rangle^+)$ into $|0_{A2}\rangle$ and $|1_N\rangle(\in |1\rangle^+)$, which can be generated optically by using a UV photon, implements another quantum NOT gate (4). Notice that the qubit substructure of the states $|0\rangle^{0,+1}$ and $|1\rangle^{0,+1}$ originated from the bipartite Hilbert space $H_1 \otimes H_2$ is irrelevant under these NOT gate operations. This implies that in order to implement the NOT gate one can work in the smaller 3-qubit Hilbert space $A_1 \otimes A_2 \otimes N$.

The quantum NOT gate (4) is also realized, within the Hamiltonian $\hat{H}_{12}{}^{switch}$ given by Eq. (3), by switching from the neutral charge state of the studied PES to its positively charged state. This latter can be prepared from the neutral state by resonant photoionization [35]. The NOT operation (4) is implemented as follows, using a consistent labeling of the target qubits assigned, e.g. to the neutral charge state:

- the neutral global minimum ("ground") state is $|0\rangle^0 \equiv |1_{A1}0_{A2}0_{H1}0_{H2}1_N\rangle$, that is Au–$(NH_3)_2{}^{II}$ maps into $|1\rangle^{+1} \equiv |1_{A1}1_{A2}0_{H1}0_{H2}0_N\rangle$, which is Au^+–$(NH_3)_2{}^{I}$; and vice versa:
- the neutral higher-energy ("excited") state Au–$(NH_3)_2{}^{I} \equiv |1\rangle^0 \equiv |1_{A1}1_{A2}0_{H1}0_{H2}0_N\rangle$ maps into Au^+–$(NH_3)_2{}^{II} \equiv |0\rangle^{+1} \equiv |1_{A1}0_{A2}0_{H1}0_{H2}1_N\rangle$.

This operation correlates the ground and excited states on both surfaces. The two-level charge-induced interchange of the conformers can occur on a timescale of a few picoseconds, which is typical for the resonant photoionization process [35]. The dynamics of such a process, $|0\rangle^0 \to |1\rangle^{+1}$ and $|1\rangle^0 \to |0\rangle^{+1}$, is monitored in real time by the change in the anchoring A–N stretch, equal to $\Delta\nu$(Au–N) $= 145$, 165 (due to the appearance of the A–N stretch doublet), and by the disappearance of the vibrational mode ν(N–H $\cdots$ N) (see Table 3) using, e.g. time-resolved picosecond UV/IR pump-probe ionization depletion spectroscopy [35].

In section 2 we described how the low-energy portion of the negatively charged PES of Au^-–$(NH_3)_2$ cluster consists of two nearly degenerate conformers Au^-–$(NH_3)_2{}^{I}$ and Au^-–$(NH_3)_2{}^{II}$. By analogy with their neutral parent states, they can be correspondingly designated by $|0\rangle^{-1}$ and $|1\rangle^{-1}$ and encoded within the 5-qubit product Hilbert space as $|0\rangle^{-1} = |0_{A1}0_{A2}1_{H1}0_{H2}1_N\rangle$ and $|1\rangle^{-1} = |0_{A1}0_{A2}1_{H1}1_{H2}0_N\rangle$. Due to quasi degeneracy of $|0\rangle^{-1}$ and $|1\rangle^{-1}$, the low-energy portion of the PES of the anionic Au^-–$(NH_3)_2$ cluster itself represents a

Table 3. The IR readout of the Au^Z–$(NH_3)_2$ cluster qubits. Frequency of a given vibrational mode in cm^{-1} and corresponding IR activity (in parentheses) in km $\cdot$ mol^{-1}.

Qubit	ν(Au-N)	ν(N- H $\cdots$ Au) ν_σ(H $\cdots$ Au)	ν(N-H $\cdots$ N) ν_σ(H $\cdots$ N)
$\|0\rangle^0 = \|1_{A1}0_{A2}0_{H1}0_{H2}1_N\rangle$	314		3291 (425) 198
$\|1\rangle^0 = \|1_{A1}1_{A2}0_{H1}0_{H2}0_N\rangle$	$\nu^{sym} = 155$ $\nu^{asym} = 175$		
$\|0\rangle^{+1} = \|1_{A1}1_{A2}0_{H1}0_{H2}0_N\rangle$	$\nu^{sym} = 459$ $\nu^{asym} = 479$		
$\|1\rangle^{+1} = \|1_{A1}0_{A2}0_{H1}0_{H2}1_N\rangle$	493		2548 (2335) 261
$\|0\rangle^{-1} = \|0_{A1}0_{A2}1_{H1}0_{H2}1_N\rangle$		3253 (467) 100	3415 (65) 118
$\|1\rangle^{-1} = \|0_{A1}0_{A2}1_{H1}1_{H2}0_N\rangle$		$\nu^{sym} = 3345(558)$ $\nu^{asym} = 3351(0)$ $\nu_\sigma^{sym} = 84$ $\nu_\sigma^{asym} = 89$	

quasi-symmetric double well, characterized by the lowest-energy eigenstates $|g_{\pm}\rangle^{-1}$ of $\hat{H}_{12}{}^{-1}$:

$$(5) \qquad |g_{\pm}\rangle^{-1} = |0\rangle^{-1} \pm |1\rangle^{-1}.$$

Therefore, the one-qubit Hadamard gate [32] can be implemented as a switch that operates within the Hamiltonian $\hat{H}_{12}{}^{\text{switch}}$ and transform its neutral-state component into the negative-state one:

$$(6) \qquad \mathbf{H} = \frac{1}{\sqrt{2}} \begin{bmatrix} 1 & 1 \\ 1 & -1 \end{bmatrix}.$$

Note that a similar Hadamard gate can be realized for the positive-negative switch that operates between the positively and negatively charged sheets of the PES of the Au^{Z}–$(NH_3)_2$ cluster. A readout of the Au^{Z}–$(NH_3)_2$ cluster qubits can be achieved in various ways, for instance by measuring its IR spectrum. Each state of the 5 qubits can be detected by IR spectroscopy – the IR identifier of each 5-qubit state is given in Table 3.

5. SUMMARY AND CONCLUSION

We have discussed how the properties of the charged states of the gold atom and the triangular gold cluster govern their reactive properties and bonding patterns with ammonia clusters. We have shown that these bonding patterns are actually a multifacet phenomenon that exhibits different characteristics. These allow to propose a scheme for encoding qubits on the PES of $Au_{1,3}{}^{Z}$–$(NH_3)_{1\leq n\leq 3}$ and implementing two simple one qubit quantum gates: the NOT and the Hadamard gates. We suggest that this way of processing quantum information on nanosize hybrid gold-organic clusters might open new horizons for the development of qubit logic-gate operations.

Acknowledgements

This work was partially funded by the European Community FET-OPEN STREP Project MOLDYNLOGIC. The computational facilities were provided by NIC (University of Liege) and by F.R.F.C. 9.4545.03F and 1.5187.05 (Belgium). One of the authors (E S K) gratefully thanks Alfred Karpfen and Camille Sandorfy for the inspiring discussions and the F.R.F.C. 2.4562.03F fellowship.

References and Notes

1. N.A. Lambropoulos, J.R. Reimers, and N.S. Hush, Binding to gold(0): accurate computational methods with application to $AuNH_3$. *J. Chem. Phys.* **116**, 10277–10286, 2002.
2. A. Antušek, M. Urban, and A.J. Sadlej, Lone pair interactions with coinage metal atoms: weak van der Waals complexes of the coinage metal atoms with water and ammonia. *J. Chem. Phys.* **119**, 7247–7262, 2003.

3. (a) J. Hrušák, R.H. Hertwig, D. Schröder, P. Schwerdtfeger, W. Koch, and H. Schwarz, Relativistic effects in cationic gold (I) complexes – a comparative study of *ab initio* pseudopotential and density-functional methods. Organometallics **14**, 1284–1291, 1995; (b) T.H. Hertwig, J. Hrušák, D. Schröder, W. Koch, and H. Schwarz, The metal-ligand bond strengths in cationic gold (I) complexes. Application of approximate density functional theory. *Chem. Phys. Lett.* **236**, 194–200, 1995; (c) D. Schröder, J. Hrušák, R.H. Hertwig, W. Koch, P. Schwerdtfeger, and H. Schwarz, Experimental and theoretical studies of gold (I) complexes $Au(L)^+$(L = H2O, CO, NH3, C3H6, C4H6, C6H6, C6F6). *Organometallics* **14**, 312–316, 1995; (d) R. Armunanto, C.F. Schwenk, and B.M. Rode, Gold (I) in liquid ammonia: *ab initio* QM/MM molecular dynamics simulation. *J. Am. Chem. Soc.* **126**, 9934–9935, 2004.
4. M. Antolovich, L.F. Lindoy, and J.R. Reimers, Explanation of the anomalous complexation of silver (I) with ammonia in terms of the poor affinity of the ion for water. *J. Phys. Chem. A* **108**, 8434–8438, 2004.
5. D.-Y. Wu, B. Ren, Y.-X. Jiang, X. Xu, and Z.-Q. Tian, Density functional study and normal-mode analysis of the bindings and vibrational frequency shifts of the pyridine-M (M = Cu, Ag, Au, Cu^+, Ag^+, Au^+, and Pt) complexes. *J. Phys. Chem. A* **106**, 9042–9052, 2002.
6. H.-C. Hsu, F.-W. Lin, C.-C. Lai, P.-H. Su, and C.-S. Yeh, Photodissociation and theoretical studies of the Au + -(C_5H_5N) complex. *New J. Chem.* **26**, 481–484, 2002.
7. (a) P. Pyykkö, Theoretical chemistry of gold I. *Angew. Chem. Int. Ed.* **43**, 4412–4456, 2004; (b) P. Pyykkö, Theoretical chemistry of gold II. *Inorg. Chim. Acta* **358**, 4113–4130, 2005.
8. (a) J.V. Burda, J. Šponer, and P. Hobza, *Ab initio* study of the interaction of guanine and adenine with various mono- and bivalent metal cations ($Li^+, Na^+, K^+, Rb^+, Cs^+$; Cu^+, Ag^+, Au^+; Mg^{2+}, Ca^{2+}, Sr^{2+}, Ba^{2+}; Zn^{2+}, Cd^{2+}, and Hg^{2+}). *J. Phys. Chem.* **100**, 7250–7255, 1996; (b) J. Šponer, M. Sabat, J.V. Burda, J. Leszcynski, P. Hobza, and B. Lippert, Metal ions in non-complementary DNA base pairs: *ab-initio* study of Cu (I), Ag (I), and Au (I) complexes with the cytosine-adenine base pair. *J. Biol. Inorg. Chem.* **4**, 537–545, 1999.
9. A. Billić, J.R. Reimers, N.S. Hush, and J. Hafner, Adsorption of ammonia on the gold (111) surface. *J. Chem. Phys.* **116**, 8981–8987, 2002.
10. (a) S.M. Hou, J.X. Zhang, R. Li, J. Ning, R.S. Han, Z.Y. Shen, X.Y. Zhao, Z.Q. Xue, and Q. Wu, First-principles calculation of the conductance of a single 4,4 bipyridine molecule. *Nanotechnology* **16**, 239–244, 2005; (b) R. Stadler, K.S. Thygesen, and K.W. Jacobsen, Forces and conductance in a single-molecule bipyridine junction. *Phys. Rev. B* **72**, 241401(R)-1 – 241401-4, 2005.
11. (a) B. Xu, X. Xiao, N.J. Tao, Measurements of single-molecule electromechanical properties. *J. Am. Chem. Soc.* **125**, 16164–16165, 2003. (b) P. Vélez, S.A. Dassie, and E.P.M. Leiva, First principles calculations of mechanical properties of 4,4(')-bipyridine attached to Au nanowires. *Phys. Rev. Lett.* **95**, 045503-1 – 045503-4, 2005.
12. (a) D.I. Gittins and F. Caruso, Spontaneous phase transfer of nanoparticule metals from organic to aqueous media. *Angew. Chem. Int. Ed.* **40**, 3001–3004, 2001; (b) V.J. Gandubert and R.B. Lennox, Assessment of 4-(dimethylamino)pyridine as a capping agent for gold nanoparticles. *Langmuir* **21**, 6532–6539, 2005.
13. (a) K. Drenck, P. Hvelplund, C.J. McKenzie, and S.B. Nielsen, Identification of the short-lived $Au(N_3)_4{}^{2-}$ dianion from its Coulomb explosion products. *Int. J. Mass Spectr.* **244**, 144–147, 2005; (b) P. Braunstein, C. Frison, N. Oberbeckmann-Winter, X. Morise, A. Messaoudi, M. Bénard, M.-M. Rohmer, and R. Welter, An oriented 1D coordination/organometallic dimetallic molecular wire with Ag-Pd metal-metal bonds. *Angew. Chem. Int. Ed.* **43**, 6120–6125, 2004; (c) S. Fukuzumi, K. Ohkubo, W.E.Z. Ou, J. Shao, K.M. Kadish, J.A. Hutchison, K.P. Ghiggino, P.J. Sintic, and M.J. Crossley, Metal-centered photoinduced electron transfer reduction of a gold (III) porphyrin cation linked with a zinc porphyrin to produce a long-lived charge-separated state in nonpolar solvents. *J. Am. Chem. Soc.* **125**, 14984–14985, 2003; (d) X. Ding, Z. Li, J. Yang, J.G. Hou, and Q. Zhu, Theoretical study of nitric oxide adsorption on Au clusters. *J. Chem. Phys.* **121**, 2558–2562, 2004; A. Fielicke, G.v. Helden, G. Meijer, B. Simard, and D.M. Rayner, Direct observation of size-dependent activation of NO on gold clusters. *Phys. Chem. Chem. Phys.* **7**, 3906–3999, 2005.

14. (a) E.S. Kryachko and F. Remacle, Complexes of DNA bases and gold clusters Au_3 and Au_4 involving nonconventional N-H $\cdots$ Au hydrogen bond. *Nano Lett.* **5**, 735–739, 2005; (b) E.S. Kryachko and F. Remacle, Complexes of DNA bases and Watson–Crick base pairs with small neutral gold clusters. *J. Phys. Chem. B* **109**, 22746–22757, 2005; (c) F. Remacle and E.S. Kryachko, Three-gold cluster as proton acceptor in nonconventional hydrogen bonds O-H $\cdots$ Au and N-H $\cdots$ Au. In *Progress in Theoretical Chemistry and Physics*, Vol. 15, J. Maruani and S. Wilson (eds.), Springer, Dordrecht, 2006, pp. 433–450.
15. (a) Q. Chen, D.J. Fraenkel, and N.V. Richardson, Self-assembly of adenine on Cu (110) surfaces. *Langmuir* **18**, 3219–3225, 2002; (b) B. Giese and D. McNaughton, Surface-enhanced Raman spectroscopic study of Uracil: the influence of the surface substrate, surface potential, and pH. *J. Phys. Chem. B* **106**, 1461–1470, 2002; (c) A.P.M. Camargo, H. Baumgärtel, and C. Donner, Coadsorption of the DNA bases thymine and adenine at the Au (111) electrode. *Phys. Chem. Chem. Phys.* **5**, 1657–1664, 2003.
16. (a) L.M. Demers, M. Östblom, H. Zhang, N.-H. Jang, B. Liedberg, C.A. Mirkin, Thermal desorption behavior and binding properties of DNA bases and nucleosides on gold. *J. Am. Chem. Soc.* **124**, 11248–11249, 2002; (b) A. Gourishankar, S. Shukla, K.N. Ganesh, and M. Sastry, Isothermal titration calorimetry studies on the binding of DNA bases and PNA base monomers to gold nanoparticles. *J. Am. Chem. Soc.* **126**, 13186–13187, 2004; (c) S. Rapino and F. Zerbetto, Modeling the stability and the motion of DNA nucleobases on the gold surface. *Langmuir* **21**, 2512–2518, 2005; (d) M. Östblom, B. Liedberg, L.M. Demers, and C.A. Mirkin, On the structure and desorption dynamics of DNA bases adsorbed on gold: a temperature-programmed study. *J. Phys. Chem. B* **109**, 15150–15160, 2005.
17. A.J. Lupinetti, S. Fau, G. Frenking, and S.H. Strauss, Theoretical analysis of the bonding between CO and positively charged atoms. *J. Phys. Chem. A* **101**, 9551–9559, 1997.
18. (a) D.H. Wells, Jr., W.N. Delgass, and K.T. Thomson, Formation of hydrogen peroxide from H_2 and O_2 over a neutral gold trimer: a DFT study. *J. Catal.* **225**, 69–77, 2004; (b) Z.-P. Liu, S.J. Jenkins, and D.A. King, Origin and activity of oxidized gold in water-gas shift catalysis. *Phys. Rev. Lett.* **94**, 196102-1–196102-4 (2005); (c) A.M. Joshi, W.N. Delgass, and K.T. Thomson, Comparison of the catalytic activity of Au_3, $Au_4{}^+$, Au_5, and $Au_5{}^-$ in the gas-phase reaction of H_2 and O_2 to form hydrogen peroxide: density functional theory investigation. *J. Phys. Chem. B* **109**, 22392–22406, 2005; and references therein.
19. The present computations were conducted with *GAUSSIAN 03* package of quantum chemical programs [20]. The Kohn-Sham self-consistent field formalism with the hybrid density functional B3LYP potential was used together with the basis sets 6-311++G (d,p) for ammonia and the energy-consistent 19-($5s^2 5p^6 5d^{10} 6s^1$) valence-electron relativistic effective core potential (RECP) for gold developed by Ermler, Christiansen, and co-workers with the primitive basis set ($5s5p4d$) [21]. All geometrical optimizations were performed with the keywords "tight" and "Int=UltraFine". The unscaled harmonic vibrational frequencies and zero-point vibrational energies (ZPVE) are also computed. Enthalpies and entropies were estimated from the partition functions calculated at room temperature (298.15 K) under a pressure of 1 atm, using Boltzmann thermostatistics and the rigid-rotor-harmonic-oscillator approximation [22]. The binding energy of the complex AB, E_b[AB], is defined as the energy difference $E_b[AB] \equiv |E[AB]-(E[A]+E[B])|$. The ZPVE-corrected binding energies E_b are reported throughout this work. For a recent computational work on ammonia clusters within the B3LYPF/6-311++G(d,p) level see Ref. [23] and references therein. For a recent review on small gold clusters see Ref. [24]. The main features of gold clusters $Au_3{}^Z (Z = 0, \pm 1)$ are gathered in note [25]. The propensity of Au_3 to behave as a nonconventional proton donor and to form nonconventional hydrogen bonds with the conventional proton donors have been computationally discovered in the works referred to at [26].
20. M.J. Frisch, G.W. Trucks, H.B. Schlegel, G.E. Scuseria, M.A. Robb, J.R. Cheeseman, J.A. Montgomery, Jr., T. Vreven, K.N. Kudin, J.C. Burant, J.M. Millam, S.S. Iyengar, J. Tomasi, V. Barone, B. Mennucci, M. Cossi, G. Scalmani, N. Rega, G.A. Petersson, H. Nakatsuji, M. Hada, M. Ehara, K. Toyota, R. Fukuda, J. Hasegawa, M. Ishida, T. Nakajima, Y. Honda, O. Kitao, H. Nakai, M. Klene, X. Li, J.E. Knox, H.P. Hratchian, J.B. Cross, C. Adamo, J. Jaramillo, R. Gomperts, R.E. Stratmann,

O. Yazyev, A.J. Austin, R. Cammi, C. Pomelli, J.W. Ochterski, P.Y. Ayala, K. Morokuma, G.A. Voth, P. Salvador, J.J. Dannenberg, V.G. Zakrzewski, S. Dapprich, A.D. Daniels, M.C. Strain, O. Farkas, D.K. Malick, A.D. Rabuck, K. Raghavachari, J.B. Foresman, J.V. Ortiz, Q. Cui, A.G. Baboul, S. Clifford, J. Cioslowski, B.B. Stefanov, G. Liu, A. Liashenko, P. Piskorz, I. Komaromi, R.L. Martin, D.J. Fox, T. Keith, M.A. Al-Laham, C.Y. Peng, A. Nanayakkara, M. Challacombe, P.M.W. Gill, B. Johnson, W. Chen, M.W. Wong, C. Gonzalez, and J.A. Pople, *GAUSSIAN 03* (Revision A.1), Gaussian, Inc., Pittsburgh, PA, USA, 2003.

21. R.B. Ross, J.M. Powers, T. Atashroo, W.C. Ermler, L.A. LaJohn, P.A. Christiansen, *Ab initio* relativistic effective potentials with spin-orbit operators. IV. Cs through Rn. *J. Chem. Phys.* **93**, 6654–6670, 1990.
22. (a) D.A. McQuarrie, *Statistical Mechanics*, Harper and Row, New York, 1976; (b) J.E. Del Bene, H.D. Mettee, M.J. Frisch, B.T. Luke, and J.A. Pople, *Ab initio* computation of the enthalpies of some gas-phase hydration reactions. *J. Phys. Chem.* **87**, 3279–3282, 1983.
23. F.M. Abu-Awad, A comparative study of the structure and electrostatic potential of H-bonded clusters of neutral ammonia $(NH_3)_n (n = 2–6)$. *J. Mol. Struct.* (THEOCHEM) **683**, 57–63, 2004, and references therein.
24. (a) F. Remacle and E.S. Kryachko, Small gold clusters $Au_{5\leq n\leq 8}$ and their cationic and anionic cousins. *Adv. Quantum Chem.* **47**, 423–464, 2004; (b) F. Remacle and E.S. Kryachko, Structure and energetics of two- and three-dimensional neutral, cationic, and anionic gold clusters $Au^Z{}_{5\leq n\leq 9} (Z = 0, \pm 1)$. *J. Chem. Phys.* **122**, 044304-1 – 044304-14 (2005).
25. Within the present computational approach, the triangular conformer of the Au_3 gold cluster is characterized by an electronic energy of −407.907290 hartree, ZPVE = 0.42 kcal · mol^{-1}, enthalpy and entropy being respectively equal to −407.900617 hartree and to 89.66 cal · K^{-1} · mol^{-1}. The geometry is determined by $r(Au_1–Au_2) = r(Au_2–Au_3) = 2.654$Å, $r(Au_1–Au_3) = 2.992$Å, and $\angle Au_1Au_2Au_3 = 68.6°$, implying the so-called "geometrical frustration" (or asymmetry) due to the Jahn-Teller distortion of the ground electronic state of the triangular conformation (see Ref. [26]).
   The chain structure $Au_3{}^{ch}$ is characterized by an electronic energy of −407.911124 hartree, ZPVE = 0.427 kcal · mol^{-1}, and an enthalpy equal to −407.904441 hartree. Its bond lengths $r(Au_1–Au_2) = r(Au_1–Au_3) = 2.619$ Å and its bond angle $\angle Au_2Au_1Au_3 = 115.2°$. The chain structure is the most stable conformer of Au_3 lying below the triangle structure by 2.4 kcal · mol^{-1} after ZPVE, which is consistent with the value of 2.3 kcal · mol^{-1} recently reported by Lee *et al.* [27] – although within a so-called DFT error (see Ref. [23] and references therein), both these clusters are nearly isoenergetic. Throughout the present work, Au_3 is identified with the triangular gold cluster since, as shown in [25b], the chain cluster $Au_3{}^{ch}$ is not relevant for binding large clusters and forming nonconventional hydrogen bonds.
   The cation $Au_3{}^+$ is characterized by an electronic energy of −407.649308 hartree, ZPVE = 0.54 kcal · mol^{-1}, enthalpy and entropy being respectively equal to −407.642608 hartree and to 86.062 cal · K^{-1} · mol^{-1}. The equilibrium geometry is given by $r(Au_1–Au_2) = r(Au_2–Au_3) = 2.685$ Å and $r(Au_1–Au_3) = 2.688$ Å.
   The anionic cluster $Au_3{}^-$ is characterized by an electronic energy of −408.040996 hartree, ZPVE = 0.45 kcal · mol^{-1}, the enthalpy and entropy being respectively equal to −408.034311 hartree and to 81.08 cal · K^{-1} · mol^{-1}. Geometrically, $Au_3{}^-$ is a triatomic chain with $r(Au_1–Au_2) = r(Au_2–Au_3) = 2.634$Å.
26. (a) E.S. Kryachko and F. Remacle, Three-gold clusters form nonconventional hydrogen bonds O–H ··· Au and N–H ··· Au with formamide and formic acid. *Chem. Phys. Lett.* **404**, 142–149, 2005; (b) E.S. Kryachko, A. Karpfen, and F. Remacle, Nonconventional hydrogen bonding between clusters of gold and hydrogen fluoride. *J. Phys. Chem. A* **109**, 7309–7318, 2005; (c) E.S. Kryachko and F. Remacle, Small gold clusters form nonconventional hydrogen bonds X-H ··· Au: gold–water clusters as example. In *Theoretical Aspects of Chemical Reactivity*, A. Torro-Labbé (ed.), *Theoretical and Computational Chemistry*, Vol. **16**, P. Politzer (ed.), Elsevier, Amsterdam, 2005.
27. (a) G. Bravo-Pérez, I.L. Garzón, and O. Novaro, *Ab initio* study of small gold clusters. *J. Mol. Struct.* (THEOCHEM) **493**, 225–231, 1999; (b) H. Grönbeck and W. Andreoni, Gold and platinum

microclusters and their anions: comparison of structural and electronic properties. *Chem. Phys.* **262**, 1–14, 2000.

28. H.M. Lee, M. Ge, B.R. Sahu, P. Tarakeshwar, and K.S. Kim, Geometrical and electronic structures of gold, silver, and gold–silver binary clusters: origins of ductility of gold and gold–silver alloy formation. *J. Phys. Chem. B* **107**, 9994–10005, 2003.
29. (a) R. Näsänen, Equilibrium in ammoniacal solution of silver nitrate. *Acta Chem. Scand.* **1**, 763–769, 1947; (b) F.A. Cotton and G. Wilkinson, *Advanced Inorganic Chemistry*, Wiley, New York, 1988; (c) R. Garner, J. Yperman, J. Mullens, L.C. van Poucke, A potentiometric study of the complexation of aliphatic acyclic monoamines with Ag (I) in 1 m nitrate. *J. Coord. Chem.* **30**, 151–164, 1993.
30. The experimental value of log $\beta_2 = 26.5$ where $\beta_2 = K_1 K_2$ the product of formation constants was measured by L.H. Skibsted and J. Bjerrum, Studies on gold complexes. I. Robustness, stability and acid dissociation of the tetramminegold (III). *Acta Chem. Scand., Ser. A* **28**, 740–746, 1974.
31. (a) C.G. Pimentel and A.L. McClellan, *The Hydrogen Bond*, Freeman, San Francisco, 1960; (b) P. Schuster, G. Zundel, and C. Sandorfy (eds.), *The Hydrogen Bond. Recent Developments in Theory and Experiments*, North-Holland, Amsterdam, 1976; (c) G.A. Jeffrey, *An Introduction to Hydrogen Bonding*, The University Press, Oxford, 1997; (d) S. Scheiner, *Hydrogen Bonding. A Theoretical Perspective*, The University Press, Oxford, 1997; (e) G.R. Desiraju and T. Steiner, *The Weak Hydrogen Bond in Structural Chemistry and Biology*, The University Press, Oxford, 1999; (f) T. Steiner, The hydrogen bond in the solid state, *Angew. Chem. Int. Ed.* **41**, 48–76, 2002.
32. M.A. Nielsen and I.L. Chuang, *Quantum Computation and Quantum Information*, The University Press, Cambridge, 2004.
33. (a) K.L. Kompa and R.D. Levine, A molecular logic gate. *Proc. Natl. Acad. Sci. USA* **98**, 410–414, 2001; (b) F. Remacle, S. Speiser, and R.D. Levine, Intermolecular and intramolecular logic gates. *J. Phys. Chem. A* **105**, 5589–5591, 2001; (c) F. Remacle and R.D. Levine, Towards molecular logic machines. *J. Chem. Phys.* **114**, 10239–10246, 2001; (d) C. Joachim, J.K. Gimzewski, and A. Aviram, Electronics using hybrid-molecular and mono-molecular devices. *Nature* **408**, 541–548, 2000; (e) C.P. Collier, E.W. Wong, M. Belohradsk, F.M. Raymo, J. F. Stoddart, P.J. Kuekes, R.S. Williams, and J.R. Heath, Electronically configurable molecular-based logic gates. Science **285**, 391–394, 1999; (f) J.R. Heath, Wires, switches and wiring, a route towards a chemically assembled electronic nanocomputer. *Pure Appl. Chem.* **72**, 11–20, 2000; (g) M.A. Reed and J.M. Tour, Computing with molecules, *Sci. Am.* **282**, 86–89 (2000); (h) J. Fiurášek, N.J. Cerf, I. Duchemin, and C. Joachim, Intramolecular Hamiltonian logic gates. *Physica E* **24**, 161–172, 2004; (i) C.M. Tesch and R. De Vivie-Riedle, Vibrational molecular quantum computing: basis set independence and theoretical realization of the Deutsch-Jozsa algorithm. *J. Chem. Phys.* **121**, 12158–12168 (2004); (k) J.H. Reina, R.G. Beausoleil, T.P. Spiller, and W.J. Munro, Radiative corrections and quantum gates in molecular systems. *Phys. Rev. Lett.* **93**, 250501-1 – 250501-4; (l) S. Giordani and F.M. Raymo, A switch in a cage with a memory. *Org. Lett.* **5**, 3559–3562, 2003; (m) E.G. Emberly and G. Kirczenow, The smallest molecular switch. *Phys. Rev. Lett.* **91**, 188301-1 – 188301-4, 2003; (n) Y.H. Jang, S. Hwang, Y.-H. Kim, S.S. Jang, and W.A. Goddard III, DFT studies [2] of the rotaxane component of the Stoddart-Heath molecular switch. *J. Am. Chem. Soc.* **126**, 12636–12645, 2004; (o) I. Duchemin and C. Joachim, A quantum digital half adder inside a single molecule. *Chem. Phys. Lett.* **406**, 167–172, 2005; (p) F. Remacle and R.D. Levine, Quasiclassical computation. *Proc. Natl. Acad. Sci. USA* **101**, 12091–12095, 2004; (q) F. Remacle, J.R. Heath, and R.D. Levine, Electrical addressing of confined quantum systems for quasiclassical computation and finite state logic machines. *Proc. Natl. Acad. Sci. USA* **102**, 5653–5658, 2005; and references therein.
34. (a) S. Wolf, G. Sommerer, S. Rutz, E. Schreiber, T. Leisner, L. Wöste, and R.S. Berry, Spectroscopy of size-selective neutral clusters: femtosecond evolution of neutral silver trimers. *Phys. Rev. Lett.* **74**, 4177–4180, 1995; (b) L.D. Socaciu-Siebert, J. Hagen, J. Le Roux, D. Popolan, M. Vaida, S. Vajda, T.M. Bernhardt, and L. Wöste, Ultrafast nuclear dynamics induced by photodetachment of Ag_2^- and $Ag_2O_2^-$: oxygen desorption from a molecular silver surface. *Phys. Chem. Chem. Phys.* **7**, 2706–2709, 2005, and references therein; (c) R. Mitrić, M. Hartmann, B. Stanca, V. Bonačić-Koutecký, and P. Fantucci, *Ab initio* adiabatic dynamics combined with Wigner distribution approach

to femtosecond pump-probe negative ion to neutral to positive ion (NeNePo) spectroscopy of Ag_2Au, Ag_4, and Au_4 clusters. *J. Phys. Chem. A* **105**, 8892–8905, 2001; (d) T.M. Bernhardt, J. Hagen, L.D. Socaciu-Siebert, R. Mitrić, A. Heidenreich, J. Le Roux, D. Popolan, M. Vaida, L. Wöste, V. Bonačić-Koutecký, and J. Jortner, Femtosecond time-resolved geometry relaxation and ultrafast intramolecular energy redistribution in Ag_2Au energy. *Chem. Phys. Chem.* **6**, 243–253, 2005.
35. S. Ishiuchi, M. Sakai, Y. Tsuchida, A. Takeda, Y. Kawashima, M. Fujii, O. Dopfer, and K. Müller-Dethlefs, Real-time observation of ionization-induced hydrophobic → hydrophilic switching. *Angew. Chem. Int. Ed.* **44**, 6149–6151, 2005.

[illegible]

POTENTIAL ENERGY SURFACES AND DYNAMICS OF HE_NBR_2 VAN DER WAALS COMPLEXES

GERARDO DELGADO-BARRIO, DAVID LÓPEZ-DURÁN, ÁLVARO VALDÉS, RITA PROSMITI, MARIA PILAR DE LARA-CASTELLS, TOMAS GONZÁLEZ-LEZANA AND PABLO VILLARREAL

Instituto de Mathemáticas y Física Fundamental, C.S.I.C., Serrano 123, 28006 Madrid, Spain

Abstract Van der Waals complexes formed by a bromine molecule and one or several He atoms are analyzed from first principles. Multidimensional potential energy surfaces and the structure and dynamics of Br_2–$(He)_N$ clusters, of increasing size *N*, are presented and discussed.

1. INTRODUCTION

For more than three decades, van der Waals (vdW) complexes have become prototypes for studying energy transfer mechanisms and weak intermolecular forces. During these years, the understanding of vdW forces has expanded dramatically. With the development of experimental techniques such as supersonic nozzle expansion, and by performing more accurate *ab initio* electronic structure calculations, it became possible to study the structure and dynamics of vdW complexes in more detail.

vdW complexes of a dihalogen molecule surrounded by several rare gas atoms have been intensely studied over the past decades by high resolution spectroscopy techniques. In fact, and depending on their size, these clusters, produced by jet supersonic expansion at extremely low temperatures, are amenable to spectroscopic studies where the diatomic molecule acts as a chromosphere. Thus, for the smaller complexes incorporating one or two rare gas atoms, spectroscopy in the visible region involving an electronic $B \leftarrow X$ transition [1, 2] offers the possibility of a detailed comparison with the theory [3, 4]. In this case, an accurate description of the driving forces through the relevant electronic potential energy surfaces (PES), and also of the photopredissociation dynamics and energy redistribution mechanisms, becomes necessary. In turn, for helium droplets doped with a variety of molecules, rotational and infrared spectroscopic studies are currently conducted, in order to answer new and challenging questions on the role of the "quantum environment" [5–7]. A number of additional

S. Lahmar et al. (eds.), Topics in the Theory of Chemical and Physical Systems, 193–202.

experiments, based on helium nanodroplet isolation techniques [8] on small and intermediate-size doped helium clusters [9, 10], have been recently performed. The first issue in this context, from the theoretical side, is how to properly describe the multidimensional PES or, in other words, at what extent the usual assumption of additive forces holds. Then, some approximate treatment to study the spatial structure, and therefore the corresponding spectroscopy, has to be used in order to deal with the many-body system.

This review is devoted to $Br_2 \cdots He_N$ clusters and is organized as follows. In section 2 we briefly overview, for $N = 1$, the photo-dissociation process $Br_2(X) \cdots He + \hbar\omega \rightarrow Br_2(B, v_B, j_B) + He$ and discuss the different PES's used, namely empirical and *ab initio* surfaces. Then, in section 3, we show, through *ab initio* calculations, how the full $Br_2(X) \cdots He_2$ surface can be accurately described by the addition of two triatomic $Br_2(X) \cdots He$ potentials plus the He–He interaction, enabling the study of larger clusters. Then in section 4, we present our Hartree approach for describing doped bosonic helium clusters and their energetics and density distributions are analyzed.

2. PHOTO-PREDISSOCIATION OF $Br_2 \cdots He$

In this process, complexes of $Br_2(X) \cdots He$ are formed in a supersonic expansion at very low temperatures, seemingly at very low vibrational and rotational states of both the diatomic and triatomic systems. In the simplest picture, a photon promotes selectively the complex towards an state in which the bromine is in an excited electronic and vibrational state. The excess of vibrational energy stored in the diatom flows to the vdW bond giving rise to its breaking up, constituting the so called vibrational predissociation (VP) process. This produces a broadening in the spectral lines which is related to the VP rate. Within the electric dipole approximation, denoting by J the quantum number associated to the total angular momentum, the selection rules (assuming that the diatomic dipole moment is not affected by complexation) are $\Delta J = 0, \pm 1, 0 \nrightarrow 0$. Thus, even at low temperatures, a large amount of rotational states have to be included in the simulations [4].

Two kind of potentials were used to describe the X and B states of this triatomic cluster: (1) The simple addition of Morse atom–atom interactions [11], and (2) An *ab initio* surface at the CCSD(T) level of the theory [12] for the X state combined with a perturbation model based on the diatomics-in-molecule approach to represent the B surface [13]. By fitting the potential parameters within the scenario (1) to reproduce the experimental blue-shifts at low diatomic v_B vibrational excitations [14], the calculations reproduce this magnitude (and also VP rates) even for moderate v_B values (≤ 30) [11]. However only the potentials of the framework (2) were able to reproduce the oscillations of such magnitudes appearing at higher v_B levels. For the $(B, v_B = 8 \leftarrow X, v_X = 0)$ transition there is also a very good agreement as regards the main band of the measured excitation spectrum [15], corresponding essentially to transitions $B \leftarrow X$ keeping a T-shaped geometry of the complex. Moreover, a weak secondary band could be mainly assigned to a transition from the collinear

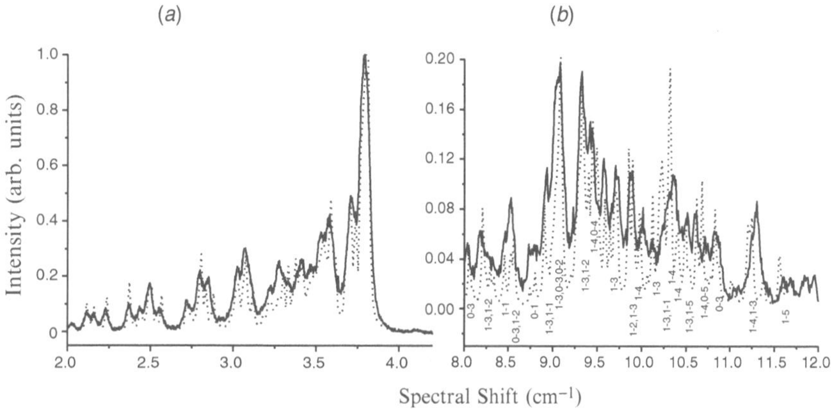

Figure 1. (a) Main and (b) secondary bands of the $B, v_B = 8 \leftarrow X, v_X = 0$ excitation spectra. *Solid lines*: experiment, *dotted lines*: simulations. The assignment of the peaks to different $n_X - n_B$ vibronic transitions are also presented.

X configuration, almost degenerated in energy with the T-shaped one, to excited bending states in the *B* state [16] by using this set of more elaborated surfaces.

Measured and simulated spectrum profiles of the main and secondary band are presented in Figure 1, showing a remarkable agreement. For the this system, laser-induced fluorescence spectra have been recently recorded at different temperature regimes [17]. Besides, the *B* state has been replaced by a further *ab initio* surface from our group [18], performing also theoretical simulations of the spectra which confirm the main conclusions of the precedent work [16]. In this context, $HeBr_2$ is the first example in which a rotational partially resolved secondary band in the spectrum has been unambiguously assigned to the linear isomer.

3. $Br_2(X) \cdots He_N$ CLUSTERS

3.1. Additivity of the PES

Electronic studies of larger species are more complex and the difficulty in the evaluation of the PES increases with their size. Four-body systems as $Br_2(X) \cdots He_2$ are now amenable for performing *ab initio* calculations with a satisfactory degree of accuracy, which then permit the testing of various models of additivity in order to describe the PES of larger $Br_2(X) \cdots He_N$ clusters. Recently, *ab initio* calculations [19] at the fourth-order Møller–Plesset (MP4) and coupled-cluster [CCSD(T)] levels of theory have been performed [20] for the above-mentioned tetra-atomic system. The surface is characterized by three minima and the minimum energy pathways through them. The global minimum corresponds to a linear He–Br_2–He configuration, while the two other ones to "police-nightstick" (one He atom in the linear configuration, and the other in the T-shaped one with respect to the bromine) and tetrahedral (with the two He atoms along a plane perpendicular to the bromine bond) structures.

Analytical representations based on a sum of pairwise atom–atom interactions and a sum of three-body $HeBr_2$ CCSD(T) potentials [12] and He–He interaction [21] were checked in comparison with the tetratomic *ab initio* results. The sum of the three-body interactions form is found to be able to accurately represent the MP4(SDTQ)/CCSD(T) data (see Figure 2). In order to extract information on nonadditive interactions in He_2Br_2 we examine its equilibrium structures based on the *ab initio* calculations [20]. By partitioning the interaction energy into components, we found a similar nature of binding in triatomic and tetratomic complexes of such type, and thus information on intermolecular interactions available for triatomic species might serve to study larger systems. For first time an analytical expression in accord with high level *ab initio* studies is proposed for describing the intermolecular interactions for such two atoms rare gas–dihalogen complexes.

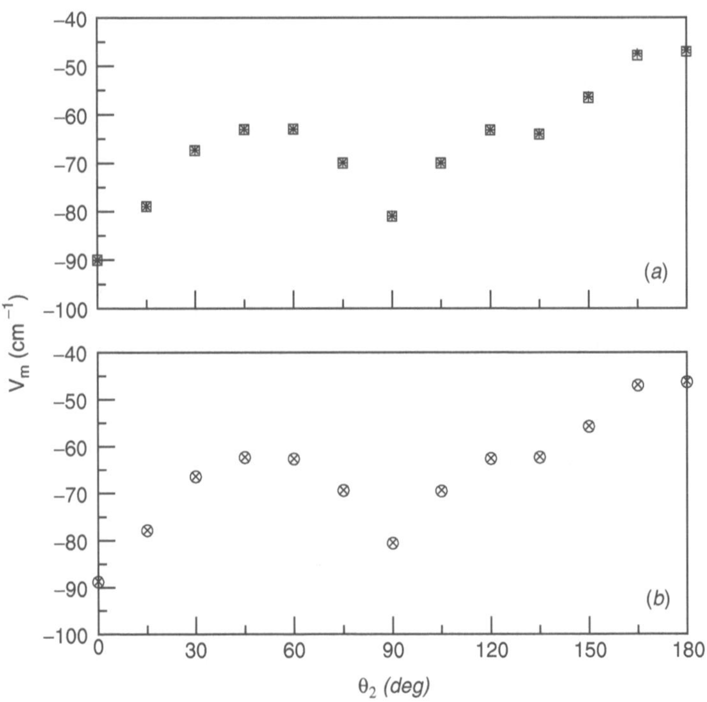

Figure 2. (a) Comparison of MP4(SDTQ) interaction energies (*squares*) and potential values for the He_2–Br_2 complex, obtained using the sum of the three-body MP4 $HeBr_2$ potentials (*stars*), at selected points along to the minimum energy path. (b) Same as Figure 1(a) for the CCSD(T) interaction energies (*circles*) and potential values of the complex, obtained using the sum of the three-body CCSD(T) $HeBr_2$ potentials (*crosses*).

Variational bound state calculation is carried out for the above surface and vdW energy levels and eigenfunctions for $J = 0$ are evaluated for He_2Br_2. Radial and angular distributions are calculated for the three lower vdW states and three different structural models, which correspond to linear, police-nightstick' and tetrahedral isomers are determined. The binding energies and the average structures for these species are computed to be D_0 = 32.240 cm^{-1} with $R^0_{1,2}$ = 4.867 Å, D_0 = 31.437 cm^{-1} with $R^0_{1,2}$ = 4.491 Å, and D_0 = 30.930 cm^{-1} with $R^0_{1,2}$ = 4.171 Å, respectively. The above values are in excellent agreement with recent LiF experimental data available [22]. This finding in combination with the very good agreement with the MP4/CCSD(T) tetratomic calculations, contribute to evaluate the present surface and justify our predictions.

We conclude, therefore, that the potential surface based on the sum of the He–Br_2(X) *ab initio* CCSD(T) potentials plus the He–He interaction, provides reliable results quantitatively comparable with the experimental observations.

4. HARTREE APPROACH: ENERGY AND DENSITY DISTRIBUTIONS OF $Br_2(X)\cdots He_N$ CLUSTERS

4.1. Overview of the methodological approach

We first define the nuclear Hamiltonian for the system as consisting of the Br_2 molecule solvated by N He atoms. Using satellite coordinates ($\mathbf{r}$, $\mathbf{R_k}$), where $\mathbf{r}$ is the vector joining the two bromine atoms and $\mathbf{R_k}$ are vectors from the diatomic center of mass to the different He atoms, this Hamiltonian can be written as

$$H^{(N)} = -\frac{\hbar^2}{2m}\frac{\partial^2}{\partial r^2} + U(r) + \frac{\mathbf{j}^2}{2mr^2} + \sum_{k=1}^{N} h_k(\mathbf{R}_k, r)$$

$$+\sum_{k<l} V_{kl}(|\mathbf{R}_k - \mathbf{R}_l|) - \frac{\hbar^2}{m_{Br_2}}\sum_{k<l} \nabla_k \cdot \nabla_l, \tag{1}$$

where the first three terms correspond to the Hamiltonian of the free diatomic molecule with m, $\mathbf{j}$, and U being the diatomic reduced mass, the angular momentum associated with $\mathbf{r}$, and the Br_2 intramolecular potential, respectively. The fourth term consists of N triatomic He–Br_2 Hamiltonians which are given by

$$h_k(\mathbf{R}_k, r) = -\frac{\hbar^2}{2\mu}\frac{\partial^2}{\partial R_k^2} + \frac{\mathbf{l}_k^2}{2\mu_\epsilon R_k^2} + W(r, R_k, \theta_k), \tag{2}$$

where μ is the reduced He–Br_2 mass, $\mathbf{l_k}$ is the angular momentum associated with $\mathbf{R_k}$, and W is the atom–diatom intermolecular interaction potential, which depends on the (r, R_k) distances and the angle between the $\mathbf{r}$ and $\mathbf{R_k}$ vectors. In the fifth term of Eq. (1), V_{kl}, represents the pair interaction potential between the kth and the lth He atoms. Finally, in the sixth term, $-\frac{\hbar^2}{m_{Br_2}}\nabla_k \cdot \nabla_l$, is the kinetic energy coupling between the kth and the lth He atoms.

Choosing a body-fixed (BF) coordinate system with the Z axis parallel to $\mathbf{r}$, and a fixed value of the intramolecular distance r, the ground-state of the bound cluster of N He atoms is obtained by solving the Schrödinger equation

$$\left[\sum_{k=1}^{N} h_k + \sum_{k<l} V_{kl} - E_{\Lambda}^{(N)}(r)\right] \Phi_{\Lambda}^{(N)}(\{\mathbf{R}_k\}; r) = 0 \tag{3}$$

in which the r-dependent eigenvalues are labeled by Λ, the projection of the orbital angular momentum $\mathbf{L} = \sum_{k=1}^{N\epsilon} \mathbf{l}_k$ on the molecular axis. Note that this representation is equivalent to the Born–Oppenheimer approximation in which the Br_2 molecule and the He atoms play the role of the nuclei and the electrons, respectively. In order to solve Eq. (3) we have developed a Hartree approach. Therefore, the wavefunction of the N bound He atoms is approximated as a symmetrized Hartree product of one-particle wavefunctions. If N_i bosons occupy the same one-particle orbital of index i, the total wavefunction of the system of $N = \sum_i^M N_i$ $(M \leq N)$ bosons can be expressed as

$$\Phi_{(N_1,\ldots,N_M)}^{(N)} = \frac{1}{\sqrt{\mathcal{N}}} \hat{S} \left(\prod_{i=1}^{N_1} \psi_1(\mathbf{R}_i; r) \prod_{j=N_1+1}^{N_1+N_2} \psi_2(\mathbf{R}_j; r) \cdots \prod_{k}^{N} \psi_M(\mathbf{R}_k; r) \right), \tag{4}$$

where $k = (N_1 + \cdots + N_{M-1}) + 1$, $\hat{S}$ is the symmetrization operator, $1/\sqrt{\mathcal{N}}$ is the normalization factor, and $\mathcal{N}$ is the number of different Hartree products obtained by interchanging the bosons occupying different orbitals,

$$\mathcal{N} = \binom{N}{N_1} \binom{N - N_1}{N_2} \cdots \binom{N - (N_1 + \cdots + N_{M-1})}{N_M}. \tag{5}$$

The energy of the N–boson system can be written as

$$E_{\Lambda}^{(N)} = \sum_{i=1}^{M} N_i \epsilon_i + \sum_{i,j=1}^{M} \frac{N_i(N_j - \delta_{ij})}{2(1 + \delta_{ij})} \left(J_{ij} + K_{ij}\right), \tag{6}$$

where

$$\epsilon_i = \int d\mathbf{R}\ \psi_i^*(\mathbf{R}; r) h(\mathbf{R}, r) \psi_i(\mathbf{R}; r) \tag{7}$$

is the average kinetic and potential energy (that of each He atom with the dopant) of a boson described by the orbital $\psi_i(\mathbf{R}; r)$. The term

$$J_{ij} = \int\int d\mathbf{R}_1 d\mathbf{R}_2\ |\psi_i(\mathbf{R}_1; r)|^2\, V_{12}'(|\mathbf{R}_1 - \mathbf{R}_2|)\, \left|\psi_j(\mathbf{R}_2; r)\right|^2 \tag{8}$$

where

$$V_{12}' = V_{12} - \frac{\hbar^2}{m_{Br_2}} \nabla_1 \cdot \nabla_2, \tag{9}$$

represents the interaction between the two bosonic clouds $|\psi_i(\mathbf{R}_1; r)|^2$ and $|\psi_j(\mathbf{R}_2; r)|^2$. It is equivalent to the *Coulomb integral* in electronic structure theory. Note that Eq. (9) explicitly incorporates the kinetic coupling. The term

$$\begin{aligned} K_{ij} &= \int\int d\mathbf{R}_1 d\mathbf{R}_2\ \psi_i^*(\mathbf{R}_1; r)\psi_j(\mathbf{R}_1; r) \\ &\quad V'_{12}(|\mathbf{R}_1 - \mathbf{R}_2|)\psi_j^*(\mathbf{R}_2; r)\psi_i(\mathbf{R}_2; r) \end{aligned} \tag{10}$$

is an analog of the exchange integral.

A finite basis set composed of products of radial and angular functions was used

$$\chi^{(n\ell m)}(\mathbf{R}; r) = g_n(R; r)Y_{\ell m}(\theta, \phi) \tag{11}$$

where $Y_{\ell m}(\theta, \phi)$ are spherical harmonics. The radial $g_n(R; r)$ functions were obtained by solving the Schrödinger equation corresponding to the triatomic He–Br_2 subsystem at different fixed orientations. The orbitals were optimized through a direct minimization tecnique [23] to ensure convergence to the global minimum.

4.2. Energy and density distributions for N = 2–60 clusters

As described in Ref. [25], the Hartree approach has been applied to get energies and density probability distributions of $Br_2(X)\cdots{}^4He_N$ clusters. The lowest energies were obtained for the value $\Lambda = 0$ of the projection of the orbital angular momentum onto the molecular axis, and the symmetric N–boson wavefunction, i.e. the "Σ_g" state in which all the He atoms occupy the same orbital (in contrast to the case of fermions). It stressed that both energetics and helium distributions on small clusters ($N \leq 18$) showed very good agreement with those obtained in "exact" DMC computations [24].

In Figure 3, total energy $E(N)$ (open squares) and energy per He atom, $E(N)/N$ (full squares), as a function of the cluster size N. The negative of the cohesive energy of the bulk ^4He is also shown. Note that the total energy and energy per He atom change continuously and monotonically with the cluster size giving no indication for shell-closure effects. The energy per atom, $E(N)/N$, increases rapidly as the cluster size increase to $N \sim 15$ and then it slowly tends to the bulk value of -4.94 cm^{-1} (see e.g Ref. [26]), which would be, obviously, attained for much larger cluster sizes than those analyzed here.

In Figure 4, helium angular density distributions around the Br_2 molecule for different cluster sizes (N = 2–40) are displayed. Note that for the smaller clusters the angular density distributions are highly anisotropic peaking at $\theta = \pi/2$. This is a consequence of the strong anisotropy in the He–Br_2 potential which favors the T-shape arrangement. The He atoms populate primarily the well associated with this arrangement up to about $N = 6$. For larger N, the increasing He–He repulsion causes the density distribution to flow from T-configuration well into the other potential regions. Indications for formation of two side peaks at $\theta = \pi/4$ and $3\pi/4$ are evident for N=12, and these peaks are clearly present in the graphs corresponding to N=16 and 18. For N=24, about 4% of the He density is found at peaks adjacent to $\theta = 0$

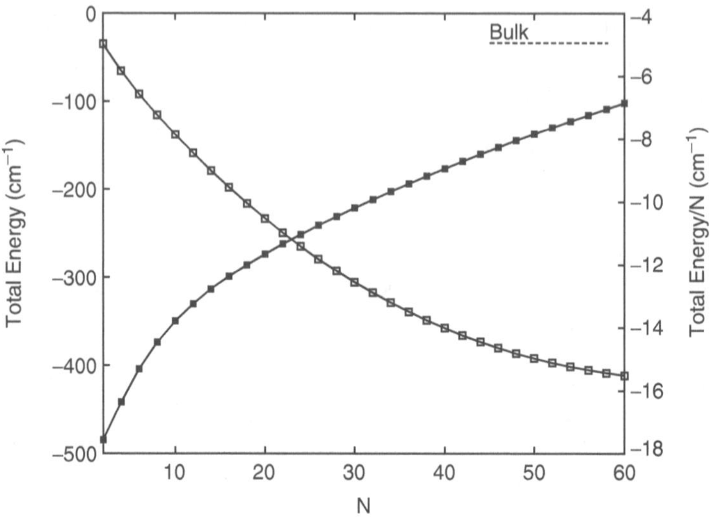

Figure 3. Total energy (*open squares*) and energy per helium atom (*solid squares*), as a function of the cluster size. The negative of the cohesive energy of the bulk ^4He is also shown.

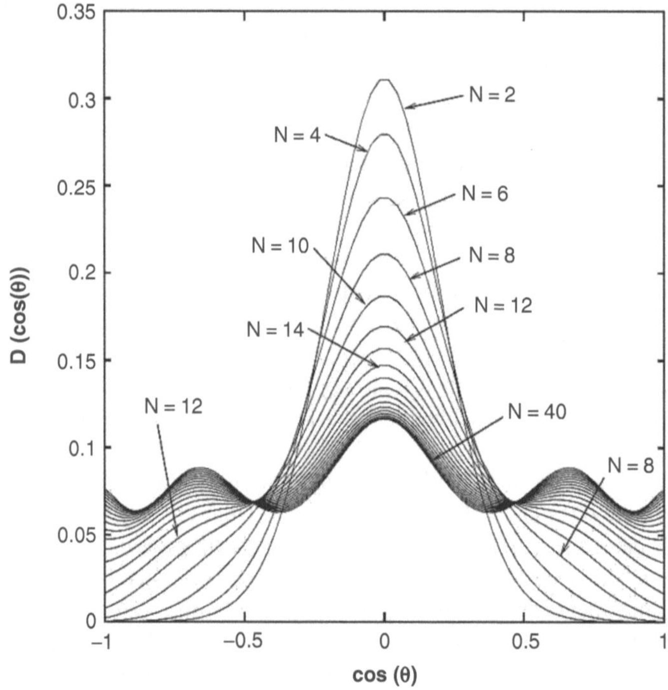

Figure 4. Angular density distributions for the He atoms surrounding the $Br_2(X)$ molecule for different cluster sizes. The distributions are normalized to 1.

and π. For $N \geq 30$, the He distributions are almost independent of the cluster size and markedly more isotropic than those for $N \leq 6$. This can be explained by taking into account that the strongly anisotropic potential is felt mainly by the He atoms that are close to the dopant molecule whereas the spatial clustering of the He atoms more distant from the impurity is driven primarily by the He–He interaction.

Finally, we stress that the quantum chemical method presented here has the advantage over DFT-based techniques that it also furnishes wavefunctions that can be used to perform computations of spectra, and therefore have a better contact with the experiment. Another advantage of this approach is that, unlike the diffusion Monte–Carlo method, it can coherently be applied to studies of fermion and mixed boson / fermion doped clusters. An example can be found in our recent work on the Raman spectra of $(He)_N$–Br_2(X) clusters [27, 28].

Acknowledgements

This work has been supported by the DGICYT Spanish Grant No. FIS2004-02461.

References

1. D.H. Levy, *Adv. Chem. Phys.* 47:323, 1981.
2. A. Rohrbacher, N. Halberstadt, and K.C. Janda, *Annu. Rev. Phys. Chem.* 51:405, 2000.
3. J.A. Beswick and J. Jortner, *Adv. Chem. Phys.* 47:363, 1981.
4. M.I. Hernández, T. González-Lezana, A.A. Buchachenko, R. Prosmiti, M.P. de Lara-Castells, G. Delgado-Barrio, and P. Villarreal, *Recent Res. Devel. Chem. Physics* 4:1, 2003.
5. S. Grebenev, J.P. Toennies, and A.F. Vilesov, *Science* 279:2083, 1998.
6. K. Nauta and R.E. Miller, *Science* 287:293, 2000.
7. J.P. Toennies and A.F. Vilesov, *Angew. Chem. Int. Ed.* 43:2622, 2004.
8. C. Callegari, K.K. Lehmann, R. Schmied, and J. Scoles, *J. Chem. Phys.* 115:10090, 2001.
9. J. Tang, Y. Xu, A.R.W. McKellar, and W. Jäger, *Science* 297:2030, 2002.
10. J. Tang and A.R.W. McKellar, *J. Chem. Phys.* 119:5467, 2003.
11. T. González-Lezana, M.I. Hernández, G. Delgado-Barrio, A.A. Buchachenko, and P. Villarreal, *J. Chem. Phys.* 105:7454, 1996.
12. R. Prosmiti, C. Cunha, P. Villarreal, and G. Delgado-Barrio, *J. Chem. Phys.* **116**, 9249, 2002; Á. Valdés, R. Prosmiti, P. Villarreal, and G. Delgado-Barrio, *Mol. Phys.* 102:2277, 2004.
13. A.A. Buchachenko, T. González-Lezana, M.I. Hernández, M.P. de Lara-Castells, G. Delgado-Barrio, and P. Villarreal, *Chem. Phys. Lett.* 318:578, 2000.
14. D.J. Jahn, S.G. Clement, and K.C. Janda, *J. Chem. Phys.* 101:283, 1994.
15. D.J. Jahn, W.S. Barney, J. Cabalo, S.G. Clement, A. Rohrbacher, T.J. Slotterback, J. Williams, K.C. Janda, and N. Halberstadt, *J. Chem. Phys.* 104:3501, 1996.
16. A.A. Buchachenko, R. Prosmiti, C. Cunha, G. Delgado-Barrio, and P. Villarreal, *J. Chem. Phys.* 117:6117, 2002.
17. D.S. Boucher, D.B. Strasfeld, R.A. Loomis, J.M. Herbert, S.A. Ray, and A.B. McCoy, *J. Chem. Phys.*, 123:104312, 2005.
18. M.P. de Lara-Castells, A.A. Buchachenko, G. Delgado-Barrio, and P. Villarreal, *J. Chem. Phys.* 120:2182, 2004.
19. Gaussian 98, Revision A.7, M.J. Frisch, G.W. Trucks, H.B. Schlegel, G.E. Scuseria, M.A. Robb, J.R. Cheeseman, V.G. Zakrzewski, J.A. Montgomery, Jr., R.E. Stratmann, J.C. Burant, S. Dapprich, J.M. Millam, A.D. Daniels, K.N. Kudin, M.C. Strain, O. Farkas, J. Tomasi, V. Barone, M. Cossi, R. Cammi, B. Mennucci, C. Pomelli, C. Adamo, S. Clifford, J. Ochterski, G.A. Petersson,

P.Y. Ayala, Q. Cui, K. Morokuma, D.K. Malick, A.D. Rabuck, K. Raghavachari, J.B. Foresman, J. Cioslowski, J.V. Ortiz, A.G. Baboul, B.B. Stefanov, G. Liu, A. Liashenko, P. Piskorz, I. Komaromi, R. Gomperts, R.L. Martin, D.J. Fox, T. Keith, M.A. Al-Laham, C.Y. Peng, A. Nanayakkara, C. Gonzalez, M. Challacombe, P.M.W. Gill, B. Johnson, W. Chen, M.W. Wong, J.L. Andres, C. Gonzalez, M. Head-Gordon, E.S. Replogle, and J.A. Pople, Gaussian, Pittsburgh PA, 1998.

20. Á. Valdés, R. Prosmiti, P. Villarreal, and G. Delgado-Barrio, *J. Chem. Phys.* 122:044305, 2005.
21. R.A. Aziz and M.J. Slaman, *J. Chem. Phys.* 94:8047, 1991.
22. R.A. Loomis (private communication).
23. J. Fernández-Rico, J.M. García-Vela, M. Paniagua, and J.I. Fernández-Alonso, *J. Chem. Phys.* 79:4407, 1983.
24. C. Di Paola, F.A. Gianturco, D. López-Durán, M.P. de Lara-Castells, G. Delgado-Barrio, P. Villarreal, and J. Jellinek, *Chem. Phys. Chem.* 6:1, 2005.
25. M.P. de Lara-Castells, D. López-Durán, G. Delgado-Barrio, P. Villarreal, C. Di Paola, F.A. Gianturco, and J. Jellinek, *Phys. Rev. A* 71:033203, 2005.
26. M.H. Kalos, M.A. Lee, P.A. Whitlock, and G.V. Chester, *Phys. Rev.* B 24:115, 1981.
27. D. López-Durán, M.P. de Lara-Castells, G. Delgado-Barrio, P. Villarreal, C. Di Paola, F.A. Gianturco, and J. Jellinek, *J. Chem. Phys.* 121:2975, 2004.
28. D. López-Durán, M.P. de Lara-Castells, G. Delgado-Barrio, P. Villarreal, C. Di Paola, F.A. Gianturco, and J. Jellinek, *Phys. Rev. Lett.* 93:053401, 2004.

THEORETICAL TREATMENT OF CHARGE TRANSFER PROCESSES: FROM ION/ATOM TO ION/BIOMOLECULE INTERACTIONS

M.C. BACCHUS-MONTABONEL[1], M. ŁABUDA[1], Y.S. TERGIMAN[1], AND J.E. SIENKIEWICZ[2]

[1]*Laboratoire de Spectrométrie Ionique et Moléculaire, CNRS et Université Lyon I, 43 bd. du 11 Novembre 1918, 69622 Villeurbanne Cedex, France*
[2]*Department of Theoretical Physics and Mathematical Methods, Gdańsk University of Technology, ul. Narutowicza 11/12, Pl-80952 Gdańsk, Poland*

Abstract A unified approach is proposed for the treatment of charge transfer processes occurring in collisions of ions on atomic or molecular targets. The theoretical treatment, including *ab initio* molecular calculations of the potential energy curves and couplings followed by a semi-classical collision dynamics, is extended in particular to polyatomic molecules of biological interest. The method is presented on the example of the important astrophysical reaction $S^{3+}(3s^23p) + H$, and extended to the charge transfer of the RNA base Uracil onto C^{q+} ions ($q = 2$–5). In that sense a simple model is proposed, correlated to experimental investigations on ionization and fragmentation processes in C^{q+}-Uracil collisions.

1. INTRODUCTION

Electron transfer is an important process in physics, chemistry, and biochemistry. Charge transfer of multiply charged ions in collision with atomic or molecular targets has been shown to be a determinant process in controlled thermonuclear fusion research, as well as in astrophysical plasmas at low collision energy [1–4]. In such processes, an electron is in general captured in an excited state of the ion followed by line emission, and observation of line intensities may provide information on electron temperature, density, and spatial distribution in the emitting region of the plasma. Such approaches, previously developed for ion/atom or ion/diatomic molecule collision systems, could be applied for systems of increasing complexity and theoretical models can be extended for charge transfer reactions between multicharged ions and

S. Lahmar et al. (eds.), Topics in the Theory of Chemical and Physical Systems, 203–214.

systems of biological interest. In that sense, a uniform treatment may be developed with an *ab initio* determination of potential energies and couplings of the states involved in the process, followed by a semi-classical or quantum collision dynamics, with regard to the collision energy range of interest. We propose to illustrate this point on two examples, taken in completely different physicochemical domains, and to show in each case which data may be extracted from the theoretical approach to be compared to the experimental measurements, leading to complementary information.

As a first example, we present a detailed study of the $S^{3+}(3s^23p) + H$ collision system [4, 5], which plays an important role in the reactions occurring in the interstellar medium [6,7]. Experimental translational energy spectroscopy experiments in the 2.4–9.0 keV energy range [8] show evidence of a dominant $S^{2+}(3s^23p3d)^3F^\circ$ capture channel over the entire range of impact energies, largely underestimated by the multi-channel Landau–Zener approach. In that case, a complete *ab initio* treatment, including all the triplet and singlet states correlated to the $^{1,3}\Sigma$ and $^{1,3}\Pi$ entry channels by means of radial or rotational coupling matrix elements, has been undertaken, leading to a quantitative comparison with measured charge transfer cross-sections. The metastable $S^{3+}(3s3p^2)^4P$ ion, evidenced in the beam, has not, however, been taken into account.

A similar method may be developed to study the action of ions on biomolecules. In that sense, we have considered the collision of C^{q+} ions on the Uracil molecule, one of the RNA bases [9]. Effectively, it is well known that interaction of ionizing radiation with biological tissue can induce severe damage to DNA and RNA [10], with single and double strand breaks which appear to be the main underlying mechanisms for cancer disease and controlled cell killing used in radiotherapy. Important damage has been shown [11] to be due to secondary particles, low-energy electrons, radicals, or singly and multiply charged ions, generated along the track, after interaction of the ionizing radiation with a biological medium. Consequently, experiments have been performed recently in order to investigate the action of these secondary particles on biological relevant molecules and a number of experimental investigations have been devoted to the action of ions on biomolecules, generally at relatively low collision energies in the range [2–150] keV. In particular, coincidence measurements between outgoing projectiles have been carried out in collisions of C^{q+} ions on Uracil [12] or Thymine [13], with analysis of the fragment mass distribution by time-of-flight spectrometry. In such collisions various processes may be considered: excitation and fragmentation of the biomolecule, ionization of the gaseous target, and also possible charge transfer from the multi-charged ion onto the biomolecule.

From the experimental point of view, the peak integration of the mass spectra as a function of collision velocity leads to an evaluation of the relative cross-sections for the different ionization and fragmentation channels, but cannot give any information on possible charge transfer, which should however be a complementary process. In a recent study on collisions of C^{q+} ions with Uracil at keV energies [8], it has been observed, in particular, a strong influence of the electronic structure and charge q (q = 1–6) of the C^{q+} ion on the ionization and fragmentation processes. At low velocities, almost complete fragmentation was observed for the C^{2+} projectile ion, whereas for all the other projectile charge states the induced fragment yield was markedly lower. A complete theoretical treatment, taking account of all degrees of freedom of the problem, cannot, of course, be handled, and a simple model has thus been proposed in order to investigate the charge transfer processes.

2. MOLECULAR CALCULATIONS

The potential energy curve calculations have been performed using the MOLPRO suite of *ab initio* programs [14], at the state average CASSCF–MRCI level. Spin–orbit coupling being negligible in the energy range of interest, we have assumed the electron spin to be conserved in the collision processes.

In the case of the $S^{3+}(3s^2 3p) + H$ system, a complete treatment has been undertaken, with consideration of all molecular states involved in the process, in both the singlet and triplet manifolds. The active space includes the 1s orbital of hydrogen and $n = 3, 4(sp)$ orbitals of sulphur. The correlation-consistent aug-cc-pVQZ basis sets of Dunning [15] with a ECP10sdf relativistic pseudo-potential to take account of the 10 sulphur core electrons [16] have been used in the calculation and lead to a good agreement with experimental atomic energy levels [17] (Table 1).

The collision system appears relatively complex, with consideration of the $\{S^{2+}(3s^2 3p3d)^3F^\circ + H^+\}$ level accessible from the $^{1,3}\Pi$ and $^{1,3}\Sigma^+\{S^{3+}(3s^2 3p)^2P^\circ + H(1s)^2S\}$ entry channel. A great number of states have been taken into account in both the singlet and triplet manifolds, including one-electron capture processes as well as re-actions of capture and excitation of the 3s core electron:

$S^{3+}(3s^2 3p)^2P^\circ + H(1s)^2S \quad {}^3\Pi, {}^3\Sigma^+ \qquad {}^1\Pi, {}^1\Sigma^+$ entry channel
$S^{2+}(3s3p^3)^1P^\circ + H^{+1}S \qquad {}^1\Pi, {}^1\Sigma^+$
$S^{2+}(3s3p^3)^3S^\circ + H^{+1}S \ {}^3\Sigma^-$
$S^{2+}(3s^2 3p3d)^3F^\circ + H^{+1}S \quad {}^3\Phi, {}^3\Pi, {}^3\Delta, {}^3\Sigma^+$
$S^{2+}(3s3p^3)^1D^\circ + H^{+1}S \qquad {}^1\Pi, {}^1\Delta, {}^1\Sigma^-$
$S^{2+}(3s3p^3)^3P^\circ + H^{+1}S {}^3\Pi, {}^3\Sigma^+$
$S^{2+}(3s3p^3)^3D^\circ + H^{+1}S \quad {}^3\Pi, {}^3\Delta, {}^3\Sigma^-$
$S^{2+}(3s^2 3p^2)^1S + H^{+1}S \qquad {}^1\Sigma^+$
$S^{2+}(3s^2 3p^2)^1D + H^{+1}S \qquad {}^1\Pi, {}^1\Delta, {}^1\Sigma^+$
$S^{2+}(3s^2 3p^2)^3P + H^{+1}S \quad {}^3\Pi, {}^3\Sigma^-$

The Σ^- levels are not correlated to the entry channel and have not been calculated.

Table 1. Comparison of atomic energy levels with experiment [17] (in eV).

	MRCI calculation	Experiment
$S^{3+}(3s^2 3p)^2P^\circ$	35.11	34.98
$S^{2+}(3s3p^3)^3P^\circ$	12.31	12.17
$S^{2+}(3s3p^3)^3D^\circ$	10.34	10.35
$S^{2+}(3s^2 3p^2)^3P$	0.0	0.0

According to statistical weight, the triplet manifold is preponderant in the charge transfer mechanism, and the main features of the potential energy curves, correlated to the $^3\Pi$ and $^3\Sigma^+\{S^{3+}(3s^23p)^2P^\circ + H\}$ entry channel, are displayed in Figure 1a,b. They show several important avoided crossings. First of all, one around $R = 9.7$ a.u., with an energy defect of 5.6 eV, between the entry channel and the $\{S^{2+}(3s^23p3d)^3F^\circ + H^{+1}S\}$ exit channel, in both $^3\Sigma^+$ and $^3\Pi$ symmetries. The energy defect appears however slightly lower than evidenced for the experimental peak A [8]. Two short range avoided crossings around $R = 6$ a.u., associated to an energy defect of 9.19 eV, are also observed with $^3\Pi$, $^3\Sigma^+\{S^{2+}(3s3p^3)^3P^\circ + H^{+1}S\}$ levels, and may contribute to the experimental peak B. Another interaction appears with the $^3\Pi\{S^{2+}(3s3p^3)^3D^\circ + H^{+1}S\}$ level around $R = 5$ a.u. which may contribute to peak C, with an energy defect of 11.16 eV.

Radial and rotational coupling matrix elements have been calculated in both the triplet and singlet manifolds. The radial coupling between all pairs of states of the same symmetry have been calculated by means of the finite difference technique:

$$g_{KL}(R) = \langle \psi_K \, |\partial/\partial R\, | \psi_L \rangle = \lim_{\Delta\to 0} \frac{1}{\Delta} \langle \psi_K(R) | \; \psi_L(R+\Delta) \rangle \,,$$

with the parameter $\Delta = 0.0012$ a.u., as previously tested, and using the three-point numerical differentiation method, for reasons of numerical accuracy. The sulphur nucleus has been chosen as the origin of the electron coordinates. They present the

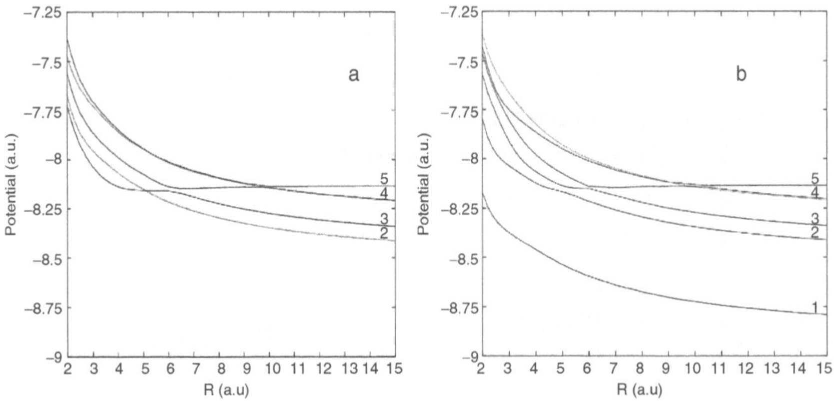

Figure 1. (a) Adiabatic potential energy curves of the $^3\Sigma^+$ and $^3\Delta$ states of the $S^{3+}(3s^23p) + H$ collisional system. — $^3\Sigma^+$ states; - - - $^3\Delta$ states. (2) $^3\Delta$ state corresponding to $\{S^{2+}(3s3p^3)^3D^\circ + H^{+1}S\}$. (3)$^3\Sigma^+$ state corresponding to $\{S^{2+}(3s3p^3)^3P^\circ + H^{+1}S\}$. (4) $^3\Sigma^+$ and $^3\Delta$ states corresponding to $\{S^{2+}(3s^23p3d)^3F^\circ + H^{+1}S\}$. (5) $^3\Sigma^+$ state corresponding to the $\{S^{3+}(3s^23p)^2P^\circ + H(1s)^2S\}$ entry channel. (b) Adiabatic potential energy curves of the $^3\Pi$ and $^3\Phi$ states of the $S^{3+}(3s^23p) + H$ collisional system. — $^3\Pi$ states; - - -$^3\Phi$ state. (1) $^3\Pi$ state corresponding to $\{S^{2+}(3s^23p^2)^3P + H^{+1}S\}$. (2) $^3\Pi$ state corresponding to $\{S^{2+}(3s3p^3)^3D^\circ + H^{+1}S\}$. (3) $^3\Pi$ state corresponding to $\{S^{2+}(3s3p^3)^3P^\circ + H^{+1}S\}$. (4) $^3\Pi$ and $^3\Phi$ states corresponding to $\{S^{2+}(3s^23p3d)^3F^\circ + H^{+1}S\}$. (5) $^3\Pi$ state corresponding to the $\{S^{3+}(3s^23p)^2P^\circ + H(1s)^2S\}$ entry channel.

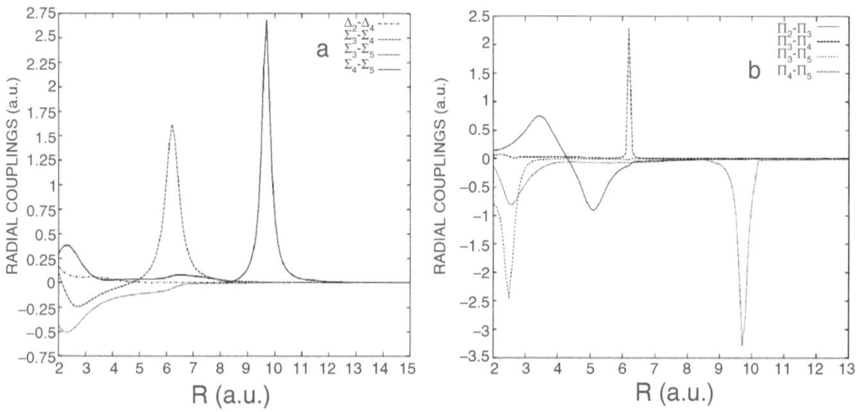

Figure 2. (a) Radial coupling matrix elements between $^3\Sigma^+$ and $^3\Delta$ states of the $S^{3+}(3s^23p) + H$ collision system (see labels in Figure 1a). (b) Radial coupling between $^3\Pi$ states of the $S^{3+}(3s^23p) + H$ collision system (see labels in Figure 1b).

same features as exhibited by the potential energy curves, with strong peaks in correspondence with the sharp avoided crossings (Figure 2a,b). In particular, a very sharp peak, about 3 a.u. high, is observed for R = 9.7 a.u., in correspondence with the strong avoided crossing between the entry channel and the $^3\Sigma^+$, $^3\Pi\{S^{2+}(3s^23p3d)^3F^\circ + H^{+1}S\}$ capture level. Radial coupling between $^3\Delta$ states remains small for all internuclear distances. The rotational coupling matrix elements between states of angular momentum $\Delta\Lambda = \pm 1$ have been calculated directly from the quadrupole moment tensor, and are also connected to the potential energy features.

Such a complete treatment cannot, of course, be considered for the collision of the C^{q+} ions on the Uracil biomolecule. A simple model of this polyatomic complex system may, however, be proposed by means of the one-dimensional reaction coordinate approximation, widely used in a number of cases [18]. Effectively, the collision may be represented as the evolution of the polyatomic C^{q+}.... Uracil complex treated, in a first approximation, as a pseudo diatomic system, the reaction coordinate corresponding to the distance between the centre-of-mass of the Uracil molecule and the colliding carbon ion. The rearrangement of the Uracil molecule when it approaches the C^{q+} ion can be taken into account by relaxating the geometry of the C^{q+}... Uracil complex along the reaction path. Such an approach does not take into account all the degrees of freedom of the complex and the internal motions of the molecule, but it seems reasonable in a very fast collision process where nuclear vibration and rotation periods are assumed to be much longer than the collision time.

The potentials of the different states of the C^{q+}-Uracil collision system may then be calculated along the reaction coordinate R for different approach angles θ, from perpendicular ($\theta = 90°$) to planar geometry ($\theta = 0°$) (Figure 3a), in order to take into account the anisotropy of the process. This requires, of course, extensive

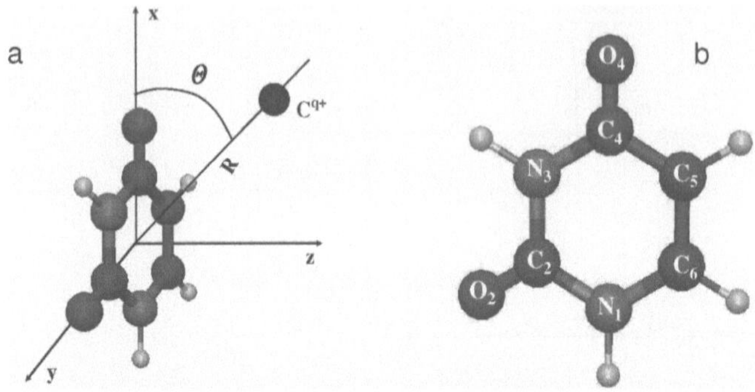

Figure 3. (a) Internal coordinates of the C^{q+}-Uracil system. (b) Geometry of the Uracil molecule.

calculations. In a first step, we have focused our attention on the planar attack ($\theta = 0°$) of the carbon ion on the Uracil molecule where calculations can be performed using the C_s symmetry group, in order to compare the capture mechanism for the different charges of the colliding carbon ion. A large number of complexes between the Uracil molecule and a series of single or double charged ions have been investigated already [19–24] with calculation of the optimized structures and attachment sites. These studies show, for the different metal ions, a minimum energy for the metal-Uracil complex, corresponding to an attachment site on oxygen O4 (Figure 3b). We have thus considered a collision attack of the carbon ion C^{q+} along the C4–O4 direction, corresponding to $\theta = 0°$, for the charges $q = (2–5)$.

Calculations were carried out for the A' states involved in the process; the rotational interaction has not been taken into account in that case. Spin symmetry being conserved, singlet states have been considered in the cases of $C^{2+}(1s^2 2s^2)$ and $C^{4+}(1s^2)$ projectile cations, and doublet states in the case of $C^{3+}(1s^2 2s)$ and $C^{5+}(1s)$ ions. The calculations have been performed in the C_s symmetry group with consideration of a planar approach along the C4–O4 chemical bond. The active space included the five highest valence orbitals as well as the n = 2,3 levels of the colliding carbon atom which might be involved in the process, the corresponding 1s orbital being treated as a frozen core. The 6-31 G** atomic basis set has been used throughout this study. The radial coupling matrix elements have been calculated, as in the previous case, by means of the finite difference technique, the centre-of-mass of the Uracil molecule being chosen, in this case, as the origin of the electron coordinates.

The geometries of the ground state as well as of the singly and doubly ionized Uracil molecules have been optimized by means of DFT calculations using the B3-LYP functional. The vertical and adiabatic first and second ionization potentials presented in Table 2 compare favourably to previous calculations [25, 26] and to experimental data [27,28].

The potential energy curves and radial coupling matrix elements have been calculated for a distance C^{q+}-O4 varying from 0.5 to 5 Å. The charge transfer levels are presented in Figure 4 for the series $q = (2–5)$. For the charges $q = (2–4)$, the

Table 2. Ionization potentials of the Uracil molecule (in eV).

	This calculation B3LYP/6-31 G**	[25] B3PW91	[26] PMP2/ 6-31++G(d,p)	Experiment
PI1 : $U \rightarrow U^+$ vertical	9.56		9.43	9.50 [27]
PI1 : $U \rightarrow U^+$ adiabatic	9.34	9.32 TZVP 9.33 6–311 ++G**	9.36	9.32 [27] 9.35 [28]
PI2 : $U^+ \rightarrow U^{2+}$ vertical	17.57			
PI2 : $U^+ \rightarrow U^{2+}$ adiabatic	17.33			

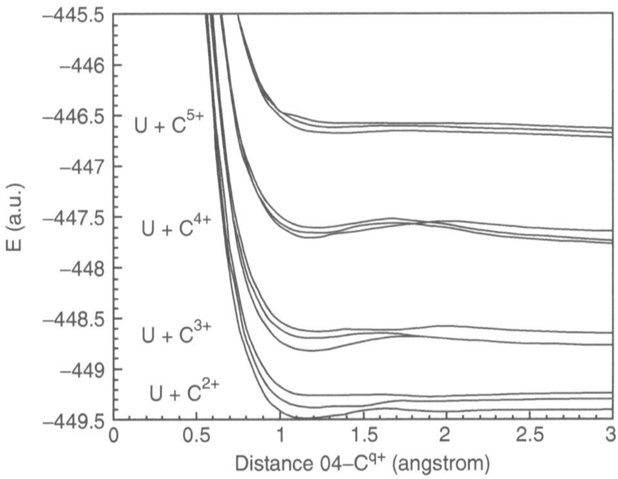

Figure 4. Adiabatic potential energy curves for the A′ charge transfer levels of the C^{q+}-Uracil collisional systems (by increasing energy for each system):

$U + C^{2+}$: $^1A'\pi_{C5C6} \rightarrow 2p_z$; $^1A'\pi_{C4O4} \rightarrow 2p_z$; $^1A'\pi_{C2O2} \rightarrow 2p_z$

$U + C^{3+}$: $^2A'\pi_{C5C6} \rightarrow 2p_z$; $^2A'\pi_{C4O4} \rightarrow 2p_z$

$U + C^{4+}$: $^1A'p^x{}_{O2} \rightarrow 2p_x$; $^1A'p^x{}_{O2} \rightarrow 2p_x, \pi_{C4O4} \rightarrow \pi_{C5C6}$; $^1A'p^x{}_{O2} \rightarrow 2p_x, \pi_{C2O2} \rightarrow \pi_{C5C6}$

$U + C^{5+}$: $^2A'p^x{}_{O4}$; $^2A'p^x{}_{O4}, \pi_{C4O4} \rightarrow \pi_{C5C6}$; $^2A'p^x{}_{O4}, \pi_{C2O2} \rightarrow \pi_{C5C6}$.

charge transfer potentials show clearly avoided crossings in the 1.5–2.0 Å range, corresponding to a strong interaction with the entry channel. Such an interaction appears at a very short range in the case of the C^{5+}-Uracil system. For the $C^{2+}(1s^2 2s^2)$-Uracil system, single charge transfer is observed, and the corresponding levels are correlated clearly to an excitation of the highest occupied molecular orbitals of the Uracil molecule to the unoccupied $2p_z$ orbital of the carbon ion. The excited states correspond respectively, by increasing energy, to excitations of the out-of-plane (a″ symmetry) π_{C5C6} and π_{C4O4}, π_{C2O2} orbitals towards the $2p_z$ orbital of the colliding carbon, leading to the $^1A'\{C^+(1s^22s^22p) + U^+\}$ molecular charge transfer levels.

Such features appear somewhat different for the higher-charged collision systems. In the case of the collision with the $C^{3+}(1s^22s)$ ion, an initial delocalization of the electrons leading to $C^{2+}(1s^22s^2)$ is clearly observed. The charge transfer levels are then attributed, similarly to the $C^{2+}(1s^22s^2)$ -Uracil system, to the out-of-plane $\pi_{C5C6} \rightarrow 2p_z$ and $\pi_{C4O4} \rightarrow 2p_z$ excitations. For the $\{C^{4+}(1s^2) + U\}$ and $\{C^{5+}(1s)+U\}$ systems, a first excitation of the in-the plane (a′ symmetry) $p^x{}_{O2}$ orbital towards the $2p_x$ orbital of the colliding carbon is observed, followed by out-of plane π–π excitations among the electron cloud of the Uracil ring. In all cases, Mulliken population analysis shows that the electronic molecular levels correspond to a charge around +1 on the colliding carbon at long range, attributed to $C^+(1s^22s^22p)$. The Uracil molecule appears as an electron reservoir which fills up the electron deficit of the colliding ion, and an initial delocalization of the electron of the Uracil ring towards the carbon projectile, to reach the $C^{2+}(1s^22s^2)$ configuration, is observed at long range, followed by a charge transfer characterized by the interaction around 2 Å.

3. COLLISION DYNAMICS

In the keV energy domain, where experiments are available on both examples, semi-classical approaches using the EIKONXS code, based on an efficient propagation method [29], may be used with a good accuracy.

For the $S^{3+}(3s^23p) + H$ collision system, calculations have been performed in the [2–8] keV energy range, in order to be compared to the experimental data. Both radial and rotational coupling matrix elements were taken into account as well as translational effects using the common translation factors [30]. The sulphur nucleus was taken as the origin of the electron coordinates. The cross-sections corresponding to the different exit channels were calculated with respect to the statistical weights 1/4 and 3/4, respectively, for the singlet and triplet manifolds, taking account of the 1/3 and 2/3 statistical weights for the Σ and Π entry channels. The results are presented in Table 3 and Figure 5. They appear to be in good agreement with experimental data.

Table 3. Partial cross-sections on the different exit channels (in 10^{-16} cm^2).

E_{lab} keV	S^{2+i} P^o	$S^{2+3}F^o$ peak A	$S^{2+1}D^o$ peak B	$S^{2+3}P^o$ peak B	$S^{2+3}D^o$ peak C	$S^{2+3}P$	σ_A/σ_B	σ_B/σ_C
2.002	0.06	17.63	0.02	9.49	6.40	0.0001	1.8	1.5
2.883	0.79	16.29	0.03	11.90	6.85	0.0004	1.3	1.7
3.924	0.63	14.57	0.05	12.19	8.08	0.0006	1.2	1.5
5.125	0.73	13.54	0.12	10.54	7.78	0.0016	1.3	1.3
6.487	0.73	13.08	0.23	9.52	7.87	0.0018	1.3	1.2
8.009	0.87	11.85	0.31	8.37	7.63	0.0036	1.3	1.1

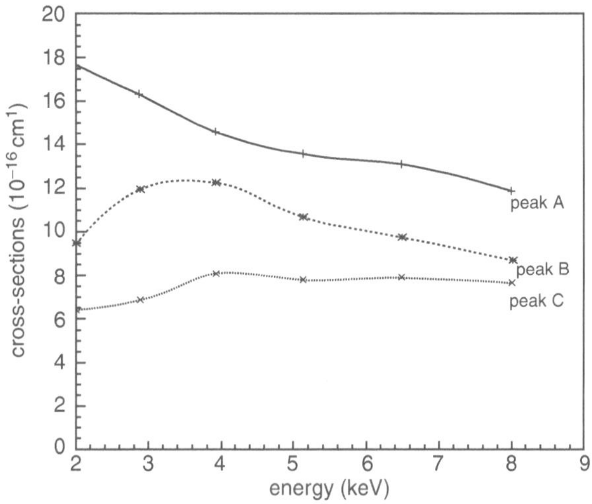

Figure 5. Calculated cross-sections on peak A, B, C (in 10^{-16} cm^2).

Effectively, the $S^{2+}(3s^2 3p3d)^3F^\circ$ capture channel attributed to peak A is clearly the dominant charge transfer channel. With regard to the energy defect, both $S^{2+}(3s3p^3)^1D^\circ$ and $S^{2+}(3s3p^3)^3P^\circ$ contribute to experimental peak B, with a dominant contribution of the $S^{2+}(3s3p^3)^3P^\circ$ channel. The ratio σ_A/σ_B of the partial cross-sections between peaks A and B is about 1.3, as observed on the experimental spectra. Both peaks A and B are shown to decrease at higher energies. The partial cross-section on peak C corresponding to the $S^{2+}(3s3p^3)^3D^\circ$ exit channel is slightly under-estimated and remains always lower than peak B; nevertheless it increases at higher energies and reaches the same order of magnitude as peak B, as observed experimentally. The partial cross-sections on the $S^{2+}(3s3p^3)^1P^\circ$ channel also increase with energy but remains much lower. The cross–sections towards the $S^{2+}(3s^2 3p^2)^1S$ and $S^{2+}(3s^2 3p^2)^1D$ exit channels are almost zero for all collision energies, and the cross-sections on the $S^{2+}(3s^2 3p^2)^3P$ capture channel remain negligible.

In the case of the collision of C^{q+} ions with Uracil, the collision dynamics was treated in the framework of the sudden approximation hypothesis, which assumes the electronic transition to occur so quickly that the rovibrational motion remains unchanged. The molecular vibration is thus ignored, and the ion-molecule collision may be visualized as an ion bumping onto an anisotropic atom. The charge transfer mechanism may be induced by very fast electronic transitions, and the Uracil molecule considered as frozen during the process. This is, of course, a first order level of approximation, but it has shown its efficiency in a number of ion-diatomic collisions [31–34] in the energy range we are dealing with.

The collision dynamics has been performed in the 0.1–0.7 a.u. collision velocity range (3–20 keV laboratory energies), corresponding to the same impact energies as the fragmentation experiments [8]. All the transitions between the A′ states, driven by radial coupling elements, have been considered for the series of colliding C^{q+} (q = 2–5) carbon ions, and the centre-of-mass of the Uracil molecule was taken as the origin of the electron coordinates. As previously stated, rotational couplings have not been taken into account in this calculation. The results are presented in Table 4 and Figure 6.

At low energies, they show clearly a strong increase of the charge transfer cross-section with increasing velocity in the case of the C^{2+}-Uracil collision system. For this system, the charge transfer cross-section is very small, of the order of 10^{-18} cm^2 at $v = 0.1$ a.u. ($E_{lab} = 3$ keV), in quite a good agreement with the experimental measurements, which assume almost complete fragmentation of the Uracil molecule in that case. The relative yield for fragmentation is experimentally shown to decrease with increasing impact velocity, which is corroborated by the first increase of the charge transfer cross-section. For velocities higher than 0.4 a.u. ($E_{lab} = 48$ keV),

Table 4. Charge transfer cross-sections for C^{q+}-Uracil, $q = (2–5)$ (in 10^{-16} cm^2).

Velocity (a.u.)	E_{lab} (keV)	$C^{2+} + U$ $^1\Sigma$ states	$C^{3+} + U$ $^2\Sigma$ states	$C^{4+} + U$ $^1\Sigma$ states	$C^{5+} + U$ $^2\Sigma$ states
0.1	3	0.0098	0.327	0.099	0.576
0.2	12	0.175	0.128	0.112	0.359
0.3	27	0.209	0.059	0.123	0.166
0.4	48	0.165	0.034	0.142	0.094
0.5	75	0.123	0.022	0.211	0.061
0.6	108	0.093	0.015	0.215	0.042
0.7	147	0.072	0.011	0.247	0.031

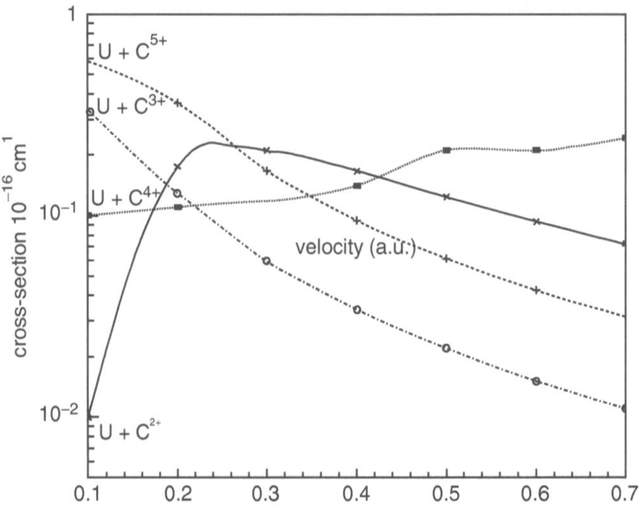

Figure 6. Charge transfer cross-sections for the C^{q+}-Uracil systems.

the charge transfer cross-section of C^{2+}-Uracil decreases smoothly with velocity to ca. 0.1 10^{-16} cm^2, in good correlation with the increase of fragmentation at $v = 0.4$ a.u.

The behaviour of the other colliding species appears quite different. The C^{4+}-Uracil collision system shows only small variations of charge transfer cross-sections with velocity, around [0.1–0.25] 10^{-16} cm^2, which is corroborated by fragmentation experiments. At higher velocities, the calculated charge transfer cross-sections are higher when compared to the case of C^{2+}-Uracil, in agreement with the lower fragmentation yield observed experimentally. The C^{3+} and C^{5+}-Uracil collision systems appear to have a very similar behaviour, with a regular decrease of the charge transfer cross-section with increasing velocity. This global variation is still in accordance with the experimental measurements which show a regular increasing yield for fragmentation in the velocity range. This discussion remains of course qualitative, as we consider only the two main processes, fragmentation and charge transfer, but do not take into account the ionization processes, which could be significant. Besides, the orientation dependence of these processes is not taken into account. Nevertheless, this shows that these processes can be considered, with quite a significant accuracy, as complementary with regard to the variation with the impact energy. This work provides also an order of magnitude of the absolute charge-transfer cross-sections which cannot be achieved experimentally.

4. CONCLUSION

This work shows, in a first approach, the possibility to extend the close-coupling method, well-adapted to the treatment of ion-atom collisions, to much more complex systems, in particular of biological interest. The model presented in the case of the C^{q+}-Uracil collision system, although very simple, could provide results compa-rable to experimental data. Such an approach, of course, does not take into account all the motions of the biomolecule; in particular, the orientation of the biomolecule with respect to the angle of attack of the ion should be considered and orientation-average calculations be performed. Nevertheless, this first approach appears very encouraging and further developments are in progress.

Acknowledgements

The support of COST action P9, *Radiation Damage in Biomolecular Systems*, is gratefully acknowledged, as well as the use of the computing facilities of IDRIS (Project n$^{\theta}$ 51566).

References

1. Y.S. Tergiman, and M.C. Bacchus-Montabonel, *Phys. Rev. A* **64**, 042721, 2001.
2. N. Vaeck, M.C. Bacchus-Montabonel, E. Baloïtcha, and M. Desouter-Lecomte, *Phys. Rev. A* **63**, 042704, 2001.

3. E. Baloïtcha, M. Desouter-Lecomte, M.C. Bacchus-Montabonel, and N. Vaeck, *J. Chem. Phys.* **114**, 8741, 2001.
4. M. Łabuda, Y.S. Tergiman, M.C. Bacchus-Montabonel, and J.E. Sienkiewicz, *Chem. Phys. Lett.* **394**, 446, 2004.
5. M. Łabuda, Y.S. Tergiman, M.C. Bacchus-Montabonel, and J.E. Sienkiewicz, *Int. J. Mol. Sci.* **5**, 265, 2004.
6. D. Péquignot, S.M.V. Aldrovandi, and G. Stasinska, *Astron. Astrophys.* **63**, 313, 1978.
7. R. McCray, C. Wright, and S. Hatchett, *Astrophys. J.* **211**, L29, 1977.
8. S.M. Wilson, T.K. McLaughlin, R.W. McCullough, and H.B. Gilbody, *J. Phys. B.* **23**, 1315, 1990.
9. M.C. Bacchus-Montabonel, M. Łabuda, Y.S. Tergiman, and J.E. Sienkiewicz, *Phys. Rev. A* **72**, 052706, 2005.
10. C. von Sonntag, *The Chemical Basis for Radiation Biology*, Taylor and Francis, London, 1987.
11. B.D. Michael and P.D. O'Neill, *Science* **287**, 1603, 2000.
12. J. de Vries, R. Hoekstra, R. Morgenstern, and T. Schlathölter, *J. Phys. B* **35**, 4373, 2002.
13. J. de Vries, R. Hoekstra, R. Morgenstern, and T. Schlathölter, *Eur. Phys. J. D* **24**, 161, 2003.
14. MOLPRO (version 2002.1) is a package of ab-initio programs written by H.J. Werner and P. Knowles with contributions from J. Almlöf, R.D. Amos, A. Berning, D.L. Cooper, M.J.O Deegan, A.J. Dobbyn, F. Eckert, S.T. Elbert, C. Hampel, R. Lindh, A. Lloyd, W. Meyer, A. Nicklass, K. Peterson, R. Pitzer, A.J. Stone, P.R. Taylor, M.E. Mura, P. Pulay, M. Schütz, H. Stoll, and T. Thorsteinsson.
15. D.E. Woon and T.H. Dunning Jr., *J. Chem. Phys.* **98**, 1358, 1993.
16. A. Nicklass, M. Dolg, H. Stoll, and H. Preuss, *J. Chem. Phys.* **102**, 8942, 1995.
17. NIST Atomic Spectra Database Levels Data: http://physics.nist.gov/cgi-bin/AtData/ main_asd
18. L. Salem, Electrons in Chemical Reactions: First Principles, Wiley Interscience, New York 1982.
19. N. Russo, M. Toscano, and A. Grand, *J. Phys. Chem. B* **105**, 4735, 2001.
20. N. Russo, M. Toscano, and A. Grand, *J. Am. Chem. Soc.* **123**, 10272, 2001.
21. N. Russo, M. Toscano, and A. Grand, *J. Phys. Chem. A* **107**, 11533, 2003.
22. N. Russo, E. Sicilia, M. Toscano, and A. Grand, *Int. J. Quant. Chem.* **90**, 903, 2002.
23. N. Russo, M. Toscano, and A. Grand, *J. Mass. Spectrom.* **38**, 265, 2003.
24. A.lM. Lamsabhi, M. Alcani, O. Mó, and M. Yanez, *Chemphyschem* **4**, 1011, 2003.
25. N. Russo, M. Toscano, and A. Grand, *J. Comput. Chem.* **21**, 1243, 2000.
26. C.E. Crespo-Hernandez, R. Arce, Y. Ishikawa, L. Gorb, J. Leszczynski, and D.M. Close, *J. Phys. Chem A* **108**, 6373, 2004.
27. V.M. Orlov, A.N. Smirnov, and Y.M. Varshavsky, *Tetrahedron Lett.* **48**, 4377, 1976.
28. B.I. Verkin, L.F. Sukodub, and I.K. Yanson, *Dolk. Akad. Nauk SSSR* **228**, 1452, 1976.
29. R.J. Allan, C. Courbin, P. Salas, and P. Wahnon, *J. Phys. B* **23**, L461, 1990.
30. L.F. Errea, L. Mendez, and A. Riera, *J. Phys. B* **15**, 101, 1982.
31. M.C. Bacchus-Montabonel, *Phys. Rev. A* **59**, 3569, 1999.
32. L.F. Errea, J.D. Gorfinkiel, C. Harel, H. Jouin, A. Macias, L. Méndez, and A. Riera, *Phys. Scr.* **T62**, 33, 1996.
33. L.F. Errea, J.D. Gorfinkiel, E.S. Kryachko, A. Macias, L. Méndez, and A. Riera, *J. Chem. Phys.* **106**, 172, 1997.
34. P.C. Stancil, B. Zygelman, and K. Kirby, in: *Photonic, Electronic, and Atomic Collisions*, ed. F. Aumayr and H.P. Winter, World Scientific, Singapore, 1998.

Part III
Excited States and Condensed Matter

SHELL EFFECTS AND HOMOTHETIC EXPRESSIONS FOR ELECTRON RELAXATION AND OTHER CORRECTIONS TO 2p-CORE IONIZATION ENERGIES AND SPIN–ORBIT SPLITTING FOR ATOMS FROM Cl TO Ba

JEAN MARUANI AND CHRISTIANE BONNELLE

Laboratoire de Chimie Physique – Matière et Rayonnement, CNRS and UPMC,
11, rue Pierre et Marie Curie – 75005 Paris, France

Abstract On the basis of numerical *ab initio*, ΔBDF and MCBDF computations, we have investigated the electron relaxation, Breit interactions, quantum electrodynamics (*qed*) and nuclear size and motion (*nuc*) effects in the 2p-core ionization energies and spin–orbit splitting of atoms from Cl to Ba, excluding transition elements. Calculations made by mixing the ground and the ionized configurations (*gsm*) yield significant improvement in the ionization energies, but not in the spin–orbit splitting, with respect to experimental data. As with the relativistic contributions to 1s- and 2s-core ionization energies investigated in earlier papers, we have identified shell effects and performed homothetic fits of corrections to the 2p-core ionization energies and spin–orbit splitting in homologous families of atoms.

Keywords: 2p-core ionization; spin–orbit splitting; electron relaxation; ground-state mixing; shell effects; homothetic fits.

1. INTRODUCTION

A longstanding problem in the interpretation of X-ray photoelectron and absorption spectra (XPS, XAS, ...) is the theoretical calculation of the core excitation and ionization energies (CIE's) and of their chemical shifts, experimentally measured in molecular systems or in condensed matter [1–3]. In computing these properties, one has to take into account various effects: (i) electron correlation, particularly strong between inner-core electrons [4, 5], which prevents the use of simple SCF schemes; (ii) electron relaxation, which increases with the number of electrons [6–9], making the use of Koopmans' theorem [10] rather inaccurate; (iii) for systems containing

S. Lahmar et al. (eds.), Topics in the Theory of Chemical and Physical Systems, 217–233.

heavy atoms, various relativistic, quantum electrodynamics (*qed*) and nuclear size and motion (*nuc*) effects, which may overcome the previous effects in inner-core excitations [8,9,11–13].

Consecutive steps of approximation for computing inner-core energies may follow various paths. The simplest step is to use Koopmans' theorem (no relaxation: *nx*), with nonrelativistic (*nt*) self-consistent field (SCF) Hartree–Fock (HF) orbital energies (no correlation: *nc*). In order to include relaxation (*yx*), one may take differences between the total SCF energies computed for the ground and excited states, this yielding ΔSCF results. This requires the use of core-hole adapted methods for avoiding "variational collapse" [14–21]. One may include correlation (*yc*) by replacing the SCF by, e.g. EHF, CI, MCSCF, CC, or DFT-type methods. If necessary, relativity may be included (*yt*) by using a Breit–Dirac–Fock (BDF) procedure, and quantum electrodynamics (*qed*) and nuclear size and motion (*nuc*) corrections by perturbation. The ultimate step would be to take differences between the corrected relativistic, correlated total energies for the ground and excited states.

In earlier papers [6–8] we have proposed a procedure for evaluating core ionization/excitation chemical shifts in molecules from computed core ionization/excitation energies for the relevant isolated atom in neutral and valence-ionized states and from computed charge transfer relative to this atom within the molecule. The atomic calculations involved relaxation, possibly correlation and, when appropriate, relativity and other effects, while in the molecule one could use any approximate method (possibly involving effective core potentials) yielding reliable charges.

In later papers [9] we have investigated relaxation and relativity effects in the 1s- [9a] and 2s- [9b] core ionization energies of neutral atoms, from Be/Mg to Xe, not including the inner-core correlation arising from ground-state mixing. We showed that an allometric function [22] can allow accurately for the relativistic corrections to the 1s- and 2s-core ionization energies [9]. However, due to shell effects [23,24], the variations of energy contributions with atomic number Z are not monotonous, but involve an oscillating irregular part. As a result, a monotonous expression such as the allometric formula could not yield a perfect fit throughout the periodic table. Therefore, we have fitted this formula separately for series of atoms whose external configuration expresses the filling of a given subshell.

More generally, in order to estimate specific corrections to uncorrected energy values for complex systems, one may compute these corrections for a set of simpler structures, then derive parameterized formulae between the corrected and uncorrected values for this set, assuming the relations derived for the simpler structures can be applied to the complex ones. In previous papers, the problem has been split into including, in this manner, either relaxation, or correlation, or relativity, in an approximation involving only the other contributions.

In this paper, we shall present a procedure for evaluating electron relaxation, inner-core correlation, and Breit, *qed*, and *nuc* corrections to 2p-core ionization energies and spin–orbit splitting in molecular systems, from tabulated results on atoms from Cl to Ba, excluding transition elements.

2. PROCEDURES

The 2p levels are split by spin–orbit coupling into sublevels labeled $j = 1/2$ and $j = 3/2$. The following relations define the correspondence between the measured sublevel energies $E_{1/2}$ and $E_{3/2}$, their weighted average $\underline{E}$, and the level splitting Λ:

(1a, 1b) $$\Lambda = E_{1/2} - E_{3/2}, \quad \underline{E} = (E_{1/2} + 2E_{3/2})/3;$$

(1c, 1d) $$E_{3/2} = \underline{E} - \Lambda/3, \quad E_{1/2} = \underline{E} + 2\Lambda/3.$$

In Table 1 we have gathered measured values of 2p-core ionization energies and spin–orbit splitting of atoms from Al through Ba [25].

Figure 1 displays the variations of these properties and their second-order finite differences (discrete derivatives) with atomic number Z. As shown earlier [24], the second derivative possesses two advantages over the first one: a stronger signal (but also noise) enhancement and a maximum located as in the function itself.

However, the second-order finite differences involve, in their computation, four values around each considered point, which may therefore be affected by their anomalies. Due to the irregularities occurring with fully filled and half-filled shells and the transition-element series, the discrete derivatives show an oscillating behavior, similar to that appearing in the 1s- and 2s-core ionization energy variations [9].

For the ionization energies (lower left), minima appear at P, As and (before) Sb, as well as at Mn and (after) Tc (half-filled shells), but also at Sc and Y (start of transition series); while maxima appear at Cl, Br, I, but also at Cu and Ag (close to full shell).

Table 1. For elements from Al (Z = 13) to Ba (Z = 56), measured 2p-core ionization energies [25b] and resulting spin–orbit splitting Λ (in eV). Energy values are given as being accurate to within 0.3–0.7 eV.

^{Z}El.	L_{II} ($2p_{1/2}$)	L_{III} ($2p_{3/2}$)	Λ	^{Z}El.	L_{II} ($2p_{1/2}$)	L_{III} ($2p_{3/2}$)	Λ	^{Z}El.	L_{II} ($2p_{1/2}$)	L_{III} ($2p_{3/2}$)	Λ
^{13}**Al**	73.3	72.9	0.4	^{31}**Ga**	1142.3	1115.4	26.9	^{49}**In**	3938.0	3730.1	207.9
^{14}**Si**	99.5	98.9	0.6	^{32}**Ge**	1247.8	1216.7	31.1	^{50}**Sn**	4156.1	3928.8	227.3
^{15}**P**	136.2	135.3	0.9	^{33}**As**	1358.6	1323.1	35.5	^{51}**Sb**	4380.4	4132.2	248.2
^{16}**S**	165.4	164.2	1.2	^{34}**Se**	1476.2	1435.8	40.4	^{52}**Te**	4612.0	4341.4	270.6
^{17}**Cl**	201.6	200.0	1.6	^{35}**Br**	1596.0	1549.9	46.1	^{53}**I**	4852.1	4557.1	295.0
^{18}**Ar**	247.3	245.2	2.1	^{36}**Kr**	1727.2	1674.9	52.3	^{54}**Xe**	5103.7	4782.2	3215
^{19}**K**	296.3	293.6	2.7	^{37}**Rb**	1863.9	1804.4	59.5	^{55}**Cs**	5359.4	5011.9	347.5
^{20}**Ca**	350.0	346.4	3.6	^{38}**Sr**	2006.8	1939.6	67.2	^{56}**Ba**	5623.6	5247.0	376.6
^{21}**Sc**	406.7	402.2	4.5	^{39}**Y**	2155.5	2080.0	75.5	^{71}**Lu**			
^{22}**Ti**	460.4	454.5	5.9	^{40}**Zr**	2306.7	2222.3	84.4	^{72}**Hf**			
^{23}**V**	520.5	512.9	7.6	^{41}**Nb**	2464.7	2370.5	94.2	^{73}**Ta**			
^{24}**Cr**	583.7	574.5	9.2	^{42}**Mo**	2625.1	2520.2	104.9	^{74}**W**			
^{25}**Mn**	651.4	640.3	11.1	^{43}**Tc**	2793.2	2676.9	116.3	^{75}**Re**			
^{26}**Fe**	720.5	707.5	13.0	^{44}**Ru**	2966.9	2837.9	129.0	^{76}**Os**			
^{27}**Co**	793.6	778.6	15.0	^{45}**Rh**	3146.1	3003.8	142.3	^{77}**Ir**			
^{28}**Ni**	871.9	854.7	17.2	^{46}**Pd**	3330.3	3173.3	157.0	^{78}**Pt**			
^{29}**Cu**	950.9	930.9	20.0	^{47}**Ag**	3523.7	3351.1	172.6	^{79}**Au**			
^{30}**Zn**	1042.8	1019.7	23.1	^{48}**Cd**	3727.0	3537.5	189.5	^{80}**Hg**			

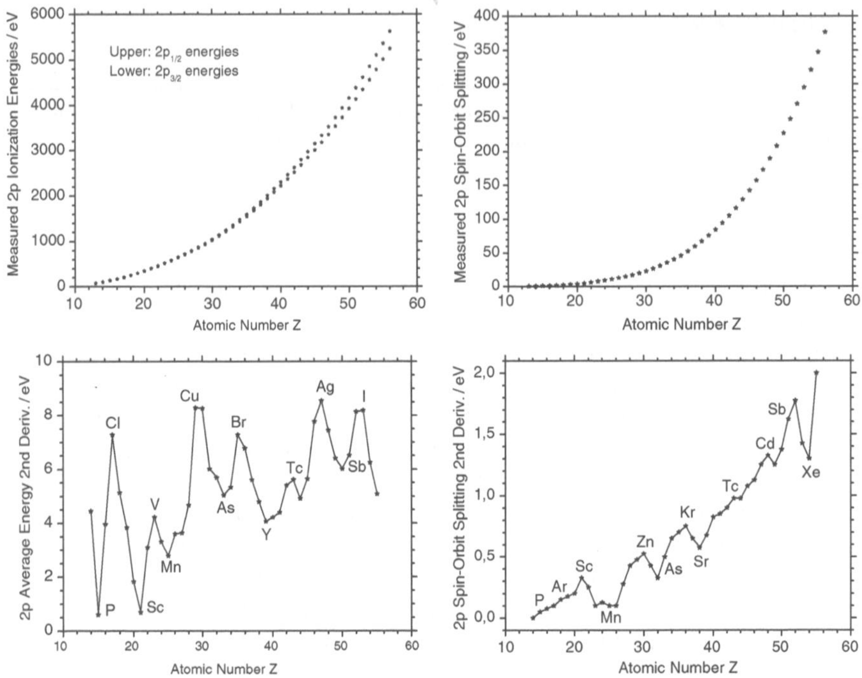

Figure 1. Experimental variations of 2p-core ionization energies (in eV) for atoms from Al(Z = 13) to Ba(Z = 56). *Upper left*: $2p_{1/2}$ and $2p_{3/2}$ energies; *lower left*: their weighted-average second-order discrete derivative, as functions of Z; *upper right*: spin–orbit splitting between the $2p_{1/2}$ and $2p_{3/2}$ levels; *lower right*: its second derivative, as functions of Z. On the derivative diagrams shell effects appear about fully filled and half-filled shells and near filling irregularities of the transition elements.

Anomalies (such as V) come from irregularities in the shell filling process around the element. For the spin–orbit splitting (lower right), maxima do not appear at P, As, Sb, or Mn and Tc (half filled shells), but do appear at Sc, Ar, Zn, Kr, Cd (full shells); while minima appear at Sr and Xe. The variations of this property present more anomalies because its values result from small differences between large values, which cumulates experimental errors and anomalies due to the finite-difference procedure.

2a. COMPUTATIONS

The computations of 2p-core ionization energies were performed using a pattern similar to that used for 1s- and 2s-core ionization energies [9]. Here again we have used Bruneau's multiconfiguration Dirac–Fock (MCDF) *ab initio* program [26–28], which is based on a numerical resolution of the Dirac equation corrected for Breit terms, vacuum polarization, and radiative (*qed*) contributions, and nuclear size and motion (*nuc*) effects.

Earlier [22] it was noticed that, for lighter elements, an artifact deviation from the real variation could occur, because some of the $2p^5$ states involved in the weighted averages did not have a well defined quantum number j. In these elements, the interaction between the 2p core vacancy and 3p valence open shell is large compared to spin–orbit coupling. The problem does not arise when the core hole level is deep enough relative to the valence open shell, and we have restricted the present investigation to these elements.

In order to assess the various energy contributions, for each element three sets of computations were performed: (1) one for the atom in the ground state; (2) one for the atom in the 2p-ionized state; (3) and one mixing the ground and ionized states (*gsm*). For each of these sets computations were made at four successive levels: (a) Dirac–Fock with no correction; (b) Dirac–Fock including the Breit terms (BDF); (c) BDF including *qed* effects, and (d) also including *nuc* effects. From level (1a) one derives the simplest relativistic, Koopmans energies, E_K. From level (3a) one derives E_R, energies that include electron relaxation and part of inner-core correlation, through *gsm*, but not Breit, *qed* and *nuc* contributions. From a combination of (1c) and (2c) one derives E_Q, energies that include electron relaxation and the Breit, *qed* and *nuc* contributions, but not *gsm* correlation. And from level (3c) one derives E_G, energies which include all effects.

In Table 2 there are displayed, as landmarks in our set of elements, the results obtained for the rare gases at the different steps of computation.

Four observations can be made: (1) Although the *nuc* contributions are not negligible and increase with Z, they are practically the same for the ground and ionized states and therefore do not contribute to the ionization energies and spin–orbit splitting. This justifies their neglect in our previous 1 s and 2 s investigations [9]. (2) The same holds true for *qed* effects in lighter atoms, but in the heavier ones their contribution to spin–orbit splitting may reach 0.5 eV, at the limit of the experimental errors. (3) The main correction comes from the Breit terms, which may decrease the ionization energies by up to 16.2 eV (for Xe) and the spin–orbit splitting by 4.3 eV. (4) Ground-state mixing (as defined earlier) increases all computed energies. For those including *fnoy* (defined earlier) the increase is less for the upper sublevel than for the ground state and still a little less for the lower sublevel (see subtable). As a result there appears a decrease of the ionization energy for $2p_{3/2}$ and an increase for $2p_{1/2}$, resulting in an increase of the spin–orbit splitting, as will be discussed later.

2b. FITTINGS

Ideally, in order to determine the ionization/excitation energies of complex systems one would like to make simple, e.g. Koopmans SCF (KSCF) computations of these energies and deduce, from the approximate results thus obtained, more accurate values by making use of some simple formula (or scaling table) where the corrective terms would appear as a parameterized (possibly nonlinear) function of the approximate value:

(2) $$E_{acc} = E_{app} + \Delta E_{cor} \text{ where } \Delta E_{cor} \cong \Delta E'_{cor} \equiv f_{zq}(\boldsymbol{k}; E_{app}).$$

Table 2. For the three rare gases investigated (Ar, Kr, and Xe), there are given the computed relativistic energies for the ground state (GS: $j = 0$), the lower 2p sublevel (LT: $j = 3/2$) and the upper 2p sublevel (UT: $j = 1/2$), with and without ground-state mixing (*gsm*). For each level, there are given the Dirac–Fock energy (label *mcdf*) and energies obtained by including the Breit terms (*hbrt*), *qed* effects (*crrd*), and *nuc* corrections (*fnoy*), and the corresponding contributions Δ. The *lower* subtable shows the effects of *gsm* on the *fnoy* values. All energies are given in eV.

with *gsm*	Ar	Δ	Kr	Δ	Xe	Δ
UT (1/2) *fnoy*	−14 127.104	+0.164	−74 085.683	+0.398	−197 267.172	0.727
UT (1/2) *crrd*	−14 127.268	+2.211	−74 086.081	+24.527	−197 267.899	+98.193
UT (1/2) *hbrt*	−14 129.479	+3.551	−74 110.608	+35.404	−197 366.092	+142.447
UT (1/2) *mcdf*	−14 132.830	Ref val.	−74 146.012	Ref val.	−197 508.539	Ref val.
LT (3/2) *fnoy*	−14 129.299	+0.164	−74 138.855	+0.398	−197 588.497	0.728
LT (3/2) *crrd*	−14 129.463	+2.208	−74 139.253	+24.439	−197 589.225	+97.720
LT (3/2) *hbrt*	−14 131.671	+3.433	−74 163.692	+36.392	−197 686.945	+146.780
LT (3/2) *mcdf*	−14 135.104	Ref val.	−74 200.084	Ref val.	−197 833.725	Ref val.
GS (0) *fnoy*	−14 377.291	+0.167	−75 817.001	+0.402	−202 371.875	+0.729
GS (0) *crrd*	−14 377.458	+2.211	−75 817.403	+24.501	−202 372.604	+98.170
GS (0) *hbrt*	−14 379.669	+3.664	−75 841.904	+39.309	−202 470.774	+158.673
GS (0) *mcdf*	−14 383.333	Ref val.	−75 881.213	Ref val.	−202 629.447	Ref val.
no gsm	*Ar*	Δ	*Kr*	Δ	*Xe*	Δ
UT (1/2) *fnoy*	−14 129.922	+0.165	−74 092.711	+0.401	−197 275.881	+0.731
UT (1/2) *crrd*	−14 130.087	+2.222	−74 093.112	+24.603	−197 276.612	+98.449
UT (1/2) *hbrt*	−14 132.309	+3.411	−74 117.715	+35.872	−197 375.061	+143.905
UT (1/2) *mcdf*	−14 135.720	Ref val.	−74 153.587	Ref val.	−197 518.966	Ref val.
LT (3/2) *fnoy*	−14 132.079	+0.165	−74 145.641	+0.401	−197 596.508	+0.732
LT (3/2) *crrd*	−14 132.244	+2.219	−74 146.042	+24.515	−197 597.240	+97.975
LT (3/2) *hbrt*	−14 134.463	+3.496	−74 170.557	+36.864	−197 695.215	+148.248
LT (3/2) *mcdf*	−14 137.959	Ref val.	−74 207.421	Ref val.	−197 843.463	Ref val.
GS (0) *fnoy*	−14 380.251	+0.166	−75 825.040	+0.400	−202 383.664	0.725
GS (0) *crrd*	−14 380.417	+2.200	−75 825.440	+24.424	−202 384.389	+97.916
GS (0) *hbrt*	−14 382.617	+3.602	−75 849.864	+38.825	−202 482.305	+157.152
GS (0) *mcdf*	−14 386.219	Ref val.	−75 888.689	Ref val.	−202 639.457	Ref val.
Δ*gsm*	Ar	δ	Kr	δ	Xe	δ
UT (1/2) *fnoy*	+2.818	+0.038	+7.03	+0.24	+8.71	+0.70
LT (3/2) *fnoy*	+2.780	−0.106	+6.79	−1.25	+8.01	−3.78
GS (0) *fnoy*	+2.886	Ref val.	+8.04	Ref val.	+11.79	Ref val.

Here we call f "homothetic" in the sense that the correction ΔE_{cor} is expressed as a parameterized function of the approximate value E_{app}. The index z labels the set of atomic numbers Z and orbitals (n, l) for the considered family of atoms, and q is the degree of (positive or negative) valence ionization for charged atoms (for neutral atoms, $q = 0$). For different z and q, f may be different. The function f being given, the label $\boldsymbol{k}$ represents the set of values of the function parameters yielded by a best-fit procedure. In Eq. (1), ΔE_{cor}, which is computed from the energies provided by quantum-mechanical calculations, designates the "exact" correction, whereas $\Delta E'_{cor}$, which is yielded by the use of the best-fit function f_{zq}, is the "estimated" correction.

The best-fit function f_{zq} is determined by computing E_{app} and E_{acc} for some ionization/excitation energy in a given family of atoms/ions of the periodic table, and then fitting their difference ΔE_{cor} to various homothetic functions:

$$\Delta E_{cor} \equiv E_{acc} - E_{app} \cong f_{zq}(\boldsymbol{k}; E_{app}). \quad (3)$$

The selected function f_{zq} is that which gives the smallest mean-square deviation δ (msd) with the optimized parameters $\boldsymbol{k}$.

We have tried the following formulae in looking for best fits of the variations of the corrective terms as functions of the approximate values. In the following, y represents ΔE_{cor}, x stands for E_{app}, and the set of optimized parameters $\boldsymbol{k} = \{x_0, y_0, b, c, \lambda, \ldots\}$ may depend on z and q:

(4a) $Linear$ (2 parameters, y_0 and b) : $y = y_0 + bx$

(4b) $Allometric$ (2 parameters, y_0 and b) : $log\, y = log\, y_0 + b\, log\, x$, or $y = y_0 x^b$

(4c) $Quadratic$ (3 parameters, $y_0, b,$ and c) : $y = y_0 + bx + cx^2$

(4d) $Exponential\text{-}decay$ (3 parameters, $y_0, b,$ and λ) : $y = y_0 + b\, exp(-x/\lambda)$

(4e) $Sigmoid$ (4 parameters, $y_0, b, x_0,$ and λ) : $y = y_0 + b/[1 + exp(x - x_0)/\lambda]$

3. RESULTS

The various approximate energies obtained for the $2p_{1/2}$ and $2p_{3/2}$ sublevels of atoms from Cl to Ba are given in Table 3. There are also given the most significant energy differences.

In Figure 2 there are displayed the variations, with atomic number Z, of our best computed values E_G (left) and their second-order finite differences (*right*). For the discrete derivative of the average energy, one observes a minimum for half-filling (As, Sb) and a large value for fulfilling (Kr, Xe), not too different from the experimental trend (Figure 1, *lower left*). For the discrete derivative of the spin–orbit splitting, there is no apparent modulation, contrary to the experimental oscillations (Figure 1, *lower right*). Therefore, one may expect to represent all theoretical values of the spin–orbit splitting in the investigated range by a single functional form.

In Figure 3 we have compared the effects of the various corrections to the Koopmans energies: relaxation and *gsm* (R), relaxation and Breit, *qed*, and *nuc* (Q), and all contributions (G). It is to be remembered that, contrary to our previous 1s- and 2s-core calculations [9], the comparisons are made in the frame of the Dirac–Fock scheme. It can be seen that just after the middle of the periodic table (Ba) the Breit, *qed*, and *nuc* corrections amount to nearly half the relaxation and *gsm* corrections for ionization energies, and are up to six times larger for spin–orbit splitting. Including *gsm* improves significantly the ionization energies (*left diagrams*) but slightly degrades the spin–orbit splitting (*right diagrams*) with respect to the experimental values. It is also seen that the differences with the measured values are smaller for rare gases, for which experiments are made on atoms.

Table 3. For elements from Cl(Z = 17) to Ba(Z = 56), there are given computed $2p_{1/2}$ and $2p_{3/2}$ ionization energies. These are obtained by applying Koopmans' theorem to the ground-state 2p-core orbitals computed relativistically but without Breit, *qed*, and *nuc* corrections (E_K); or by difference between ground and excited state energies computed together (*gsm*) to account for relaxation and inner-core correlation (E_R); or by two separate calculations (not including *gsm*) including Breit, *qed*, and *nuc* corrections (E_Q); or with all corrections included (E_G). In col. 1, the index + stands for 2p-core vacancy and in col. 2, the second line gives the numbers of states in neutral atoms and in $j = 1/2$ and 3/2 2p-core vacancy ions. Columns 7 and 8 give differences between our best computed values and Koopmans or experimental energies. The last columns give the K and G computed spin–orbit splittings Λ and their departures from the experimental values X. All energies are given in eV.

Elt	Term	$E_K(2p_{1/2})$ $E_K(2p_{3/2})$	$E_R(2p_{1/2})$ $E_R(2p_{3/2})$	$E_Q(2p_{1/2})$ $E_Q(2p_{3/2})$	$E_G(2p_{1/2})$ $E_G(2p_{3/2})$	$\Delta_{KG(1/2)}$ $\Delta_{KG(3/2)}$	$\Delta_{GX(1/2)}$ $\Delta_{GX(3/2)}$	Λ_K Λ_G	$\delta\Lambda_{KX}$ $\delta\Lambda_{GX}$
Cl	$^2P_{3/2}$	220.879	210.419	210.157	210.172	10.707	8.572	1.762	0.162
Cl_+	2/4/6	219.117	208.132	208.100	207.893	11.224	7.893	2.279	0.679
Ar	$^1S_0$	262.099	250.504	250.329	250.187	11.912	2.887	2.310	0.210
Ar_+	1/1/1	259.789	248.229	248.172	247.992	11.787	2.792	2.195	0.095
K	$^2S_{1/2}$	315.614	303.365	303.157	302.976	12.638	6.676	2.980	0.280
K_+	1/2/2	312.634	300.425	300.355	300.133	12.501	6.533	2.843	0.143
Ca	$^1S_0$	373.657	360.933	360.664	360.462	13.195	10.462	3.787	0.187
Ca_+	1/1/1	369.870	357.191	357.085	356.837	13.033	10.437	3.625	0.025
Cu	$^2S_{1/2}$	988.323	960.181	958.978	958.374	29.949	7.474	20.795	0.795
Cu_+	1/2/2	967.528	939.511	938.869	938.132	29.396	7.232	20.242	0.242
Zn	$^1S_0$	1081.389	1053.338	1051.940	1051.299	30.090	8.499	24.191	1.091
Zn_+	1/1/1	1057.198	1029.285	1028.514	1027.728	29.470	8.028	23.571	0.471
Ga	$^2P_{1/2}$	1182.076	1153.308	1152.160	1151.466	30.610	9.166	27.992	1.092
Ga_+	2/4/6	1154.084	1125.465	1125.000	1124.163	29.921	8.763	27.303	0.403
Ge	$^3P_0$	1288.054	1259.380	1257.650	1256.904	31.150	9.104	32.233	1.133
Ge_+	5/8/13	1255.821	1227.310	1226.355	1225.435	30.386	8.735	31.469	0.369
As	$^4S_{3/2}$	1399.374	1370.462	1368.407	1367.605	31.770	9.005	36.948	1.448
As_+	5/10/18	1362.426	1333.692	1332.495	1331.502	30.924	8.402	36.103	0.603
Se	$^3P_2$	1516.026	1486.739	1484.425	1483.564	32.462	7.364	42.173	1.773
Se_+	5/8/13	1473.853	1444.760	1443.392	1442.325	31.528	6.525	41.239	0.839
Br	$^2P_{3/2}$	1638.009	1608.302	1605.820	1604.784	32.225	8.784	47.945	1.845
Br_+	2/4/6	1590.064	1560.569	1559.137	1557.868	32.196	7.968	46.916	0.816
Kr	$^1S_0$	1765.334	1735.201	1732.330	1731.318	33.004	4.118	54.303	2.003
Kr_+	1/1/1	1711.031	1681.129	1679.400	1678.146	31.631	3.246	52.930	0.872
Rb	$^2S_{1/2}$	1905.247	1874.766	1871.593	1870.490	34.757	6.590	61.288	1.788
Rb_+	1/2/2	1843.959	1813.727	1811.803	1810.441	33.518	6.041	60.049	0.549
Sr	$^1S_0$	2051.089	2020.409	2016.894	2015.713	35.376	8.913	68.944	1.744
Sr_+	1/1/1	1982.145	1951.732	1949.580	1948.121	34.024	8.521	67.592	0.392
Ag	$^2S_{1/2}$	3579.146	3540.710	3532.890	3530.788	48.358	7.088	176.220	3.620
Ag_+	1/2/2	3402.926	3364.954	3359.925	3357.340	45.586	6.240	173.448	0.848
Cd	$^1S_0$	3783.002	3744.423	3735.961	3733.737	49.265	6.737	193.399	3.899
Cd_+	1/1/1	3589.603	3551.516	3546.056	3543.323	46.280	5.823	190.414	0.914
In	$^2P_{1/2}$	3995.461	3956.530	3947.397	3945.037	50.424	7.037	211.855	3.955
In_+	2/4/6	3783.606	3745.194	3739.288	3736.390	47.216	6.290	208.647	0.747

Sn	$^3P_0$	4214.663	4175.441	4165.591	4163.099	51.564	6.999	231.661	4.361
Sn_+	5/8/13	3983.002	3944.330	3937.943	3934.882	48.120	6.082	228.217	0.917
Sb	$^4S_{3/2}$	4440.780	4401.255	4390.648	4388.019	52.761	7.619	252.887	4.687
Sb_+	5/10/18	4187.893	4148.946	4142.054	4138.825	49.068	6.625	249.194	0.994
Te	$^3P_2$	4673.899	4634.051	4622.646	4619.874	54.025	7.874	275.610	5.010
Te_+	5/8/13	4398.289	4359.051	4351.626	4348.223	50.066	6.823	271.651	1.051
I	$^2P_{3/2}$	4914.097	4873.905	4861.669	4858.739	55.358	6.639	299.908	4.908
I_+	2/4/6	4614.189	4574.640	4566.664	4563.071	51.118	5.971	295.668	0.668
Xe	$^1S_0$	5161.454	5120.908	5107.784	5104.704	56.750	0.704	325.862	4.862
Xe_+	1/1/1	4835.592	4795.722	4787.156	4783.378	52.214	0.378	321.326	0.326
Cs	$^2S_{1/2}$	5421.760	5380.844	5366.804	5363.550	58.210	4.150	353.552	6.052
Cs_+	1/2/2	5068.208	5027.999	5018.828	5014.884	53.324	2.984	348.666	1.166
Ba	$^1S_0$	5689.577	5648.436	5633.415	5629.999	59.578	6.399	383.076	6.476
Ba_+	1/1/1	5306.501	5266.104	5256.284	5252.101	54.400	5.101	377.898	1.298

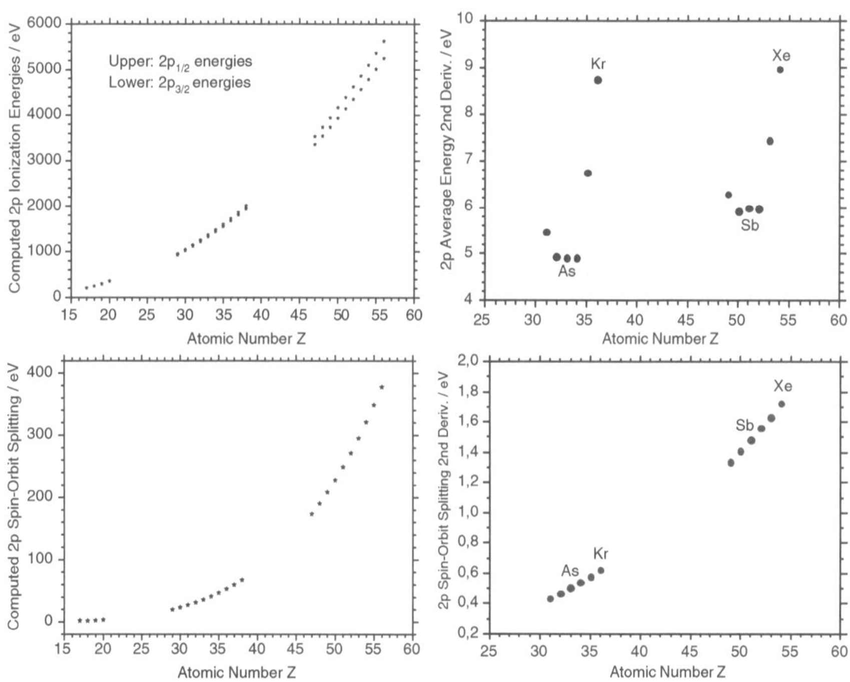

Figure 2. 2p ionization energies computed including relaxation and ground-state mixing (*gsm*) as well as quantum-electrodynamics (*qed*) and nuclear size and motion (*nuc*) effects. *Upper left*: $2p_{1/2}$ and $2p_{3/2}$ energies; *lower left*: resulting spin–orbit splitting; *upper right*: second-order finite difference of the weighted average of *upper left* energies; *lower right* second-order finite difference of *lower left* energy splitting.

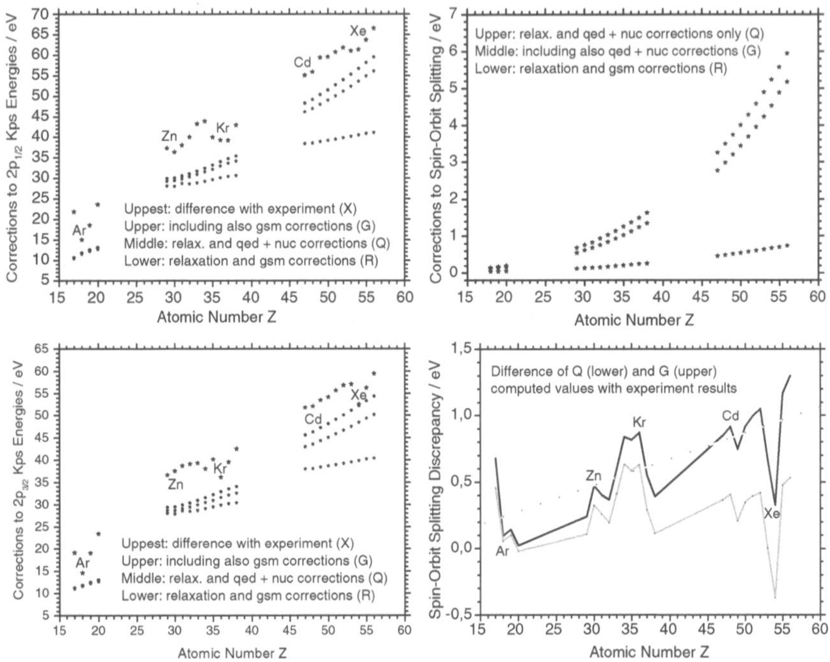

Figure 3. Corrections to 2p Koopmans ionization energies when including: (R) relaxation and *gsm*; (Q) relaxation and *qed* plus *nuc* terms; (G) relaxation and *gsm* plus *qed*, and *nuc* terms. *Upper left*: $2p_{1/2}$ corrections; *lower left*: $2p_{3/2}$ corrections; *upper right*: spin–orbit splitting corrections to Koopmans values; *lower right*: spin–orbit splitting discrepancies with respect to experimental values.

The fact that ground-state mixing has an effect smaller than electron relaxation or Breit, *qed*, and *nuc* effects confirms that in these systems, inner-core correlation is overcome by electron relaxation and, for heavy systems, relativity effects [6, 8]. However, it has been seen earlier (Table 2), the closer the ionized level is to the ground state, the less it is increased by *gsm*.

This entails an increase in spin–orbit splitting. As all computed values are larger than the measured ones this will appear as a decrease of the correction to the Koopmans values (Figure 3, *upper right*), and therefore will increase the discrepancy with the experiment (Figure 3, *lower right*).

Table 3 and Figure 3 show that, while the R, Q, and G corrections to the K energies tend to increase with Z, the difference between our best computed (G) and the measured (X) values remains about 5–10 eV – slightly smaller for the lower ($2p_{3/2}$) excited level, with a tendency to decrease while Z increases and some irregularities around filled shells (Ar, Zn, Kr, Cd, Xe; Ca, and Sr do not appear in the diagrams). The lower-right diagram shows that, for spin–orbit splitting, the discrepancy of the Q-computed values with the measured values remains between -0.5 and $+0.5$ eV, which is well within experimental errors.

It should be remembered that most of the measured data reported in Table 1 do not refer to ionization to the continuum, as in our calculations, but to excitation to the Fermi level (for conductors) or to the lowest unoccupied level (for the other elements). This makes the measured values systematically lower than the computed values, by about 3–4 eV. In addition, our calculations were performed on a single spectroscopic term of the isolated atom (Table 2), whereas the measurements were made on single atoms for rare gases only: for some nonmetals they were made on molecules and for most other elements, including metals, on the solid. Due to the chemical environment, and atomic bonding, this induces an additional decrease, of about 2–6 eV, in the measured values. These two effects account nearly perfectly for the discrepancies, of about 5–10 eV, between the theoretical and experimental ionization/excitation energies. However, as these effects cancel out in the difference between the $2p_{1/2}$ and $2p_{3/2}$ levels, it is not surprising that the computed splittings reproduce nearly perfectly the measured values, within experimental errors.

The splitting Λ between the 2p levels, when it is solely due to spin–orbit interaction, is believed to increase roughly as Z^4, due to its quantum mechanical perturbative formulation. This is actually the trend appearing on the second discrete derivative of measured values (Figure 1, *lower right*). Here we have investigated the accurate Z variation of the components of Λ. In a polynomial fit involving only powers of Z^4, one has to go up to Z^{28} (8 terms) to reach a fit accuracy ($R^2 = 1$; Sd $= 0.030$) comparable to that obtained with terms up to Z^6 (4 terms) in a formula involving also powers of Z^2($R^2 = 1$; Sd $= 0.032$). The adjunction of a Z^2 term to a fit of the form $A + C\,Z^4$ already increases the correlation coefficient R^2 from .9995 to .9999 and reduces the square deviation Sd from 4.163 to 1.285. Our best bet is to use a 4-parameter polynomial of Z^2 (where the last term may be neglected for the lighter elements). Detailed results are reported in Table 4, which is the translation of the observations made on Figure 3 (*right diagrams*).

As explained earlier, in order to transfer atomic results to molecular systems, it is useful to search for homothetic relations between the corrective terms and the non-corrected values. This is because the dependence of energy properties on the

Table 4. Best bets for polynomial fits of Koopmans (K) and experimental (X) spin–orbit splittings. Similar fits are used for the increments due to relaxation and *gsm* (ΔR) plus *qed* and *nuc* (ΔG) corrections.

$P(Z^2)$	$K(Z^2)$	$+\Delta R(Z^2)$	$+\Delta G(Z^2)$	$X(Z^2)$
A	0.752	−0.014	−0.058	1.200
	±0.040	±0.003	±0.011	±0.301
$+B\,Z^2$	−0.631e-2	1.305e-4	3.156e-4	−0.819e-2
	± 0.009e-2	± 0.068e-4	± 0.239e-4	± 0.069e-2
$+C\,Z^4$	3.387e-5	0.485e-7	3.036e-7	3.416e-5
	± 0.006e-5	± 0.046e-7	± 0.160e-7	± 0.046e-5
$+D\,Z^6$	2.238e-9	−0.419e-11	1.712e-11	2.121e-9
	± 0.012e-9	± 0.091e-11	± 0.319e-11	± 0.091e-9
R^2; Sd	1.0000; 0.0316	0.9999; 0.0023	1.0000; 0.0082	1.0000; 0.2350

atomic number Z is not molecular dependent, whereas the corrective terms and the non-corrected values depend both critically on the charge transferred to or from the excited/ionized atom in the molecule. In Figure 4 there are displayed, as functions of the Koopmans values, the total G-corrections (relaxation, *gsm*, Breit, *qed*, and *nuc*) for the $2p_{1/2}$ and $2p_{3/2}$ ionization energies (*left*) and the spin–orbit splitting (*upper right*) for atoms from Cl to Ba (excluding transition elements). There are also shown the deviations of the Koopmans values from measured ones (*lower right*). On these diagrams it appears clearly that, as with our earlier 1s and 2s results [22], any functional fitting of the variations must be made separately on the three homologous families Cl–Ca, Cu–Sr, and Ag–Ba, at least for the ionization energies (Figure 5). For the spin–orbit splitting, a single functional form can be used (Figure 4). Solid lines in Figures 4 and 5 are computed using the functional forms and best-fit parameters listed in Table 5.

In Table 5 we have gathered the optimized parameters (and mean deviations) for the best-fit functions representing the total G-corrections to the Koopmans $2p_{1/2}$ and

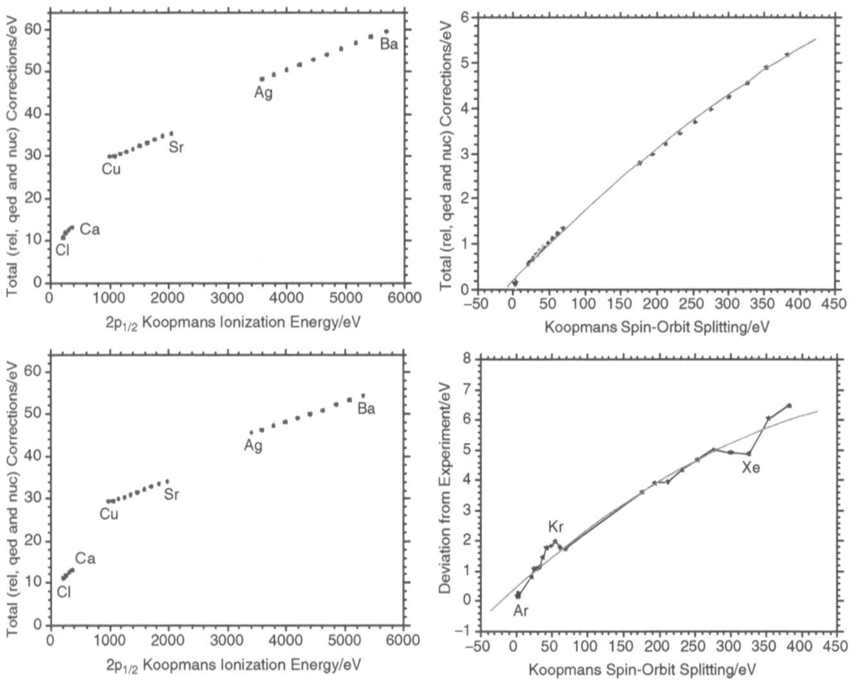

Figure 4. Total corrections to the Koopmans values, as functions of these same values, for the $2p_{1/2}$ (*upper left*) and $2p_{3/2}$ (*lower left*) ionization energies and the spin–orbit splitting (*upper right*) of atoms from Cl to Ba, excluding transition elements. There are also shown the deviations of the Koopmans values from measured ones (*lower right*). *Solid lines* are computed using the functional forms and best-fit parameters listed in Table 5.

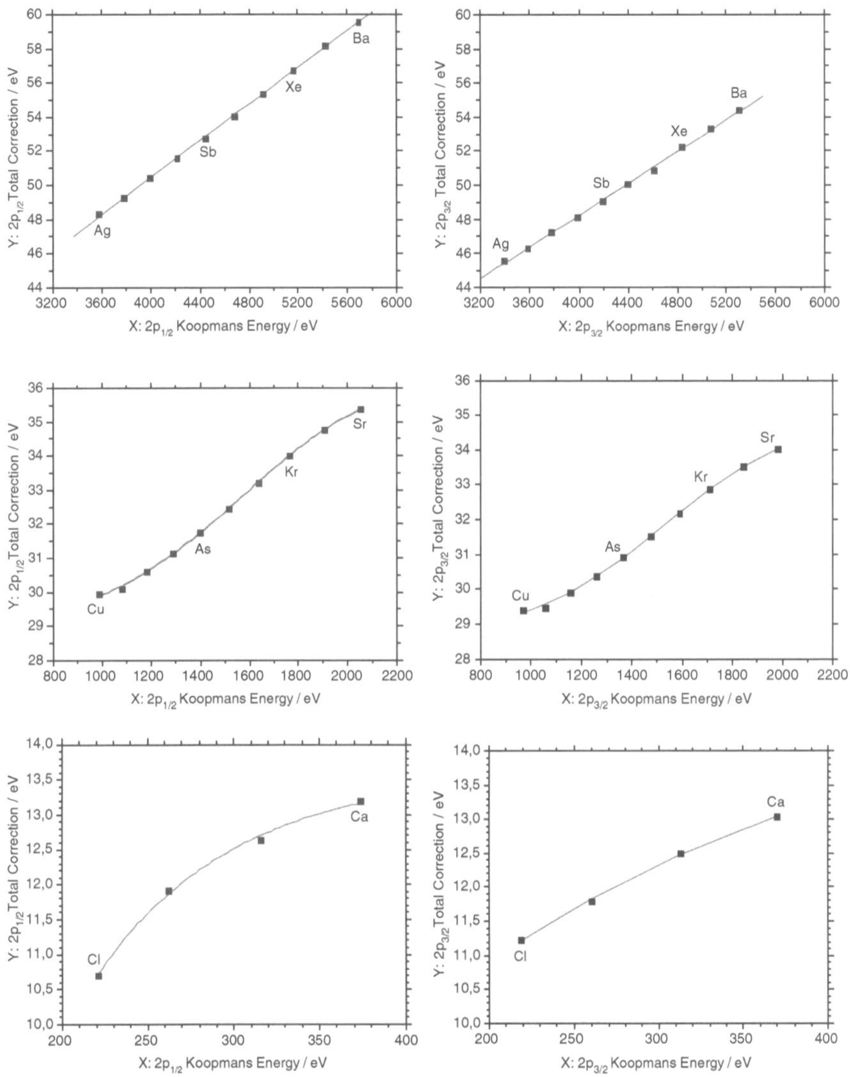

Figure 5. Total corrections (Y axis) to the Koopmans values as functions of these same values (X axis), for the $2p_{1/2}$ (*left*) and $2p_{3/2}$ (*right*) ionization energies of the homologous series of atoms Cl–Ca, Cu–Sr, and Ag–Ba. Elements corresponding to half or full filling of a (sub) shell are shown. *Solid lines* are computed using the functional forms and best-fit parameters listed in Table 5.

$2p_{3/2}$ ionization energies and spin–orbit splitting for the atoms in the investigated set. There are shown only the results obtained from the functions yielding the best fits (Eq. 3a–e). The allometric function appears less appropriate here than it was for relativity corrections to 1s- and 2s-core ionization energies [22]. There are displayed:

Table 5. Optimized parameters and standard deviations for best-fit functions representing corrections to Koopmans $2p_{1/2}$ and $2p_{3/2}$ ionization energies and spin–orbit splitting for the investigated atoms. There are shown the results obtained from the functions giving the best fits: linear, quadratic, exponential-decay, or sigmoid. The solid lines shown in Figures 4 and 5 are computed using these functions.

Level	Series	B/eV	Y_0/eV	X_0/eV	λ or C/eV	χ^2/eV^2	R^2/eV^2
$2p_{1/2}$	Cl-Ba (expdec)	−56.042 ±1.838	63.378 ±2.280		2555.827 ±273.354	3.762	0.9859
$2p_{1/2}$	Cl-Ca (expdec)	−42.419 ±19.759	13.619 ±0.265		82.305 ±16.502	0.008	0.9978
$2p_{1/2}$	Cu-Sr (sigmoid)	−7.959 ±0.274	36.760 ±0.307	1565.130 ±16.482	312.144 ±27.316	0.003	0.9995
$2p_{1/2}$	Ag-Ba (linear)	0.00538 ±0.00004	28.927 ±0.173				0.9998
$2p_{3/2}$	Cl-Ba (expdec)	−49.785 ±1.481	56.741 ±1.745		2152.075 ±215.109	3.233	0.9849
$2p_{3/2}$	Cl-Ca (expdec)	−9.725 ±0.582	15.190 ±0.928		244.940 ±76.881	0.001	0.9992
$2p_{3/2}$	Cu-Sr (sigmoid)	−6.487 ±0.234	35.041 ±0.264	1518.819 ±16.744	275.618 ±26.090	0.003	0.9993
$2p_{3/2}$	Ag-Ba (linear)	0.00468 ±0.00006	29.525 ±0.261				0.9994
Λ	Δ G/K (quadratic)	0.01617 ±0.00045	0.2087 ±0.0254		$-0.872*10^{-5}$ $\pm0.126*10^{-5}$	0.066	0.9986
Λ	Δ X/K (quadratic)	0.02091 ±0.00198	0.4654 ±0.1091		$-0.167*10^{-4}$ $\pm0.057*10^{-4}$	0.303	0.9791

1. for the $2p_{1/2}$ (line 2) and $2p_{3/2}$ (line 6) ionization energies of atoms from Cl to Ba, the overall parameters B, Y_0, and λ (and standard deviations) of exponential-decay best fits;

2. for the $2p_{1/2}$ (lines 3–5) and $2p_{3/2}$ (lines 8–9) ionization energies in the homologous series Cl–Ca, Cu–Sr, and Ag–Ba, the specific parameters (and standard deviations) corresponding to exponential-decay, sigmoid, and linear best fits, respectively (see Figure 5, *left and right*);

3. for the spin–orbit splitting of atoms from Cl to Ba, the parameters B, Y_0, and C (and standard deviations) of quadratic best fits of the variations of the G-corrections to Koopmans values as a function of these latter (see Figure 4, *upper right*) and of the X-deviations from them (see Figure 4, *lower right*).

The following observations can be made on the diagrams shown in Figures 4 and 5 and on the coefficients given in Table 5.

(i) The best overall fits for the $2p_{1/2}$ and $2p_{3/2}$ energies in the Cl–Ba range are obtained by using exponential-decay functions (Eq. 3d). However, the uncertainties are of the order of 3% for the parameters B and Y_0 and of 10% for the parameter λ, and the *msd* is much larger than 1 eV.

(ii) The best fits in the Cl–Ca series are obtained with exponential-decay functions, while in the Cu-Sr series one must use sigmoid functions (Eq. 3e) and in the Ag–Ba series simple linear functions (Eq. 3a) give the best fits.

(iii) The fits with linear functions are more accurate than those with sigmoid or exponential-decay functions. The series Cl–Ca, which involves only four elements, expectedly gives less reliable results than the series Cu–Sr or Ag–Ba, which involve ten elements. However, in all the fits with homologous elements the *msd* is much smaller than 1 eV.

(iv) The best overall fits for spin–orbit splitting in the Cl–Ba range, both for the G corrections and for the X deviations from K values, as functions of these values, are obtained by using quadratic functions (Eq. 3c). Although relative uncertainties on parameters C and Y_0 are rather large (while those on the linear factor B are 3–4 times smaller), the reliability of the fit is rather good even for the more irregular X deviations (Figure 4, *right diagrams*).

4. CONCLUSION

In this paper we have investigated the variations of the corrections to Koopmans 2p-core ionization energies in atoms as functions of the atomic number Z or of uncorrected Koopmans values. These latter were computed using Bruneau's numerical *ab initio*, relativistic program, in order to get consistent results for the spin–orbit splitting. Corrections involved electron relaxation obtained through a ΔDF procedure, inner-core correlation yielded by ground-state mixing, the second-order Breit terms, quantum-electrodynamics effects, and nuclear size and motion effects, introduced by perturbation. The present investigation differs from our previous ones [9] on 1s- and 2s-core ionization energies, where the stress was on the variations of the relativistic corrections and where nuclear effects were neglected.

The results described in the previous section provide a rational ground for using homothetic formulas in representing various corrections to uncorrected atomic calculations. Due to shell effects, such formulas ought to be derived, in principle, for homologous families of atoms. Contrary to the relativistic corrections to the 1s- and 2s-core ionization energies [9], the allometric formula does not show very relevant here, and various 2, 3, or 4-parameter functions have to be used to obtain best fits. The sets of optimized parameters gathered in Table 5 are those of the best functions representing the total corrections to the Koopmans $2p_{1/2}$ and $2p_{3/2}$ ionization energies and spin–orbit splitting.

The interest of the procedure described in this paper for incorporating specific corrections to approximate values resides in the possibility that our results may be transferable to complex molecules. Indeed, for heavy atoms there exists a variety of computer programs taking all the effects into account, especially the MCBDF and coupled-cluster four-component relativistic codes used in the study of superheavy elements [13]. In contrast, for complex molecules there are no available programs allowing computations of core-hole excitations including relativity, relaxation, and correlation together. This is the reason why we choose to fit the total corrections to the uncorrected energies rather than to the atomic number Z. Although these latter are related in the atomic case, in the molecular case shifts induced by the chemical environment of the ionized atom, mainly through the relative electronegativities of this latter and the bonded chemical groups [1, 7], make Z partly irrelevant.

As the relativistic and *qed* corrections essentially depend on the velocities of the involved electrons, which in turn critically depend on the charge transferred onto or from the core-ionized atom, and as the relaxation and correlation corrections also depend, in a different way, on charge transfer [6–8], one may conjecture that in a molecule all significant corrections to Koopmans energies depend on charge transfer about the ionized atom, and thus on chemically shifted ionization energies, in the same way as for a bare atom they depend on the uncorrected ionization energy.

Tests of this methodology on hemoglobin and phthalocyanine-like molecules are in progress.

Acknowledgements

One of the authors (J M) expresses his thanks to Pr Alexander Wang for hosting him at UBC, Vancouver, Canada, in July 2005, and to Pr Delano Chong for stimulating discussions about allometric fits. Dr Stéphane Carniato is acknowledged for his critical reading of the manuscript.

References

1. K. Siegbahn, C. Nordling, G. Johansson, *et al. ESCA Applied to Free Molec-ules* North-Holland, Amsterdam, 1971.
2. W.L. Jolly, K.D. Bomben, and C.J. Eyermann, Atomic Data and Nuclear Data Tables **31**, 433, 1984, and references therein.
3. S. Hüfner, in M. Cardona, P. Fulde, K. von Klitzing, and H.-J. Queisser (eds.), *Photoelectron Spectroscopy*, Series in Solid-State Physics **82**, Springer Verlag, Berlin, 1995, pp. 31 ff.
4. C.L. Pekeris, *Phys. Rev.* **112**, 1649, 1958; **115**, 1216, 1959; **126**, 1470, 1962, and references therein.
5. J. Rychlewski (ed.), *Explicitly Correlated Wave Functions in Chemistry and Physics*, Progress in Theoretical Chemistry and Physics **13**, Kluwer, Dordrecht, 2003, and references therein.
6. J. Maruani, M. Tronc, and C. Dezarnaud, *C. R. Acad. Sci.* Paris II **318**, 1191, 1994, and references therein.
7. A. Khoudir, J. Maruani, and M. Tronc, in A. Hernandez-Laguna, J. Maruani, R. McWeeny and S. Wilson (eds.), *Quantum Systems in Chemistry and Physics,* vol. 2, Progress in Theoretical Chemistry and Physics **3** Kluwer, Dordrecht, 2000, pp. 57–89.
8. J. Maruani, A. Khoudir, A. Kuleff, M. Tronc, G. Giorgi, and C. Bonnelle, *Adv. Quant. Chem.* **39**, 307, 2001.
9. J. Maruani, A.I. Kuleff, Ya.I. Delchev, and C. Bonnelle: (a) in E.J. Brändas and E.S. Kryachko (eds.), *The Fundamental World of Quantum Chemistry* Kluwer, Dordrecht, 2003, vol.1, pp. 639–656; (b) Israel *J. Chem.* **44** 2004, 71–82.
10. T. Koopmans, *Physica* **1**, 104, 1933. The proof given there holds for ionizations; for excitations, a proof is given in: R. Daudel, G. Leroy, D. Peeters, and M. Sana, *Quantum Chemistry* Wiley, New York, 1983, Section 4.2.
11. G.L. Malli (ed.), *Relativistic Effects in Atoms, Molecules and Solids*, NATO ASI Series B Physics Plenum Press, New York, 1981.
12. P. Pyykkö, *Relativistic Theory of Atoms and Molecules* Springer, Berlin: vol. 1 1986; vol.2 1993; vol3 2000, and web site: http://www.csc.fi/lul/rtam.
13. U. Kaldor and S. Wilson (ed.), *Theoretical Chemistry and Physics of Heavy and Superheavy Elements* (Progress in Theoretical Chemistry and Physics) **11**, Kluwer, Dordrecht, 2003.

14. (a) E. Clementi and A. Routh, *Int. J. Quant. Chem.* **6**, 525, 1972. (b) B. Lévy, Ph. Millié, J. Ridard, and J. Vinh, *J. Elec. Spectr. & Rel. Phen.* **4**, 13, 1974, and references therein.
15. H. Hsu, E.R. Davidson, and R.M. Pitzer, *J. Chem. Phys.* **65**, 609, 1976; A.R. Rossi and E.R. Davidson, *J. Chem. Phys.* **96**, 10682, 1992.
16. D. Firsht and R. McWeeny, *Molec. Phys.* **32**, 1637, 1976; C. Amovilli and R. McWeeny, in M. Defranceschi and Y. Ellinger (eds.), *Strategies and Applications in Quantum Chemistry* Kluwer, Dordrecht, 1996, p. 165.
17. C. Kozmutza, E. Kapuy, M.A. Robb, R. Daudel, and I.G. Csizmadia, *J. Comp. Chem.* **3**, 14, 1982; C. Kozmutza and E.T first, private communication.
18. (a) A. Filipponi, E. Bernieri, and S. Mobilio, *Phys. Rev. B* **38**, 3298, 1988. (b) S. Bodeur, P. Millié and I. Nenner, *Phys. Rev. A* **41**, 252, 1990.
19. H.J.Aa. Jensen, P. Jorgensen, and H. Agren, *J. Chem. Phys.* **87**, 451, 1987; H.J.Aa. Jensen, H. Agren and J. Olsen, in E. Clementi (ed.), *Modern Techniques in Computational Chemistry*, Escom, Leiden, 1990, chap. 8; see also: H. Agren and H.J.Aa. Jensen, *Chem. Phys.* **172**, 45, 1993.
20. D.P. Chong, *Chem. Phys. Lett.* **232**, 486, 1995; *J. Chem. Phys.* **103**, 1842, 1995; Can. *J. Chem.* **74**, 1005, 1996; D.P. Chong, C.H. Hu, and P. Duffy, *Chem. Phys. Lett.* **249**, 491, 1996; C. Bureau and D.P. Chong, *ibid.* **264**, 186, 1997.
21. S. Carniato and P. Millié, *J. Chem. Phys.* **116**, 3521, 2002; S. Carniato and Luo Yi, *J. Elec. Spectrosc. & Rel. Phenom.* **142**, 163, 2005.
22. J. Maruani, A.I. Kuleff, D.P. Chong, and C. Bonnelle, *Int. J. Quant. Chem.* **104**, 2005, 397–410.
23. B-G. Englert and J. Schwinger, *Phys. Rev. A* **32**, 26, 36, 47, 1985, and references therein.
24. Ya.I. Delchev, R.L. Pavlov, K.A. Pavlova, L.P. Marinova, and J. Maruani, *Int. J. Quant. Chem.* **52**, 1349, 1994.
25. (a) Y. Cauchois: *J. de Physique & Le Radium* **16**, 253, 1955. (b) A. Bearden and A.F. Burr: *Rev. Mod. Phys.* **39**, 125, 1967. (c) C.E. Moore, *Nat. Bur. of Stand. Ref.* Data Series **34**, 1970. (d) Gwyn P. Williams: *Rev. Sci. Instr.* **67**, 681, 1996, and data on web site: http://xdb.lbl.gov/Section1/Sec_1-1.html.
26. J. Bruneau, *J. Phys B* **16**, 4135, 1983; CEA/CEL MCDF Technical Report, 1995.
27. C. Bonnelle, P. Jonnard, C. Barré, G. Giorgi, and J. Bruneau, *Phys. Rev. A* **55**, 3422, 1997.
28. C. Bonnelle, *Electron distributions in heavy elements by x-ray and electron spectroscopy*, in U. Kaldor and S. Wilson (eds.), *Theoretical Chemistry and Physics of Heavy and Superheavy Elements* Kluwer, Dordrecht, 2003.

ON THE ROLE OF ELECTRONIC MOLECULAR STATES OF HIGH SPIN MULTIPLICITY

S. BEN YAGHLANE[1], A. BEN HOURIA[1], AND M. HOCHLAF[2]

[1] *Laboratoire de Spectroscopie Atomique et Moléculaire et Applications (LSAMA), Université de Tunis El-Manar, Tunis, Tunisia*
[2] *Groupe de Chimie Théorique, Université de Marne-la-Vallée, Champs-sur-Marne, F-77454, Marne-la-Vallée Cedex 2, France*

Abstract Several examples are presented in order to illustrate the crucial role of high spin multiplicity electronic states on the formation and decomposition of the corresponding molecular systems. For instance, these states are good candidates where electronically excited, metastable negative ions can be found. Moreover, they are needed in order to explain fully unimolecular and bimolecular reaction pathways. During these reactions, the importance of the couplings between these states, such as vibronic and Renner–Teller, and with the states of lower spin multiplicity, such as spin–orbit, are pointed out.

1. INTRODUCTION

In this review, we are considering the case of molecular species where the electronic ground state is of low spin multiplicity (for instance, singlet, doublet, or triplet) and where high spin multiplicity can be found in electronically excited states. These latter are less known: generally, they are hardly accessible by standard spectroscopic techniques. For example, when photoionizing a singlet neutral state only the doublet cationic states can be reached in the experimental spectra: higher spin multiplicities are not observed. Moreover, when the spin multiplicity increases it is hard to find deep potential wells for the corresponding electronic states, where conventional spectroscopic techniques can be used for their identification. Finally, the bound rovibrational levels of the electronic states of interest here are mostly located relatively far from the Franck–Condon regions reached by these techniques.

Our methodology is the following: large *ab initio* calculations are performed in order to compute the potential energy curves for these electronic states and their

S. Lahmar et al. (eds.), *Topics in the Theory of Chemical and Physical Systems*, 235–247.

spin–orbit coupling with the close lying electronic states. These calculations are performed using large basis sets (e.g. cc-pVQZ, cc-pV5Z Dunning's basis sets [1]) and interaction configuration methods, such as the complete active space self-consistent field (CASSCF) approach [2], followed by internally contracted multi-reference configuration interaction (MRCI) [3], and the coupled cluster approach with perturbative treatment of triple excitations (RCCSD(T)) [4]. These methods have the advantage of taking into account a large part of the electronic correlation, which is needed for better describing that kind of electronic states.

In the following, four examples will be investigated. The first one, considering the CS^- negative ion, illustrates that CS^- metastable electronically excited states can be formed only in the high angular momentum states. However, the low CS^- angular momentum states undergo rapid autoionization, hence reducing their lifetime. Then we are presenting the unimolecular decomposition of a diatomic ion ($SO^+(\tilde{b}\,^4\Sigma^-)$) and that of a triatomic ion ($CO_2{}^+(\tilde{C}\,^2\Sigma_g{}^+)$). The quartets (in both cases) and the sextets (for SO^+) are needed to fully describe the corresponding reactive processes. Finally, the $S^- + N \rightarrow SN^-$ bimolecular reactive processes are detailed in the last section, where the electronically excited bound quintet state of SN^- is found to be the main intermediate.

2. ELECTRONICALLY EXCITED METASTABLE NEGATIVE IONS: CS^-

Stable and metastable states have been characterized for several negatively charged atoms and molecules both in their ground and/or electronically excited states. Long-lived electronically excited molecular systems, where the anionic electronic ground state is not bound, do exist and have been characterized experimentally. A detailed discussion and a full presentation of the examples known are reviewed in Refs. [5,6]. In this section, we are treating the case of the CS^- anion.

Below the $C(^3P) + S(^3P)$ asymptote there are two asymptotes, $C^-(^4S) + S(^3P)$ followed by $C(^3P) + S^-(^2P)$, with bound states of the C^- and S^- atoms. The lowest doublets ($\tilde{X}\,^2\Pi$, $^2\Sigma^+$, $^2\Sigma^-$, $^2\Delta$) and quartets ($\tilde{a}\,^4\Sigma^-$, $\tilde{b}\,^4\Pi$, $\tilde{c}\,^4\Pi$, $\tilde{d}\,^4\Sigma^+$, $\tilde{e}\,^4\Delta$) and the two sextets ($\tilde{a}\,^6\Pi$ and $\tilde{b}\,^6\Sigma^+$) of the CS^- negative ion correlate to these two dissociation limits. Therefore all these states will be bound at long internuclear separation with respect to electron detachment, before reaching the molecular regions of the electronic states of the neutral CS, related to short-lived core excited resonances. Only the vibrational ground level of $CS^-(\tilde{X}\,^2\Pi)$ is lying below the autodetachment threshold, i.e. the energy of $CS(\tilde{X}\,^1\Sigma^+)v' = 0$ (cf. Figure 1). The $CS^-(\tilde{X}\,^2\Pi)v > 0$ levels can autodetach rapidly to $CS(\tilde{X}\,^1\Sigma^+)$. The $CS(\tilde{X}\,^1\Sigma^+)$ is also the parent state of the $CS^-(^2\Sigma^+)$ state. The $^2\Delta$ and $^2\Sigma^-$ anionic states are formed by binding an extra electron to the $CS(\tilde{a}\,^3\Pi)$ state. Our calculations show that these doublets are located, in the molecular region, above their respective parent states, thus reducing their lifetimes, and only their long range parts are predicted to be bound.

Hence, electronically excited metastable CS^- ions should be found only in the higher spin multiplicities. Three conditions should be fulfilled by these electronic states in order to exist: first, they should possess positive electron affinity with

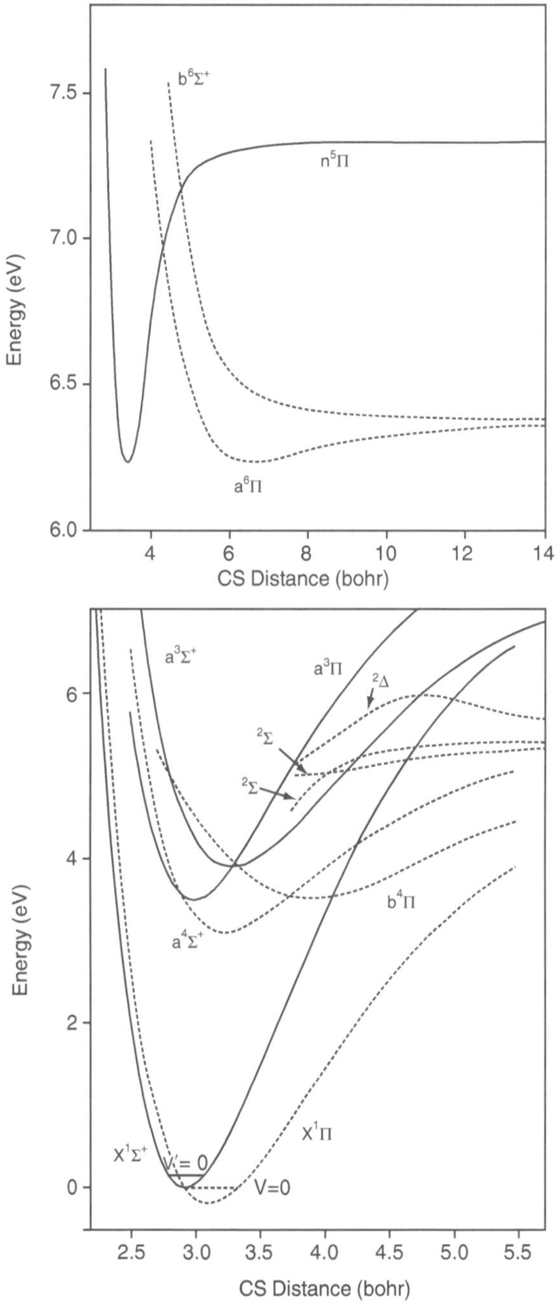

Figure 1. CASSCF-CI potential energy curves of CS($\tilde{X}^1\Sigma^+$, $\tilde{a}'^3\Sigma^+$, $\tilde{a}^3\Pi$, and $\tilde{a}^5\Pi$) and of CS$^-$ ($\tilde{X}^2\Pi$, $\tilde{a}^4\Sigma^-$, $\tilde{b}^4\Pi$, $^2\Sigma^+$, $^2\Sigma^-$, $^2\Delta$, $\tilde{a}^6\Pi$, and $\tilde{b}^6\Sigma^+$) along the CS distance. These curves are given relative to the equilibrium ground-state energy of CS (from [7]).

respect to their parent neutral state; secondly, they should exhibit slow depletion by spin-forbidden autodetachment for at least one fine-structure component and by radiative depletion; third, their wavefunctions should undergo weak interaction with the electron continuum wave.

The CS($\tilde{a}\,^3\Pi$) state has a leading $5\sigma^2 6\sigma^2 7\sigma^1 2\pi^4 3\pi^1$ configuration and is the parent state of the $\tilde{a}\,^4\Sigma^-$ state. The CS$^-$($\tilde{b}\,^4\Pi$) state has a leading configuration $5\sigma^2 6\sigma^2 7\sigma^2 2\pi^3 3\pi^2$ (for $R_{CS} > 1.6$ Å) by adding an electron to the CS($\tilde{a}'\,^3\Sigma^+$) state of $5\sigma^2 6\sigma^2 7\sigma^2 2\pi^3 3\pi^1$ leading configuration. Figure 1 shows that these two anionic states are lying definitely below their respective parent neutral states. For $\tilde{a}\,^4\Sigma^-$, however, the angular momentum of the fine-structure component (i.e. $J = 3/2$) has its counterpart in the low-lying CS$^-$($\tilde{X}\,^2\Pi$) state and will probably autodetach rapidly to the final electronic ground state of CS. Similarly, the CS$^-$($\tilde{b}\,^4\Pi$)$J = 1/2$ and 3/2 fine-structure components have their counterparts in the CS$^-$($\tilde{X}\,^2\Pi_{1/2,3/2}$) state. In contrast, the $\tilde{a}\,^4\Pi J = 5/2$ fine-structure component has no lower-lying counter-part and it lies below its parent state. Our calculations presented in Ref. [7] show that at least four vibrational states are expected to exist below CS($\tilde{a}'\,^3\Sigma^+$)$v' = 0$.

Long-lived CS$^-$ ions may also be formed at large internuclear distances in the shallow potential wells of CS$^-$($\tilde{a}\,^6\Pi$). Indeed, the sextet is located well below its parent neutral state [i.e. CS($\tilde{a}\,^5\Pi$)] at these distances. Here again, the $J = 7/2$ fine component of this sextet has no counterpart in the lower electronic states of the CS$^-$ anion. This potential well is due to polarization effects. More than ten vibrational states were calculated to be bound there.

3. UNIMOLECULAR DECOMPOSITION OF DIATOMICS: SPIN–ORBIT INDUCED PREDISSOCIATION OF SO$^+$($\tilde{b}\,^4\Sigma^-$) LEADING TO S$^+$(^4S_u) AND O(^3P_g)

The predissociative nature of the SO$^+$($\tilde{b}\,^4\Sigma^-$)$v \geq 7$ vibrational levels forming the S$^+$ and O species in their electronic ground states has been established 20 years ago in Cosby's SO$^+$ photofragmentation study [8]. Several attempts have been made in order to propose an explanation to this predissociation [9–11]. The proposed mechanisms are spin–orbit based induced predissociation pathways involving the repulsive $1\,^4\Sigma^+$, which leads directly to the S$^+$(^2P_u) + O(^3P_g) asymptote. However, the potential energy curves of the electronic states of SO$^+$ lying at these energy ranges and computed at the cc-pV5Z/MRCI level of theory (cf. Figure 2), show that the crossing between the $\tilde{b}\,^4\Sigma^-$ and the $1\,^4\Sigma^+$ states is occurring close to $\tilde{b}\,v = 9$ level, i.e. far from SO$^+$($\tilde{b}\,^4\Sigma^-$, $v = 7$) corresponding to the onset of this predissociation. Ornellas and Borin [11] in their *ab initio* investigations of the quartet states of SO$^+$, have suggested the involvement of the $2\,^4\Pi$ state crossing the $\tilde{b}\,^4\Sigma^-$ electronic state close to $v = 13$. Finally, Bissantz *et al.* [10] have confirmed the rapid predissociation of the $\tilde{b}\,^4\Sigma^-$($v \geq 13$) levels already stressed out by Cosby based on their later photofragment spectroscopic study. The analysis of the corresponding data has permitted to deduce the shape of the “predissociating” electronic state, which has been found to be naturally repulsive. Bissantz *et al.* have wrongly identified this state to $1\,^4\Sigma^+$.

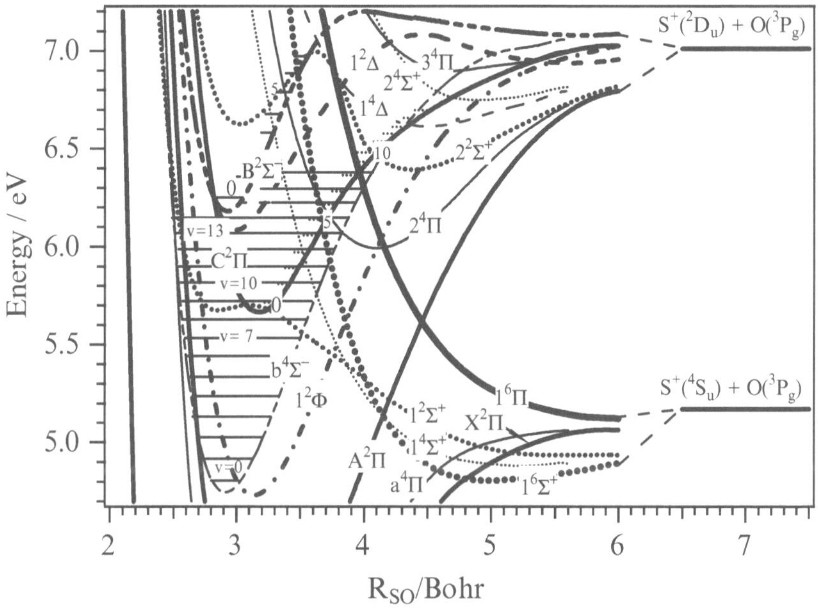

Figure 2. MRCI potential energy curves of the electronic states of SO^+ correlating to $S^+(^4S_u) + O(^3P_g)$ and $S^+(^2D_u) + O(^3P_g)$ dissociation limits. These states are positioned relative to the minimum of $SO^+(\tilde{X}\,^2\Pi)$ together with the vibrational levels of $\tilde{b}\,^4\Sigma^-$, $\tilde{C}\,^2\Pi$ and $\tilde{B}\,^2\Sigma^-$ states (see [12] for more details).

Recently, Ornellas and Borin found that the short range part of this experimental potential curve is coinciding with the short range potential curve of $2^4\Pi$. No further details are given in the literature about the nature of the long range part of this experimentally determined curve. Moreover, the predissociation onset at $\tilde{b}$ v = 7 remained unexplained. In Ref. [12] a detailed discussion is given on the predissociation mechanisms proposed in light of the potential energy curves of Figure 2 and their mutual spin–orbit coupling functions. In the present review, only the role of the $1\,^6\Sigma^+$ and $1\,^6\Pi$ states will be enlightened.

Figure 2 shows that the $\tilde{b}$ v $\geq$ 16 vibrational levels are located above the crossing of $\tilde{b}\,^4\Sigma^-$ with $1\,^6\Pi$ state. Hence, they may be predissociated via this sextet. Indeed, the spin–orbit conversion of the $SO^+(\tilde{b}\,^4\Sigma^-)$ ions is allowed: the $i < \tilde{b}\,^4\Sigma$, $m_s = 3/2|\mathbf{L_x S_x}|1^6\Pi_x$, $m_s = 5/2 >$ integral is calculated to be $-122\,\mathrm{cm}^{-1}$ at $R_{SO} \sim 3.8$ B, corresponding to the crossing of the quartet and the sextet. This coupling is high enough to permit such a conversion. Then, the $SO^+\tilde{b} \rightarrow S^+ + O$ reaction follows the potential energy curve of the repulsive $1^6\Pi$ state. However, the direct participation of the dissociative $1\,^6\Sigma^+$ state should be ruled out here despite its crossing of the $\tilde{b}$ state close to v $\sim$ 11. At these internuclear ranges, the $i < \tilde{b}\,^4\Sigma^-$, $m_s = 3/2|L_z S_z|1\,^6\Sigma^+$, $m_s = 5/2 >$ integral is computed to be close to zero (cf. Ref. [12]).

Figure 2 reveals that the $1^6\Pi$ state is crossing the $2^4\Pi$ state, which has already been proposed by Ornellas and Borin [11] as the starting point for the predissociation of the $SO^+(\tilde{b}\,^4\Sigma^-, v \geq 13)$ levels. The $SO^+(2^4\Pi)$ ions hence prepared may be predissociated later by the $1^6\Pi$ state after spin–orbit interactions. Indeed, the $i < 2^4\Pi_y, m_s = 3/2|\mathbf{L_zS_z}|1^6\Pi_x, m_s = 3/2 >$ spin–orbit integral is computed to be $21.5\,cm^{-1}$ at this crossing ($R_{SO} \sim 4.17$ Bohr), allowing efficient conversion. Moreover, the experimentally determined potential energy curve is found corresponding to that of the $1^6\Pi$ for long range internuclear separations. Similarly, the $1^6\Sigma^+$ is crossing the $2^4\Pi$ for $R_{SO} \sim 3.63$ Bohr, where the $i < 2^4\Pi_y$, $m_s = 3/2|\mathbf{L_xS_x}|1^6\Sigma^+, m_s = 5/2 >$ integral is evaluated to be $135.5\,cm^{-1}$. Accordingly, the $SO^+(2^4\Pi) \rightarrow SO^+(1\,^6\Sigma^+) \rightarrow S^+ + O$ reactive pathway can also be proposed. The involvement of the sextet states in the predissociation of $SO^+(\tilde{b}\,^4\Sigma^-)$ has never been considered before, and not yet checked for the predissociation of the important isovalent O_2^+ cation.

4. UNIMOLECULAR DECAY PATHWAYS OF A TRIATOMIC MOLECULAR SPECIES: THE CASE OF THE $CO_2^+(\tilde{C}^2\Sigma_g{}^+)$ VIBRONIC LEVELS

The ground vibrational level of the quartet $CO_2{}^+(\tilde{C}\,^2\Sigma_g{}^+)$ is located above the first dissociation limit $\{O^+(^4S_u) + CO(\tilde{X}\,^1\Sigma^+)\}$. The $CO_2{}^+(\tilde{C}\,^2\Sigma_g{}^+)$ excited vibrational levels are positioned above the first and second asymptotes $\{CO^+(\tilde{X}\,^2\Sigma^+)+O(^3P_g)\}$. Hence they are expected to be predissociated to these limits. The dissociative photoionization of CO_2 has been extensively studied experimentally [13]. These experimental works have provided evidence that $CO_2{}^+(\tilde{C}\,^2\Sigma_g{}^+)$ selected in its ground vibrational level dissociates completely to $O^+(^4S) + CO(\tilde{X}\,^1\Sigma^+, v'' = 0, 1)$ and that the excited vibrational bands of the $\tilde{C}$ state are also fully predissociated, forming both the O^+ and the CO^+ products. The direct formation of $O^+(^4S)+CO(\tilde{X}\,^1\Sigma^+)$ by dissociation of $CO_2{}^+(\tilde{C}\,^2\Sigma_g{}^+)$ is a spin-forbidden process according to the Wigner–Witmer correlation rules [14]. The involvement of a quartet state was first suggested in the theoretical work of Praet *et al.* [15]. This quartet either dissociates to $O^+(^4S) + CO(\tilde{X}\,^1\Sigma^+)$ or undergoes a fast intersystem crossing to the ground state $CO_2{}^+(\tilde{X}^2\Pi_g)$ and then dissociates to $CO^+(\tilde{X}\,^2\Sigma^+)+O(^3P)$. However, the role of this high spin multiplicity electronic state during the decomposition of $CO_2{}^+(\tilde{C}\,^2\Sigma_g{}^+)$ ions was not fully explained. Therefore, we have performed a high-level *ab initio* study on the potential energy functions (PEF) of $CO_2{}^+(\tilde{C}\,^2\Sigma_g{}^+)$ and of the lower doublet and quartet states of $CO_2{}^+$, along with their spin–orbit interactions. These calculations were performed at the CASSCF/MRCI level, using the *spdf* cc-pVQZ subset basis of Dunning [1]. The results are depicted in Figures 3–5. These potential curves provide insight into the possible predissociation pathways of excited $CO_2{}^+$ prepared in the internal energy range of 5.2–6.2 eV.

Close examination of Figure 3 reveals that only the repulsive $\tilde{a}\,^4\Sigma^-$ state correlates adiabatically to the lowest dissociation limit $O^+(^4S) + CO(\tilde{X}\,^1\Sigma^+)$. As a result, each

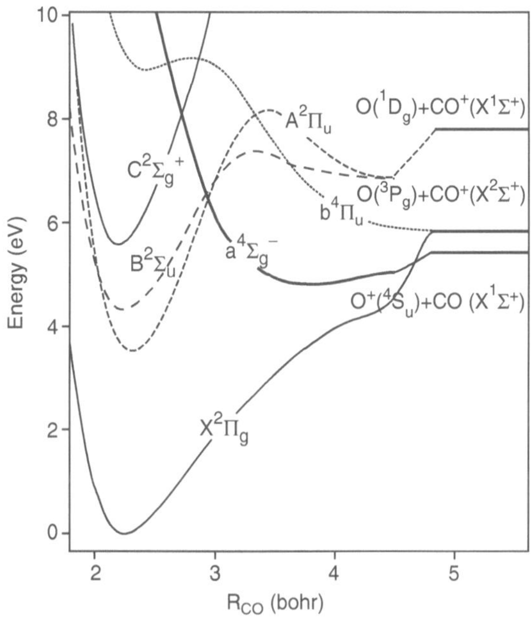

Figure 3. Collinear CASSCF one-dimensional cuts of the three-dimensional potential energy functions for the doublet and quartet states of $CO_2{}^+$ possibly involved in the predissociation of $CO_2{}^+(\tilde{C}\,{}^2\Sigma_g{}^+)$. The other R_{CO} distance is set to the equilibrium geometry of the neutral molecule (2.2 Bohr). The energies of the dissociation limits and electronic states are shifted to known experimental values. Strictly speaking, the g-u symmetry is only applicable for $R_{CO} = 2.2$ Bohr (from [13]).

predissociation process forming the O^+ ions in their electronic ground states should involve this quartet.

The $\tilde{b}\,{}^4\Pi_u$ state correlates at long range to the second dissociation limit, forming the $CO^+\tilde{X}$ ions. Figure 3 shows that these two quartets cross the lower doublet states for collinear configurations, where spin–orbit couplings can take place. By bending the $CO_2{}^+$ ion, the $\tilde{b}\,{}^4\Pi_u$ state splits into two components, due to the well-known Renner–Teller effect and that this quartet possesses a strongly bent component (the ${}^4B_1$), which also crosses the lower electronic states of $CO_2{}^+$ for bent geometries (cf. Figure 4). In contrast to the example detailed in the previous section, the unimolecular decomposition of the $CO_2{}^+$ triatomic ions may occur for both linear and bent structures. Generally the selection rules are relaxed for non-linear geometries where additional couplings may take place, favouring the mixings between these electronic states.

Let us explicit the role of the $\tilde{a}\,{}^4\Sigma_u{}^-$ and $\tilde{b}\,{}^4\Pi_u$ states during the formation of the two lowest dissociation channels of interest. Despite the fact that $\tilde{a}\,{}^4\Sigma^-$ is crossing the $\tilde{C}$ state (Figure 3), the direct predissociation of $CO_2{}^+(\tilde{C}\,{}^2\Sigma_g{}^+)$ ions via the $\tilde{a}\,{}^4\Sigma_g{}^-$ state is not possible, because the spin–orbit coupling integral for these states, $< \tilde{a}\,{}^4\Sigma_g{}^-, m_s = 3/2|\mathbf{L}.\mathbf{S}|\tilde{C}\,{}^2\Sigma_g{}^+, m_s = 1/2 >$, gives no contribution for $\Delta m_s = 1$.

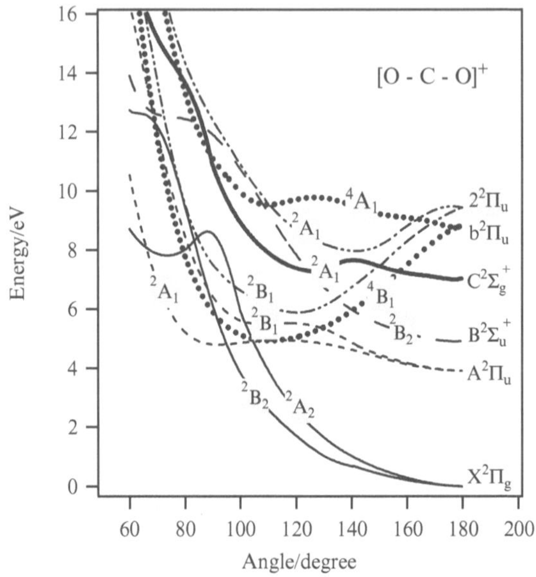

Figure 4. CASSCF potential energy curves of the quartet states of CO_2^+ along the bending coordinate. Both R_{CO} distances are fixed at 2.2 Bohr. These curves are given with respect to the potential minimum of the ground $CO_2^+(\tilde{X}\,^2\Pi_g)$ state. The energies of the electronic states have been shifted to the known experimental values (from [16]).

Nevertheless, multistep mechanisms can be proposed where both quartet states are the main intermediate in order to explain the formation of both O^+ and CO^+ ions. For the $\tilde{C}\,^2\Sigma_g{}^+(0,0,0)$ level undergoing predissociation only to the first dissociation limit $\{O^+(^4S_u) + CO(\tilde{X}\,^1\Sigma^+)\}$, Figures 3 and 5 show that the $\tilde{C}\,^2\Sigma_g{}^+(0,0,0)$ ions formed can be converted first into the $CO_2^+(\tilde{A}\,^2\Pi_u)$ ions after vibronic couplings, followed by spin–orbit interaction between the $CO_2^+(\tilde{A}\,^2\Pi_u)$ and $\tilde{a}\,^4\Sigma$-states, which leads directly to this asymptote. Readers are referred to Figures 3 and 5 and Table 1 for the evaluation of the corresponding spin–orbit couplings and the determination of the crossings between these states. The efficient dissociation pathway $CO_2^+(\tilde{C}\,^2\Sigma_g{}^+; 0,0,0) \rightarrow CO_2^+(\tilde{A}\,^2\Pi_u) \rightarrow CO_2^+(\tilde{a}\,^4\Sigma_g{}^-) \rightarrow O^+(^4S_u) + CO(\tilde{X}\,^1\Sigma^+)$ may account for the absence of fluorescence from $\tilde{C}\,^2\Sigma_g{}^+(0,0,0)$ to $A\,^2\Pi_u$, which has never been observed experimentally. Concerning the $\tilde{b}\,^4\Pi_u$ state, Figure 3 shows that for linear configurations this quartet is crossing the $\tilde{C}$ state for internal energies of ~ 9 eV. However, the bent component of this quartet, i.e. $^4B_1$, is crossing the $\tilde{X}\,^2\Pi_g$, $\tilde{A}\,^2\Pi_u$, $\tilde{B}\,^2\Sigma_u{}^+$, $\tilde{C}\,^2\Sigma_g{}^+$ states of CO_2^+ for bent geometries (cf. Figure 4) at lower internal energies (even lower than the energy of $CO_2^+(\tilde{C}\,^2\Sigma_g{}^+, 0,0,0)$. Accordingly, a second multistep reactive pathway can be proposed via the $\tilde{b}\,^4\Pi_u$ state. First, the $CO_2^+(\tilde{C}\,^2\Sigma_g{}^+)$ vibrational levels can be predissociated by this quartet since the spin–orbit coupling between the doublet and the

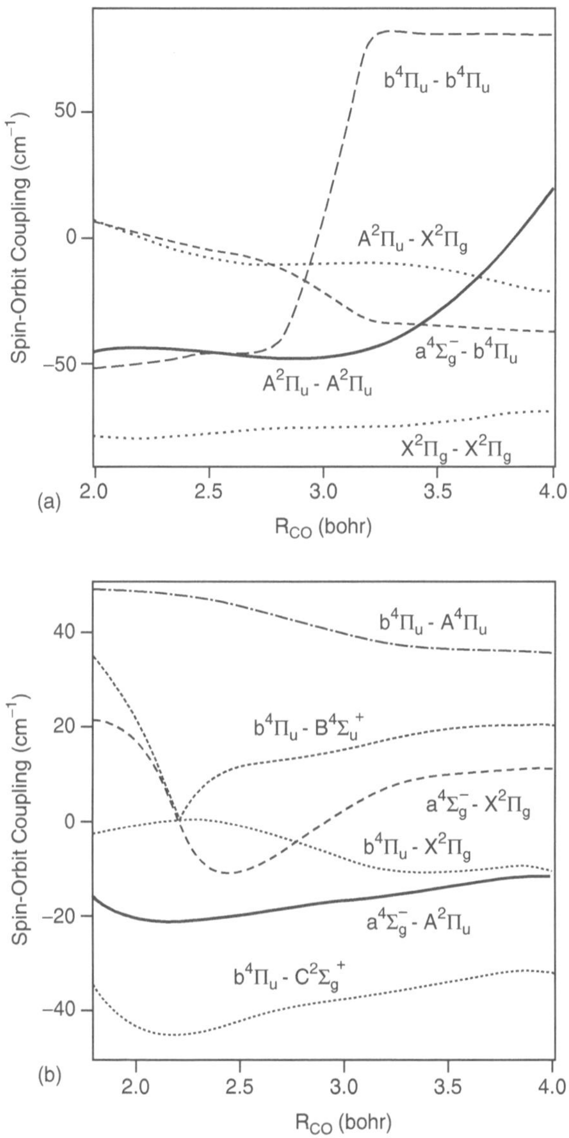

Figure 5. CASSCF collinear evolution of spin–orbit couplings between: (a) doublet–doublet, quartet–quartet and (b) doublet–quartet electronic states of CO_2^+, where the other CO distance is kept fixed at the equilibrium geometry of the neutral molecule (i.e. 2.2 Bohr). See Table 1 for the definition of these terms. Strictly speaking, the g–u symmetry is only applicable for $R_{CO} = 2.2$ Bohr.

Table 1. spin–orbit integrals between doublet–doublet, quartet–quartet, and doublet–quartet states of $CO_2{}^+$. Between parentheses there are given the schematic representations used in Figure 5.

$i < \tilde{X}\,{}^2\Pi_{g,x}, m_s = 1/2 \vert \mathbf{L_z S_z} \vert \tilde{X}\,{}^2\Pi_{g,y}, m_s = 1/2 >= (\tilde{X}\,{}^2\Pi_g - \tilde{X}\,{}^2\Pi_g)$
$i < \tilde{A}\,{}^2\Pi_{u,x}, m_s = 1/2 \vert \mathbf{L_z S_z} \vert \tilde{A}\,{}^2\Pi_{u,y}, m_s = 1/2 >= (\tilde{A}\,{}^2\Pi_u - \tilde{A}\,{}^2\Pi_u)$
$i < \tilde{b}\,{}^4\Pi_{u,x}, m_s = 1/2 \vert \mathbf{L_x S_x} \vert \tilde{a}\,{}^4\Sigma_g{}^-, m_s = 3/2 >= (\tilde{b}\,{}^4\Pi_u - \tilde{a}\,{}^4\Sigma_g{}^-)$
$i < \tilde{A}\,{}^2\Pi_{u,x}, m_s = 1/2 \vert \mathbf{L_z S_z} \vert \tilde{X}\,{}^2\Pi_{g,y}, m_s = 1/2 >= (\tilde{A}\,{}^2\Pi_u - \tilde{X}\,{}^2\Pi_g)$
$i < \tilde{b}\,{}^4\Pi_{u,x}, m_s = 3/2 \vert \mathbf{L_z S_z} \vert \tilde{b}\,{}^4\Pi_{u,y}, m_s = 3/2 >= (\tilde{b}\,{}^4\Pi_u - \tilde{b}\,{}^4\Pi_u)$
$i < \tilde{C}\,{}^2\Sigma_g{}^+, m_s = 1/2 \vert \mathbf{L_y S_y} \vert \tilde{b}\,{}^4\Pi_{u,x}, m_s = 3/2 >= (\tilde{C}\,{}^2\Sigma_g{}^+ - \tilde{b}\,{}^4\Pi_u)$
$i < \tilde{A}\,{}^2\Pi_{u,x}, m_s = 1/2 \vert \mathbf{L_x S_x} \vert \tilde{a}\,{}^4\Sigma_g{}^-, m_s = 3/2 >= (\tilde{A}\,{}^2\Pi_u - \tilde{a}\,{}^4\Sigma_g{}^-)$
$i < \tilde{b}\,{}^4\Pi_{u,x}, m_s = 1/2 \vert \mathbf{L_z S_z} \vert \tilde{X}\,{}^2\Pi_{g,y}, m_s = 1/2 >= (\tilde{b}\,{}^4\Pi_u - \tilde{X}\,{}^2\Pi_g)$
$i < \tilde{a}\,{}^4\Sigma_g{}^-, m_s = 3/2 \vert \mathbf{L_x S_x} \vert \tilde{X}\,{}^2\Pi_{g,x}, m_s = 1/2 >= (\tilde{a}\,{}^4\Sigma_g{}^- - \tilde{X}\,{}^2\Pi_g)$
$i < \tilde{B}\,{}^2\Sigma_u{}^+, m_s = 1/2 \vert \mathbf{L_y S_y} \vert \tilde{b}\,{}^4\Pi_{u,x}, m_s = 3/2 >= (\tilde{B}\,{}^2\Sigma_u{}^+ - \tilde{b}\,{}^4\Pi_u)$
$i < \tilde{A}\,{}^2\Pi_{u,y}, m_s = 1/2 \vert \mathbf{L_z S_z} \vert \tilde{b}\,{}^4\Pi_{u,x}, m_s = 1/2 >= (\tilde{A}\,{}^2\Pi_u - \tilde{b}\,{}^4\Pi_u)$

quartet ($\sim$40 cm^{-1} at their crossing, cf. Figures 3 and 5) is high enough to allow such conversion. The $CO_2{}^+(\tilde{b}\,{}^4\Pi_u)$ ions can also be obtained after spin–orbit conversion of the $CO_2{}^+(\tilde{A}\,{}^2\Pi_u)$ ions formed in the mechanism given above. At large internuclear distances, the $CO_2{}^+(\tilde{b}\,{}^4\Pi_u)$ ions may form directly the CO^+ ions, because $\tilde{b}\,{}^4\Pi_u$ correlates adiabatically to the second limit, or be predissociated by $\tilde{a}\,{}^4\Sigma_g{}^-$ after spin–orbit coupling between both quartets, forming the O^+ ions in their electronic ground state. Finally, the $\tilde{b}\,{}^4\Pi_u$ state, via the bent ${}^4B_1$ component, can undergo an intersystem crossing to the ground $\tilde{X}\,{}^2\Pi_g$ state, by more efficient spin–orbit interaction [15]. It is worth noting that, close to the barrier to linearity of the $\tilde{b}\,{}^4\Pi_u$ state, the Renner–Teller effect in this quartet should also play a role, mixing rather more the wavefunctions of the electronic states involved during the mechanisms proposed above.

The mechanisms suggested here have been confirmed, using state-of-the-art experimental techniques where the dynamics of state-selected vibronic levels of the $\tilde{C}$ state were investigated. More details can be found in Ref. [13].

5. BIMOLECULAR REACTIVE COLLISIONS BETWEEN $S^-({}^2P_u)$ AND $N({}^4S_u)$

In Figure 6 there are depicted the *spdfg* aug-cc-pV5Z/MRCI+Q potential energy curves of the $SN(\tilde{X}\,{}^2\Pi$ and $\tilde{a}\,{}^4\Pi)$ electronic states and those of the anionic states correlating to the unique bound asymptote of $SN^-[S^-({}^2P_g) + N({}^4S_u)]$. In addition to the bound electronic states of SN^- (i.e. $\tilde{X}\,{}^3\Sigma^-$ and ${}^1\Delta$) already known, our large MRCI+Q computations reveal that the ${}^3\Delta$ state is lying energetically below its quartet parent neutral state ($\tilde{a}\,{}^4\Pi$). The depletion of the J = 3 component of $SN^-({}^3\Delta)$ will mainly occur via weak interactions with the electron continuum wave. At large internuclear distances, $SN^-({}^5\Pi)$ state is predicted to possess a shallow polarization minimum supporting long-lived SN^- ions. Further details can be found in Ref. [17]. In the following, we concentrate on the role of the ${}^5\Pi$ state in the formation of this negative ion from the S^- and N species taken in their ground electronic state. These

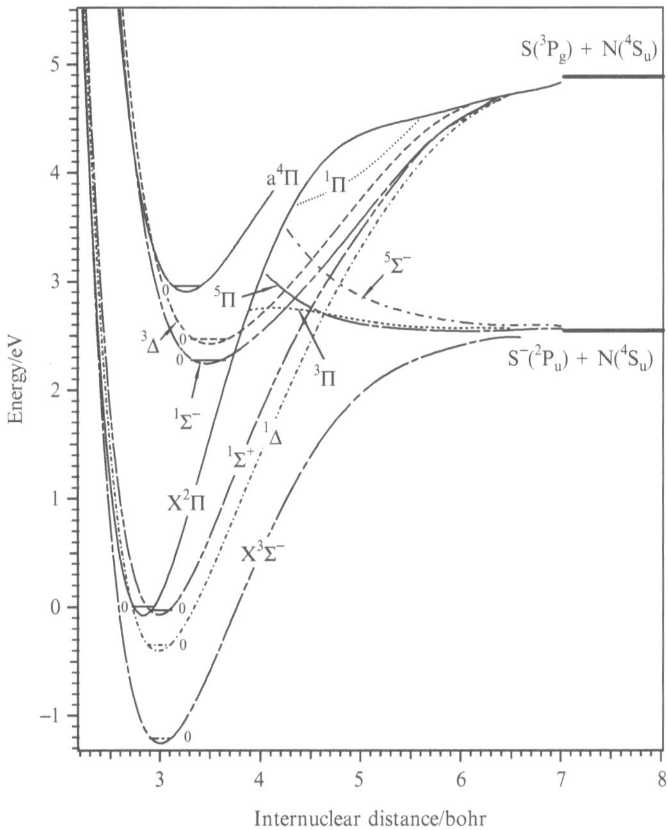

Figure 6. MRCI+Q potential energy curves (PECs) for the electronic states of SN^- together with those for SN ($\tilde{X}\,^2\Pi$ and $\tilde{a}\,^4\Pi$) (solid lines). These curves are given with respect to the energy of SN ($\tilde{X}\,^2\Pi$)v$' =$ 0. The *ab initio* computed anionic PECs were shifted by -0.162 eV to match the experimental EA of SN^- ($\tilde{X}\,^3\Sigma^-$). The horizontal lines correspond to the positions of the ground vibrational levels (from [17]).

reactive processes are occurring along the potential energy curves of SN^- presented in Figure 6 and their mutual interactions. Full discussion of all reactive pathways can be found in Ref. [17].

When the S^- and N in their electronic ground states are colliding together, the low vibrational states in the shallow minimum of SN^-($^5\Pi$) are formed directly. The relative stability of the corresponding rovibrational levels may contribute to the efficiency of this reaction. Hence, electronically excited SN^- ions are surely formed in this high spin multiplicity electronic state. The quintet may also be involved as an intermediate state to form the SN^- ions in the bound $^1\Delta$, $^1\Sigma^+$ states and in the metastable $^3\Delta$, J = 3 component via the $^3\Pi$ state. Indeed, the formation of SN^-($^1\Delta$, $^1\Sigma^+$, $^3\Delta$) ions can take place first along the $^3\Pi$ potential energy curve at long internuclear separations. In the molecular region, the SN^-($^3\Pi$) ions may be converted either to $SN^-(^1\Delta, ^1\Sigma^+, ^3\Delta)$ ions at the crossings between $^3\Pi$ and these electronic states directly, or following spin–orbit conversion of the $SN^-(^5\Pi)$ ions formed after reaction between the $S^-(^2P_u)$ and $N(^4S_u)$, as illustrated above. Indeed, the quintet is crossing the triplet for $R_{SN} \sim 4.67$ Bohr, where the $i < {^3\Pi_y}; m_s = 1|\mathbf{L_z S_z}|{^5\Pi_x}; m_s = 1 >$ integral is evaluated to be $\sim$126.1 cm^{-1}, which is high enough to allow this conversion. These reactive pathways are permitted by spin–orbit selection rules. One is referred to Ref. [17] for the values of the corresponding integrals and for the identification of the respective crossings.

However, the $^5\Sigma^-$ state is viewed to play a minor role here in spite of the fact that it crosses the $(^1\Delta, ^1\Sigma^+, ^3\Delta)$ potential energy curves (Figure 6). The off-diagonal spin–orbit coupling terms between this quintet and these electronic states are computed to be equal to zero (for $^1\Delta, ^1\Sigma^+$) or close to zero (for $^3\Delta$) [17].

6. CONCLUSION

The dynamics of the reactions presented in this review are found to be largely determined by the pattern of curve crossings of the relevant electronic states and their mutual couplings. We have treated examples of positively and negatively charged molecular species. It is worth noting that similar processes can be also found for the neutral molecules [14]. Generally, when the system size increases the symmetry restrictions are relaxed and additional couplings may show up, such as vibronic couplings, Renner–Teller effects and intramolecular isomerization processes involving these high spin multiplicities, together with spin–orbit coupling. They are expected to mix further the wavefunctions of the electronic states having the same and /or different spin multiplicities, complicating even more the reactive pathways. Hence, the lifetimes of corresponding rovibrational levels are reduced.

Acknowledgements

M.H. would like to thank the NERSC (UC Berkeley, USA) for computation time, and the people working with him on these topics for their interest.

References

1. T.H. Dunning, *J. Chem. Phys.* **90**, 1007, 1989. D.E. Woon and T.H. Dunning, Jr., *J. Chem. Phys.* **98**, 1358, 1993.
2. P.J. Knowles, and H.-J. Werner, *Chem. Phys. Lett.* **115**, 259, 1985.
3. H.-J. Werner, and P.J. Knowles, *J. Chem. Phys.* **89**, 5803, 1988. P.J. Knowles, and H.-J. Werner, *Chem. Phys. Lett.* **145**, 514, 1988.
4. C. Hampel, K.A. Peterson, and H.-J. Werner, *Chem. Phys. Lett.* **190**, 1, 1992.
5. T. Sommerfeld, and L.S. Cederbaum, *Phys. Rev. Lett.* 80, 3220, 1998.
6. A. Dreuw, T. Sommerfeld, and L.S. Cederbaum, *J. Chem. Phys.* **116**, 6039, 2002 and references therein.
7. M. Hochlaf, G. Chambaud, P. Romsus, T. Andersen, and H.J. Werner, *J. Chem. Phys.* **110**, 11835, 1999 and references therein.

8. P.C. Cosby, *J. Chem. Phys.* **81**, 1102, 1984.
9. L. Andrić, H. Bissantz, and F. Linder, *Z. Phys. D* **13**, 147, 1989.
10. H. Bissantz, L. Andrić, Ch. Hertzler, H.-J. Foth, and F. Linder, *Z. Phys. D* **22**, 727, 1992.
11. F.R. Ornellas, and A.C. Borin, *J. Chem. Phys.* **109**, 2202, 1998.
12. A. Ben. Houria, Z. Ben Lakhdar, and M. Hochlaf, *J. Chem. Phys.* 2006 in press.
13. J. Liu, W. Chen, M. Hochlaf, X. Qian, C. Chang, and C.-Y. Ng, *J. Chem. Phys.* **118**, 149, 2003.
14. G. Herzberg, *Molecular Spectra and Molecular Structure III, Electronic Spectra and Electronic Structure of Polyatomic Molecules,* van Nostrand, Princeton, 1966.
15. M.T. Praet, J.C. Lorquet, and G. Raseev, *J. Chem. Phys.* **77**, 4611, 1982.
16. R. Polák, M. Hochlaf, M. Levinas, G. Chambaud, and P. Rosmus, *Spectrochim. Acta* A 55, 447, 1999.
17. S. Ben Yaghlane, S. Lahmar, Z. Ben Lakhdar, and M. Hochlaf, *J. Phys. B* **38**, 3395, 2005.

AB INITIO CHARACTERIZATION OF ELECTRONICALLY EXCITED METASTABLE STATES OF S_2^-

B. EDHAY[1], S. LAHMAR[1], Z. BEN LAKHDAR[1], AND M. HOCHLAF[2]

[1] *Laboratoire de Spectroscopie Atomique et Moléculaire et Applications (LSAMA), Université de Tunis El-Manar, Tunis, Tunisia*
[2] *Groupe de Chimie Théorique, Université de Marne-la-Vallée, Champs-sur-Marne, F-77454, Marne-la-Vallée Cedex 2, France*

Abstract Accurate *ab initio* calculations of the potential energy curves (PEC) of the electronic states of the neutral S_2 molecule and the S_2^- negative ion, correlating to the bound $S^-(^2P_u) + S(^3P_g)$ asymptote, reveal that the depletion of the $J = 7/2$ component of the $S_2^-(1^4\Delta_g)$ state occurs mainly via weak interactions with the electron continuum wave. This quartet is found to possess several long-lived rovibrational levels which are located below the $S_2(c^1\Sigma_u^-)$ parent neutral state. Those levels may be populated, at least, during low-energy collisions between $S^-(^2P_u)$ and $S(^3P_g)$.

1. INTRODUCTION

Several conditions must be fulfilled for an anionic electronic state to exist: (i) it should possess positive electron affinity with respect to its parent neutral state; (ii) it should exhibit slow depletion by spin-forbidden autodetachment for at least one fine-structure component and by radiative depletion; and (iii) its wave-function should undergo weak interaction with the electron continuum wave. Such stable and metastable states have been identified for several negatively charged atoms and molecules, in both ground and electronically excited states. Long-lived electronically excited molecular systems, where the anionic ground state is not bound, do exist and have been observed experimentally. For a detailed presentation of the examples already known is referred in Refs. [1–3].

The S_2^- anion possesses a unique bound asymptote $[S^-(^2P_u) + S(^3P_g)]$, located at ~2.08 eV below the neutral $S(^3P_g) + S(^3P_g)$ dissociation limit [4]. Photoelectron spectroscopy has allowed to deduce the electron affinity of disulfur (measured as 1.670 ± 0.015 eV [5–7]) and an estimate of the spectroscopic parameters for the

S. Lahmar et al. (eds.), Topics in the Theory of Chemical and Physical Systems, 249–259.

$S_2^-(X^2\Pi_g)$ ground state, including the equilibrium distance ($R_e = 2.005 \pm 0.015$ Å) and the harmonic frequency ($\omega_e = 570 \pm 100\,cm^{-1}$). The vibrational structure of $S_2^-(X^2\Pi_g)$ trapped in solid matrices has been also characterized by using Raman spectroscopy [8]. The optically-allowed $X^2\Pi_g \leftarrow A^2\Pi_u$ emission (at ~20000 cm^{-1} for the 0–0 transition) has been observed for S_2^- the ions doped in alkali iodides, chlorides, and bromides after laser excitation [9, 10]. The lifetime of the emitting $A^2\Pi_u$ rovibrational levels are probably lengthened in matrices since the expected gas phase rapid nonradiative autodetachment of the $J = 1/2$ and $J = 3/2$ fine components of the $A^2\Pi_u$ state to their corresponding X state counterparts is quenched there. Therefore, no experimental evidence was made previously for the existence of electronically excited metastable S_2^- ions in the gas phase. Theoretically and in addition to relatively small scale calculations on S_2^- at the DFT and MP2 levels of theory [11–13], extensive MRCI calculations on the lowest doublets and the first quartet state (i.e. $1^4\Sigma_u^-$) of S_2^- were performed by Heinemann *et al.* [14]. These authors have deduced the spectroscopic properties of these electronic states and their dipole allowed transition elements. Close examination of these theoretical results shows that only the $S_2^-(X^2\Pi_g$ both $J = 1/2$ & 3/2) components are expected to be long-lived in the doublets. However, nothing is known yet about the higher quartet states which may present metastable components.

Recently, metastable negative ions, such as SN^- and CS^-, have been predicted to exist in electronic states having high-spin multiplicity [1, 2, 15]. Such high-spin electronic states are good candidates for metastable negative ions. In the present work we will consider the case of the S_2^- anion by calculating the PEC for all its doublet and quartet electronic states correlating to the $S^-(^2P_u) + S(^3P_g)$ dissociation limit, together with the potentials of their S_2 respective parent states. These calculations are carried out at the highest achievable level of theory, by using highly correlated *ab initio* methods and a large basis set of aug-cc-pV5Z quality [16]. Reliable predictive properties for the quartet states of S_2^- can be deduced from these computations.

2. ELECTRONIC STRUCTURE CALCULATIONS: METHODS

The calculations were performed using state-of-the-art *ab initio* methods, including the complete active space self-consistent field (CASSCF) approach [17], followed by internally contracted multi-reference configuration interaction, and the Davidson correction (MRCI + Q) method [18, 19], both implemented in the MOLPRO program [20]. Sulfur atoms were described using a large basis set (*spdfgh* aug-cc-pV5Z [16]) comprising 262 contracted Gaussian functions. In these calculations, the CASSCF active space included all configurations (configuration state functions (CSF)) obtained after excitation of all valence electrons in valence orbitals. For MRCI calculations, all configurations in the CI expansion of the CASSCF wavefunctions were taken as a reference, resulting in more than 5×10^6 CSFs to be treated in the D_{2h} point group. All valence electrons were correlated. The nuclear motion problem was solved by using the method of Cooley [21]. The spectroscopic constants were computed from the derivatives of the potentials at their respective minima, and the spin–orbit couplings were evaluated over the CASSCF wavefunctions by using the one-electron term of the Breit–Pauli Hamiltonian as implemented in the MOLPRO program.

3. ELECTRONIC STRUCTURE CALCULATIONS: RESULTS

In Figure 1 there are depicted the MRCI + Q PEC for all 24 electronic states of S_2^- (*thin lines*) correlating to the bound asymptote $S^-(^2P_u) + S(^3P_g)$, together with those of the neutral S_2 ($X^3\Sigma_g^-$, $c^1\Sigma_u^-$, $B'^3\Pi_u$, $B''^3\Pi_g$, $1^5\Sigma_u^-$, $1^5\Delta_g$, $1^5\Pi_u$, and

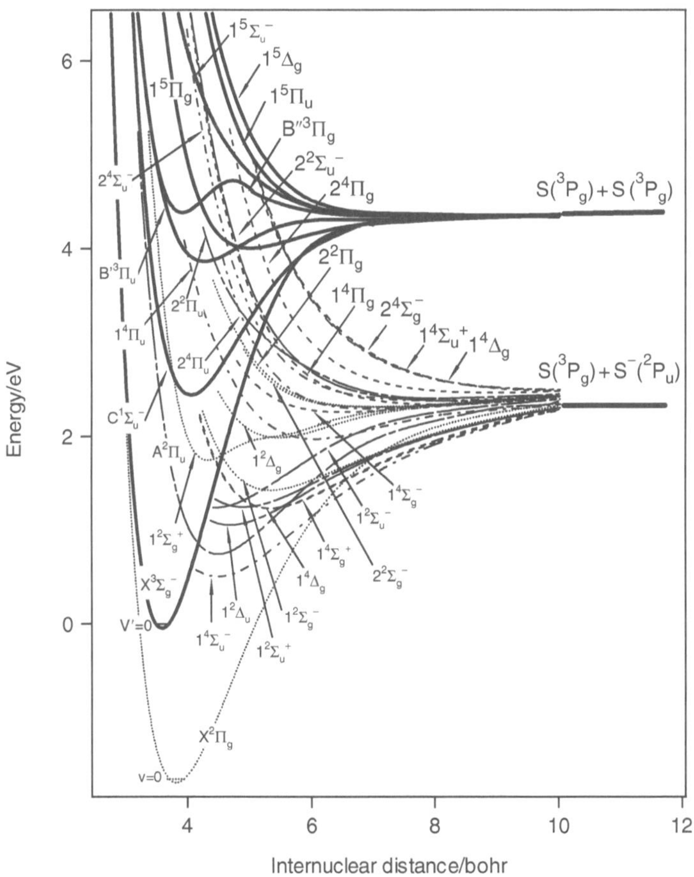

Figure 1. MRCI + Q PEC of the electronic states of S_2^- (*thin lines*) together with those of S_2 ($X^3\Sigma_g^-$, $c^1\Sigma_u^-$, $B'^3\Pi_u$, $B''^3\Pi_g$, $1^5\Sigma_u^-$, $1^5\Delta_g$, $1^5\Pi_u$, and $1^5\Pi_g$) (*thick lines*). These curves are given with respect to the energy of $S_2(X^3\Sigma_g^-)v' = 0$. The *ab initio* computed anionic PEC were shifted to lower energies by 0.23 eV in order to match the experimental electron affinity of $S_2(X^3\Sigma_g^-)$ (see text). The horizontal lines correspond to the positions of the ground vibrational levels.

$1^5\Pi_g$) states (*thick lines*). Our curves for the anionic lowest doublets and for the $S_2^-(1^4\Sigma_u^-)$ state are consistent with those of Heinemann *et al.* [14] and complete this previous theoretical work. All the S_2^- electronic states are bound at large internuclear distances since they are located below their respective parent states. The situation, however, is quite different in the molecular region, and these electronic states should be considered one by one. Here we are presenting only the bound parts of our anionic potentials, since our approach is not valid to describe accurately the negative ion resonances above the autodetachment threshold [1–3]. We noticed that our initially computed PECs for S_2^- were shifted to lower energies (by 0.23 eV) in order to match the experimental electron affinity of $S_2(X^3\Sigma_g^-)$ [5, 6]. Such a procedure has already been used in previous theoretical studies dealing with CS^- and SN^- [1,2]. Figure 1 shows the high density of electronic states for this anion favoring their mutual couplings by vibronic and spin–orbit, and complicating the computations for this molecular system. For better clarity, we are depicting in Figure 2 these anionic PECs together with those of their respective parent state separately, useful for judging the metastability of the S_2^- states (see below). These figures show that some of the anionic states are lying so close in energy that one cannot clearly distinguish them. For instance, this is the case for the $1^4\Delta_g$ and the $1^4\Sigma_g^+$ states, for which the insert in Figure 2B shows that these two states are surely close but do correspond to two different PECs.

By examining Figures 1 and 2, we see that several electronic states possess relatively deep potential wells. This is the case for almost all the states located below the anionic dissociation limit. The upper ones are mostly repulsive. Table 1 gives the spectroscopic constants for $S_2(X^3\Sigma_g^-)$ and for the S_2^- states possessing potential wells together with their comparison with previous works. These spectroscopic properties include the adiabatic excitation energies before shifting the S_2^- PECs (T_0, including zero point vibrational energy correction computed variationally), equilibrium distances (R_e), the harmonic wavenumbers (ω_e), the anharmonic terms ($\omega_e x_e$, $\omega_e y_e$), and the rotational constants (B_e, α_e, and γ_e). The agreement for the neutral spectroscopic parameters and with previous works for S_2^- makes possible reliable predictions for the unknown anionic states. Indeed, for $S_2(X^3\Sigma_g^-)$ our MRCI + Q internuclear distance of 1.898 Å differs by less than 0.01 Å from both the experimental (1.889 Å [26]) and earlier MRCI + Q (1.907 Å [14]) values, and the harmonic wavenumber (716.8 cm^{-1}) compares rather well with the values (725.65 cm^{-1} and 734 cm^{-1}) given in [26] and [14], respectively. For $S_2^-(X^2\Pi_g)$, there is also a good agreement between our R_e value (2.021 Å) and previous experimental (2.005 ± 0.015 Å [5]) and theoretical (2.018 Å [14]) determinations. The harmonic wavenumber $\omega_e = 555.5\ cm^{-1}$ for $S_2^-(X^2\Pi_g)$ is within photoelectron spectroscopy error bars (570 ± 100 cm^{-1} [5]) and in close accord with the value of 584 cm^{-1} derived from studies of S_2^- trapped in silicate solid matrices [23]. However, an ω_e of 600.8 cm^{-1} given in Ref. [24] is out of the range of these determinations. Finally, Table 1 lists the equilibrium spin–orbit constant ($A_{so,e}$) for $S_2^-(X^2\Pi_g, A^2\Pi_u, 1^2\Delta_u, 1^4\Delta_g, 1^2\Delta_g,$ and $1^4\Pi_u$). For the X state, our computed value, −355.8 cm^{-1}, compares reasonably well with the experimental determinations: −410 cm^{-1} [5], 420 cm^{-1} [22, 23],

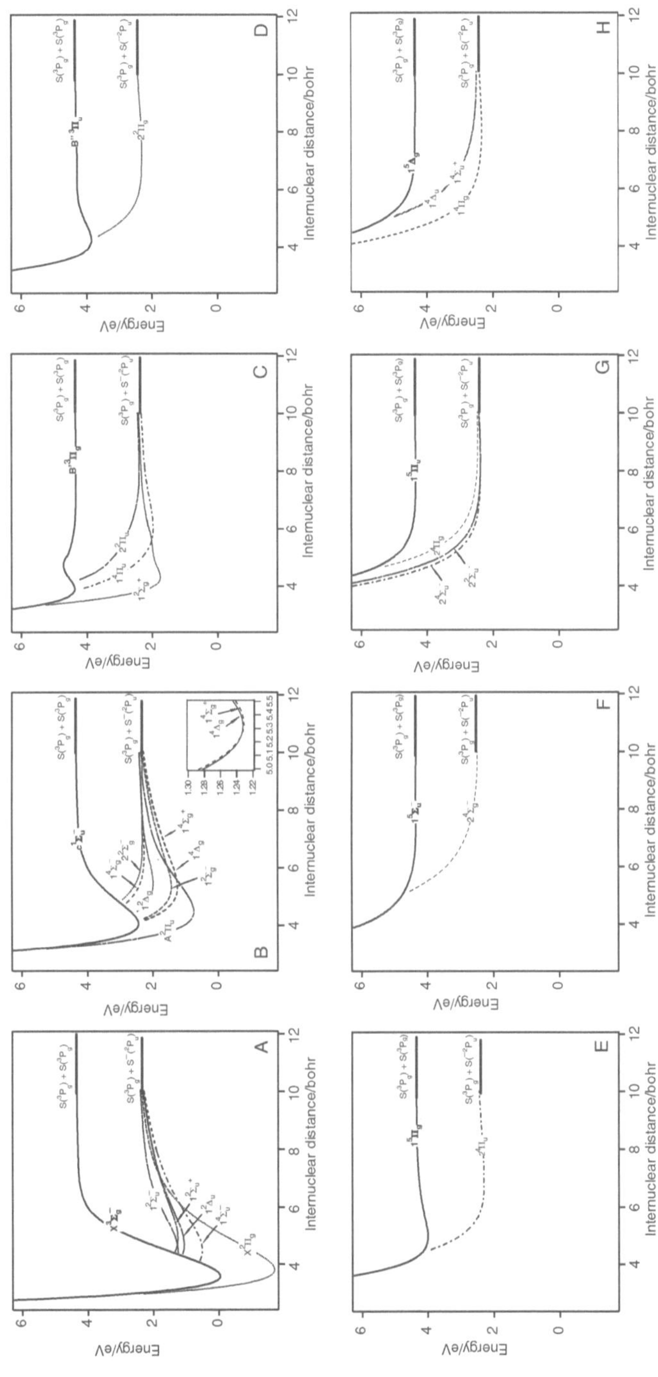

Figure 2. Details of Figure 1, where the anionic states are given separately with their respective parent state. We are using the same convention for the line types as in Figure 1. For the insert in Figure 2B see the text.

Table 1. MRCI + Q adiabatic transition energy (T_0, including zero-point vibrational energy correction computed variationally, in eV), equilibrium distances (R_e in Å), spin–orbit constant at R_e ($A_{SO,e}$ in cm^{-1}) and the spectroscopic parameters (in cm^{-1}) of S_2^- and of $S_2(X^3\Sigma^-_g)$.

State	T_0	R_e	ω_e	$\omega_e x_e$	$\omega_e y_e$	G_0[b]	B_e	α_e	γ	$A_{so,e}$[f]
$S_2(X^3\Sigma^-_g)$	0.00[a]	1.898	716.7	2.08	0.066	358.0	0.292	0.00153	0.00001	
		1.907[c]	734[c]				0.300[c]			
		1.889[d]	725.65[d]				0.295[d]			
$S_2^-(X^2\Pi_g)$	−1.469	2.021	555.5	0.16	0.169	278.5	0.257	0.00120	0.00003	355.8
		2.018[c]	582[c]				0.267[c]			416[c]
			600.8[d]	2.16[d]						420[g]
		2.005 ± 0.015[e]	570 ± 100[e]							410[e]
										440[h]
										373.6[i]
$S_2^-(1^4\Sigma^-_u)$	0.670	2.390	258.6	9.94	1.856	126.6	0.184	0.00348	0.00103	
		2.386[c]	270[c]				0.185[c]			
$S_2^-(A^2\Pi_u)$	0.933	2.373	338.0	1.55	0.052	168.8	0.186	0.00155	0.0000	301.3
		2.364[c]	340[c]				0.192[c]			
			364.2[d]	2.00[d]						

$S_2^-(1^2\Delta_u)$	1.252	2.484	226.1	1.03	0.119	112.7	0.170	0.00075	0.00004	2.0
		2.496[c]	236[c]				0.170[c]			
$S_2^-(1^4\Delta_g)$	1.398	2.834	254.9	15.34	1.944	137.4	0.131	0.00488	0.00199	108.6
$S_2^-(1^2\Sigma^+{}_u)$	1.442	2.594	210.0	1.83	0.077	104.3	0.156		0.00001	
		2.613[c]	205[c]				0.154[c]	0.00089		
$S_2^-(1^4\Sigma^+{}_g)$	1.430	2.815	223.0	1.71	0.015	111.0	0.132	0.00113	0.00002	
$S_2^-(1^2\Sigma^-{}_g)$	1.587	2.528	211.7	2.39	0.141	105.1	0.131	0.00116	0.00002	
$S_2^-(1^2\Sigma^+{}_g)$	1.972	2.372	293.0	1.64	0.029	146.3	0.187	0.00166	0.00001	
$S_2^-(1^2\Delta_g)$	2.182	2.897	202.8	4.62	0.182	99.5	0.125	0.00068	0.00009	
$S_2^-(1^4\Pi_u)$	2.203	3.191	126.1	1.56	0.025	62.7	0.103	0.00134	0.00001	112.9

[a] Used as reference.
[b] Zero-point vibrational energy calculated variationally.
[c] MRCI calculations. Ref. [14].
[d] Experiment. Ref. [24].
[e] Experiment. Ref. [5].
[f] Calculated using the one-electron term of the Breit–Pauli Hamiltonian (see ref. [29] for details).
[g] Ion trapped in matrices studies. Refs. [22, 23].
[h] Experiment. Ref. [25].
[i] MCSCF computations. Ref. [28].

and 440 cm^{-1} [25]. The differences are within our estimation of ∼5% accuracy for this property [1, 25–27]. This agreement suggests that the experimentally unknown spin–orbit constants for the S_2^- electronic states should be of similar accuracy. We refer to Table 1 for the corresponding fine-structure constants.

4. DISCUSSION

In the present section we focus on discussing the metastability of the electronic states of S_2^- in the light of their PEC (cf. Figures 2 and 3), their electronic configurations (cf. Table 2), and the spin–orbit coupling functions (depicted in Figure 3). It is worth noting again here that their long-range parts are bound, and we will consider only their behavior in the molecular region.

Table 2 lists the dominant electronic configurations of the quartets and the doublets of S_2^- together with the dominant electronic configurations of their neutral parent states. The anionic states of interest are obtained by attaching an extra electron onto either the $5\sigma_u$ or the $2\pi_g$ orbitals of S_2. The $X^2\Pi_g$, $1^4\Sigma_u^-$, $1^2\Delta_u$, $1^2\Sigma_u^+$, and $1^2\Sigma_u^-$ anionic electronic states have the $S_2(X^3\Sigma_g^-)$ ground state as the parent neutral state (cf. Figure 2A). The $S_2^-(A^2\Pi_u$, $1^4\Sigma_g^+$, $1^2\Sigma_g^-$, $1^4\Sigma_g^-$, $2^2\Sigma_g^-$, $1^2\Delta_g$, and $1^4\Delta_g)$

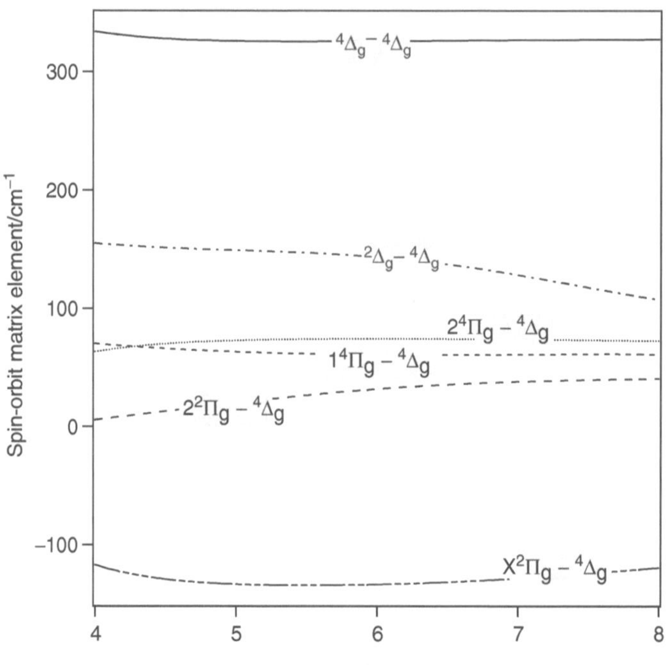

Figure 3. Evolution of the spin–orbit couplings involving the $S_2^-(^4\Delta_g)$ state along the internuclear distance. See Table 3 for the definition of these terms.

Table 2. Dominant electronic configurations for the electronic states of S_2^- and S_2 investigated. These configurations are quoted for an internuclear separation of 5.5 bohr.

S_2	S_2^-		S_2^-
$X^3\Sigma_g^-$	$(5\sigma_g)^2(5\sigma_u)^0(2\Pi_u)^4(2\Pi_g)^2$	$X^2\Pi_g$	$(5\sigma_g)^2(5\sigma_u)^0(2\Pi_u)^4(2\Pi_g)^3$
$c^1\Sigma_u^-$	$(5\sigma_g)^2(5\sigma_u)^0(2\Pi_u)^3(2\Pi_g)^3$	$1^4\Sigma_u^-$	$(5\sigma_g)^2(5\sigma_u)^1(2\Pi_u)^4(2\Pi_g)^2$
$1^5\Pi_g$	$(5\sigma_g)^2(5\sigma_u)^1(2\Pi_u)^3(2\Pi_g)^2$	$A^2\Pi_u$	$(5\sigma_g)^2(5\sigma_u)^0(2\Pi_u)^3(2\Pi_g)^4$
$B''^3\Pi_u$	$(5\sigma_g)^2(5\sigma_u)^1(2\Pi_u)^4(2\Pi_g)^1$	$1^2\Delta_u$	$(5\sigma_g)^2(5\sigma_u)^1(2\Pi_u)^4(2\Pi_g)^2$
$B'^3\Pi_g$	$(5\sigma_g)^1(5\sigma_u)^0(2\Pi_u)^4(2\Pi_g)^3$	$1^4\Delta_g$	$(5\sigma_g)^2(5\sigma_u)^1(2\Pi_u)^3(2\Pi_g)^3$
$1^5\Sigma_u^-$	$(5\sigma_g)^1(5\sigma_u)^1(2\Pi_u)^4(2\Pi_g)^2$	$1^4\Sigma_g^+$	$(5\sigma_g)^2(5\sigma_u)^1(2\Pi_u)^3(2\Pi_g)^3$
$1^5\Pi_u$	$(5\sigma_g)^2(5\sigma_u)^1(2\Pi_u)^2(2\Pi_g)^3$	$1^2\Sigma_u^+$	$(5\sigma_g)^2(5\sigma_u)^1(2\Pi_u)^4(2\Pi_g)^2$
$1^5\Delta_g$	$(5\sigma_g)^1(5\sigma_u)^1(2\Pi_u)^3(2\Pi_g)^3$	$1^2\Sigma_g^-$	$(5\sigma_g)^2(5\sigma_u)^1(2\Pi_u)^3(2\Pi_g)^3$
		$1^2\Sigma_u^-$	$(5\sigma_g)^2(5\sigma_u)^1(2\Pi_u)^4(2\Pi_g)^2$
		$1^2\Sigma_g^+$	$(5\sigma_g)^1(5\sigma_u)^0(2\Pi_u)^4(2\Pi_g)^4$
		$1^2\Delta_g$	$(5\sigma_g)^2(5\sigma_u)^1(2\Pi_u)^3(2\Pi_g)^3$
		$1^4\Pi_u$	$(5\sigma_g)^1(5\sigma_u)^1(2\Pi_u)^4(2\Pi_g)^3$
		$1^4\Sigma_g^-$	$(5\sigma_g)^2(5\sigma_u)^1(2\Pi_u)^3(2\Pi_g)^3$
		$2^2\Sigma_g^-$	$(5\sigma_g)^2(5\sigma_u)^1(2\Pi_u)^3(2\Pi_g)^3$
		$2^2\Pi_g$	$(5\sigma_g)^2(5\sigma_u)^2(2\Pi_u)^4(2\Pi_g)^1$
		$2^4\Pi_u$	$(5\sigma_g)^2(5\sigma_u)^2(2\Pi_u)^3(2\Pi_g)^2$
		$2^2\Pi_u$	$(5\sigma_g)^1(5\sigma_u)^1(2\Pi_u)^4(2\Pi_g)^3$
		$2^4\Sigma_u^-$	$(5\sigma_g)^2(5\sigma_u)^1(2\Pi_u)^2(2\Pi_g)^4$
		$1^4\Pi_g$	$(5\sigma_g)^1(5\sigma_u)^1(2\Pi_u)^3(2\Pi_g)^4$
		$2^2\Sigma_u^-$	$(5\sigma_g)^2(5\sigma_u)^1(2\Pi_u)^2(2\Pi_g)^4$
		$2^4\Pi_g$	$(5\sigma_g)^2(5\sigma_u)^2(2\Pi_u)^2(2\Pi_g)^3$
		$2^4\Sigma_g^-$	$(5\sigma_g)^1(5\sigma_u)^2(2\Pi_u)^4(2\Pi_g)^2$
		$1^4\Delta_u$	$(5\sigma_g)^1(5\sigma_u)^2(2\Pi_u)^3(2\Pi_g)^3$
		$1^4\Sigma_u^+$	$(5\sigma_g)^1(5\sigma_u)^2(2\Pi_u)^3(2\Pi_g)^3$

states are formed by binding an extra electron to $S_2(c^1\Sigma_u^-)$ (cf. Figure 2B). The $S_2(B''^3\Pi_u)$ and $S_2(B'^3\Pi_g)$ are the parent states of the $S_2^-(2^2\Pi_g)$ (Figure 2D) and $S_2^-(1^4\Pi_u$, $1^2\Sigma_g^+$, and $2^2\Pi_u)$ (Figure 2C) states, respectively. The remaining S_2^- electronic states of Figure 1 present quintets as parent states (Figures 2E, 2F, 2G, 2H).

According to our theoretical results, it is worth noting that among all the anionic states depicted in Figures 2 and 3 only the $X^2\Pi_g$ and $1^4\Delta_g$ states are long-lived although several S_2^- states are located well below their neutral parent state. Indeed, Figure 2A shows that the low vibrational levels of $S_2^-(X^2\Pi_g)$ are lying well below their autodetachment threshold, i.e. the energy of $S_2(X^3\Sigma_g^-)$ $v' = 0$. However, the other states in this Figure are lying above it, where fast electron loss is expected to occur. For all the S_2^- states given in Figure 2B, with the exception of the $J = 7/2$ fine structure component of $1^4\Delta_g$, the angular momentum of the fine structure components (i.e. $J = 1/2$ and 3/2) have their counterparts in the low-lying $X^2\Pi_g$ state. Similarly, the $J = 5/2$ component of the $1^4\Delta_g$ state may couple with its counterpart in $1^2\Delta_u$ which, however, may autodetach rapidly to $S_2(X^3\Sigma_g^-)$. Nevertheless, $1^4\Delta_g$ ($J = 7/2$) does not possess any low-lying counterpart. Moreover, the spin–orbit coupling functions involving $S_2^-(1^4\Delta_g)$ (cf. Figure 3 and Table 3) reveal

Table 3. Definition of the spin–orbit matrix elements involving the $S_2^-(1^4\Delta_g)$ state, together with our schematic representation given in Figure 3.

$< X^2\Pi_g\|H^{SO}\|^4\Delta_g >$	$= X^2\Pi_g - {}^4\Delta_g$
$< 2^2\Pi_g\|H^{SO}\|^4\Delta_g >$	$= 2^2\Pi_g - {}^4\Delta_g$
$< 1^4\Pi_g\|H^{SO}\|^4\Delta_g >$	$= 1^4\Pi_g - {}^4\Delta_g$
$< 2^4\Pi_g\|H^{SO}\|^4\Delta_g >$	$= 2^4\Pi_g - {}^4\Delta_g$
$<^2 \Delta_g\|H^{SO}\|^4\Delta_g >$	$= {}^2\Delta_g - {}^4\Delta_g$
$<^4 \Delta_g\|H^{SO}\|^4\Delta_g >$	$= {}^4\Delta_g - {}^4\Delta_g$

that couplings involving the $1^4\Delta_g$ state with close-lying states are small or null: for instance, the $<^2\Sigma^+{}_g|H^{SO}|^4\Delta_g>$, $<^4\Sigma^+{}_g|H^{SO}|^4\Delta_g>$, $<^4\Delta_g|H^{SO}|1^4\Sigma^-{}_g>$, $<^4\Delta_g|H^{SO}|2^4\Sigma^-{}_g>$, $<^4\Delta_g|H^{SO}|2^2\Sigma^-{}_g>$, $<^4\Delta_g|H^{SO}|1^2\Sigma^-{}_g>$ matrix elements are null due to spin–orbit selection rules. Hence, the depletion of the metastable $J = 7/2$ component of this quartet should occur via weak interactions. Finally, regarding the states located above the $1^4\Delta_g$ state their fine structure components are expected to undergo rapid intersystem conversion into the lower counterparts, thus reducing their lifetime even for those lying below their respective neutral state. As an illustration, we can mention the lowest $S_2^-(1^2\Sigma_g^+)$ rovibrational levels, which are lying definitely well below the $S_2(B'^3\Pi_g)$ neutral parent state (Figure 2C) and which can couple, at least, with the $S_2^-(X^2\Pi_g; J = 1/2)$ component.

5. CONCLUSION

Accurate *ab initio* computations were performed on the S_2^- electronic states correlating to the unique bound asymptote of this molecular system and to those of their respective neutral parent states, together with some diagonal and off-diagonal spin–orbit matrix elements. In the light of these calculations, the stability of the electronic states of this anion has been studied and, in addition to the well-known bound-ground state of S_2^- (i.e. the $X^2\Pi_g$ state), the $J = 7/2$ component of the $S_2^-(1^4\Delta_g)$ state is predicted to be long-lived in the gas phase. The present theoretical predictions are helpful for discussing the rapid decomposition reaction pathways that may be undergone by S_2^- and occur in competition with the autodetachment process (leading to $S_2 + e^-$) and also for proposing the reaction pathways followed when the S and S^- species in their ground states collide together. This theoretical work should motivate state-of-the-art experiments dealing with this negatively charged ion, to check the assumptions discussed above using, for instance, heavy ion storage rings.

Acknowledgements

M.H. would like to acknowledge a visiting fellowship at the University of Tunis from the Tunisian Ministry of Higher Education and Research.

References

1. M. Hochlaf, G. Chambaud, P. Rosmus, T. Andersen, and H.J. Werner, *J. Chem. Phys.* **110**, 11835, 1999, and references therein.
2. S. Ben Yaghlane, S. Lahmar, Z. Ben Lakhdar, and M. Hochlaf, *J. Phys. B* **38**, 3395, 2005.
3. A. Dreuw, T. Sommerfeld, and L.S. Cederbaum, *J. Chem. Phys.* **116**, 6039, 2002, and references therein.
4. http://webbook.nist.gov.
5. S. Moran and G.B. Ellison, *J. Phys. Chem.* **92**, 1794, 1988.
6. R.J. Celotta, R.A. Bennett, and J.L. Hall, *J. Chem. Phys.* **60**, 1740, 1974.
7. S. Hunsicker, R.O. Jones, and G. Gantefor, *J. Chem. Phys.* **102**, 5917, 1995.
8. G.-G. Lindner, Ph D thesis, Shaker, Aachen, 1994. R.J.H. Clark, T.J. Dines, and M. Curmoo, *Inorg. Chem.* **22**, 2766, 1983.
9. M. Ikezawa and J. Rolfe, *J. Chem. Phys.* **58**, 2024, 1974.
10. C.A. Sawicki and D.B. Fitchen, *J. Chem. Phys.* **65**, 4497, 1976.
11. F.A. Cotton, J.B. Harmon, and R.M. Hedges, *J. Am. Chem. Soc.* **98**, 1417, 1976.
12. F. Ramondo, N. Sanna, and L. Bencivenni, *J. Mol. Struc.* THEOCHEM, **258**, 361, 1992.
13. O. Hübner and J. Sauer, *Phys. Chem. & Chem. Phys.* **4**, 5234, 2002.
14. C. Heinemann, W. Koch, G.-G. Lindner, and D. Reinen, *Phys. Rev. A* **52**, 1024, 1995.
15. S. Ben Yaghlane, A. Ben Houria, and M. Hochlaf, *On The Role of Electronic Molecular States of High Spin Multiplicity*, this volume, 2006.
16. D.E. Woon and T.H. Dunning Jr, *J. Chem. Phys.* **98**, 1358, 1993.
17. P.J. Knowles and H.-J. Werner, *Chem. Phys. Lett.* **115**, 259, 1985.
18. H.-J. Werner and P.J. Knowles, *J. Chem. Phys.* **89**, 5803, 1988.
19. P.J. Knowles and H.-J. Werner, *Chem. Phys. Lett.* **145**, 514, 1988.
20. MOLPRO is a package of *ab initio* programs written by H.J. Werner and P.J. Knowles. Further details can be found at www.tc.bham.ac.uk/molpro.
21. J.W. Cooley, *Math. Comput.* **15**, 363, 1961.
22. L.E. Vannotti and J.R. Morton, *Phys. Rev.* **161**, 282, 1967.
23. G.J. Vella and J. Rolfe, *J. Chem. Phys.* **61**, 41, 1974.
24. K.P. Huber and G. Herzberg, in *Molecular Spectra and Molecular Structure IV. Constants of Diatomic Molecules*, Van Nostrand, New York, 1979, pp. 564–567.
25. S. Koseki, M.W. Schmidt, and M.S. Gordon, *J. Phys. Chem.* **96**, 10768, 1992.
26. A. Ben Houria, Z. Ben Lakhdar, and M. Hochlaf, *J. Chem. Phys.* **124**, 054313, 2006.
27. F. Khadri, H. Ndome, S. Lahmar, Z. Ben Lakhdar, and M. Hochlaf, *J. Mol. Spectrosc.* 2006, in press.
28. S. Koseki, M.S. Gordon, M.W. Schmidt, and N. Matasunaga, *J. Phys. Chem.* **99**, 12764, 1995.
29. H. Lefebvre-Brion and R.W. Field, *The Spectra and Dynamics of Diatomic Molecules*, Academic, Amsterdam, 2004, pp. 180–203.

References

AN EXTENSIVE STUDY OF THE PROTOTYPICAL HIGHLY SILICON DOPED HETEROFULLERENE $C_{30}SI_{30}$

MASAHIKO MATSUBARA[1] AND CARLO MASSOBRIO[2]

[1]*Laboratory of Physics, Helsinki University of Technology, P.O.Box 1100, 02015 HUT, Finland*
[2]*Institut de Physique et Chimie des Matériaux de Strasbourg, UMR 7504 ULP-CNRS, BP43, 23 rue du Loess, 67034 Strasbourg Cedex 2, France*

Abstract We provide information on structural and electronic properties of highly doped silicon fullerenes, obtained by replacing as much as 30 carbon atoms by an equivalent number of silicon atoms. A large number of isomers results from our optimization study based on the first-principles molecular dynamics approach within density functional theory. We are able to rationalize the topology taken by these clusters on the basis of preferential segregated arrangements, in which groups of atoms of the same nature form homogeneous regions on the cage.

1. INTRODUCTION

Fullerenes attract a great deal of interest since their discovery [1], stimulating the synthesis of doped derivatives enhancing chemical reactivity. Also, the search of a connection between molecular and material science has prompted investigations on a large number of doped fullerenes, expected to be building blocks of fullerene-based new nanomaterials. Among the doping methods, substitutional doping is regarded as the best method to create a reactive site in the cage and to modify the electronic properties of the fullerene itself. To date, various types of substitutional heterofullerenes have been reported, where B, N, and transition metal atoms, such as Fe, Co, Ni, Rh, Ir have been successfully incorporated into the fullerene cage [2–8]. However, the interest for silicon as a dopant atom is largely predominant, due to a striking sequence of similarities and differences among these two elements. We remind that carbon and silicon have the same number of valence electrons. Therefore, at first glance, the replacement of carbon with silicon appears trivial. However, a deeper analysis shows that the bonding nature of carbon and silicon differs notably. Silicon prefers to form multi-directional sp^3 single bonds. These are expected not to be easily adjustable

S. Lahmar et al. (eds.), Topics in the Theory of Chemical and Physical Systems, 261–270.

into the carbon cage where carbon is bonded through sp^2 threefold connections. The existence of silicon doped heterofullerenes was confirmed by ion mobility and mass spectroscopy experiments [9–12], where at least two silicon atoms are doped into the cage. A combination of mass spectroscopy and photofragmentation experiments showed the existence of $C_{60-m}Si_m$ with $m = 12$ as an upper limit of the number of doped silicon atoms [13, 14].

In addition to these experimental results, intensive calculations began on the theoretical side. First principles studies have been performed in the case of one or two silicon atoms doping C_{60} ($C_{59}Si$, $C_{58}Si_2$) [12, 15]. Then, the number of silicon atoms has been gradually increased up to $m = 12$ [16–21]. Very recently we showed the possibility of silicon-doped fullerenes with $m > 12$, such as $C_{40}Si_{20}$, $C_{36}Si_{24}$, and $C_{30}Si_{30}$ [22, 23]. These studies have made possible a remarkable breakthrough in the area of doped fullerenes, obtained via a combination of a careful analysis of charge topology and an observation of the fragmentation mechanism at high temperatures [24]. We were able to show that highly Si-doped fullerenes are thermally stable as long as Si atoms neighbors of C atoms in segregated regions (the outer atoms) are predominant over inner Si atoms. Due to its polar character, the Si–C interaction is able to stabilize the cage and offsets the energetic costs of sp^2 interactions for a number of dopant atoms smaller than 20. We demonstrated that beyond this threshold, repulsive interactions among inner Si atoms cause fragmentation. From the electronic point of view, a clear localization of the HOMO (highest occupied molecular orbital) is the chemical bonding fingerprint of this behavior [24].

We have pursued our search of structural isomers for the largest number of Si atoms considered. This allows to complement effectively our description of the $C_{30}Si_{30}$ isomers, thereby exemplifying the complexity of their classification in terms of binding energy. The present paper is intended to attain this goal. In this contribution we provide an extensive analysis of 15 silicon doped heterofullerenes $C_{30}Si_{30}$ by focusing on their structural and electronic properties. Accurate data are given on bonding distances. The distribution of the charges as well as the topology of the relevant electronic localization concur to describe the correlation among the doping content and changes in the electronic properties. Part of the information made available is obtained for isomers unexplored before, bringing complementary and novel insight on these new nanosystems. Indeed, nine of these isomers have been the object of a previous preliminary report, where we claimed that our research of the most stable structures had been essentially exhaustive [23]. By insisting on the isomer search, we are now in a position to present additional information, with an even larger number of isomers lying very close in energy. A view of the fragmentation mechanism characterizing the dynamics at high temperatures is also provided.

2. COMPUTATIONAL METHODS

The calculations are performed within the framework of the Carr–Parrinello molecular dynamics [25, 26]. Our approach is based on the density functional theory with generalized gradient approximations after Becke for the exchange energy [27] and Lee, Yang, and Parr for the correlation energy [28]. Troullier–Martins type norm-conserving pseudopotentials are used to describe the core–valence interaction. Periodic boundary conditions have been adopted and a face-centered cubic cell with an edge of 21.17 Å provides a system size much larger than the diameter of the optimized structures. Wavefunctions are expanded in plane waves with an energy cutoff of 40 Ry. We considered 15 different isomers. In order to select the appropriate initial configurations we have taken advantage of the fact that these heterofullerenes are more stable when the Si atoms lie close together in the cage [16, 19, 20, 22]. Therefore, our starting configurations for $C_{30}Si_{30}$ are obtained from $C_{36}Si_{24}$ by further replacing six C atoms. Then we optimized self-consistently the structures by minimizing the atomic forces. Optimization is allowed to proceed until the largest force component is less than 5×10^{-4} a.u. and the average force is one order of magnitude smaller. An analysis of electronic properties is performed by projecting Kohn–Sham (KS) orbitals on atomic orbitals centered on each atom and calculating the atomic populations and the probability densities. A population analysis is also performed to obtain Mulliken charges on each atom [29]. Concerning molecular dynamics, more details on the simulation parameters are given elsewhere [24].

3. RESULTS

3.1. Structural properties

In Figure 1 we give a two-dimensional schematic representations of 15 different isomers of $C_{30}Si_{30}$. Interatomic distances are listed in the Table 1. In the table, ph is the bond between a pentagon and a hexagon, while hh is the bond between two hexagons. The total number of occurrences for each bond is also given in the parenthesis. For stretched Si–Si distances, the notion of bond has to be intended as indicative of an interatomic distance, customary Si–Si bond lengths being significantly smaller than some of the Si–Si distances recorded in this study.

In all these isomers C–C bond lengths are almost the same as those of C_{60}, meaning that the conjugated pattern of C_{60} is preserved in the carbon region despite the large number of doping Si atoms. In some of the isomers, Si–Si bond distances are highly stretched (as much as 2.66 Å), suggesting that these connections will be the first to break as a reaction to thermal motion. While this preliminary conclusion is intuitively correct, we have demonstrated that these weak interactions are not necessarily a precursor of unavoidable fragmentation [19]. The different isomers have in common an unmistakable relationship between their energetic stability and the formation of homogeneous regions made by Si atoms well segregated from C atoms. This behavior goes along with the peculiar shape taken by these clusters. With increasing number of silicon atoms, two half portions can be clearly distinguished (see Figures 3 and 4 for representative cases). One, occupied by C atoms, is surprisingly similar to the undoped C_{60} cage. The second, hosting Si atoms, has a larger diameter and reflects the attempt to restore a more favorable sp^3 bonding configuration. We have calculated the average angle taken by Si–Si–Si triads, by obtaining a value very close to 109.5°, the tetrahedral angle for sp^3 bonding. At a first glance, one might wonder

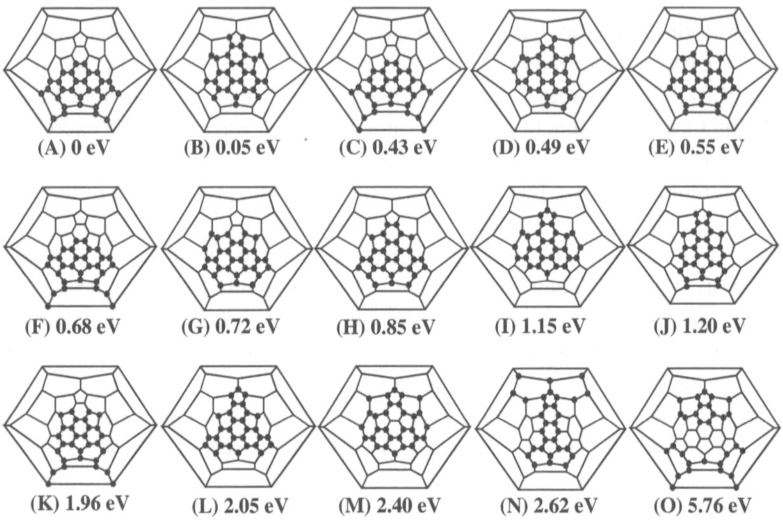

Figure 1. Two-dimensional schematic representations of 15 different isomers of $C_{30}Si_{30}$. Si atoms are denoted by the black dots. The difference of the total energies relative to the energy of isomer **A** is also given.

whether the stability of such deformed structures is real or simply resulting from a "fake" local minimum, i.e. a saddle point in the configurational space. Vibrational analysis reveals that most of these isomers are indeed vibrationally stable, thereby standing small displacements in the harmonic regime. Also, differences in zero-point energy among isomers (typically 0.002 eV) are much smaller than the differences in total energies. Therefore, no effect is found in the isomers energy scale.

3.1.1. Electronic properties

To better understand the origins of this enhanced and unexpected structural stability and see how they are deeply rooted into the electronic structure, we provide here information on the electronic properties after doping the fullerene cage with a large number of silicon atoms. By focusing on the five most stable isomers (from **A** to **E**), which are found within 0.01 eV/atom, we consider first the values of the Mulliken charges (see Figure 2).

Non negligible negative charges on the C atoms are found when C atoms have nearest-neighbor Si atoms. Conversely, when C atoms have no Si nearest-neighbors, the charges are vanishingly small. In the case of Si atoms, substantial positive charges appear when the Si atoms are nearest neighbors to C atoms. Positive charges tend to vanish whenever a Si atom is bound to three neighbors of the same kind. In this case, small negative charges can also be found on the Si atoms. One notice that the residual charges on the Si atoms not neighboring C atoms take alternate signs within a specific Si-made ring, as in an attempt to reduce repulsive effects. However, the presence of odd-membered rings (pentagons) precludes systematic sequences of alternate charges

Table 1. Bond lengths of $C_{30}Si_{30}$ calculated in this work. The smallest and the largest values and the total number of occurrences for each bond are given.

Isomers	Si–Si (Å)		Si–C (Å)		C–C (Å)
A	ph(26)	2.35–2.49	ph(8)	1.89–2.01	ph 1.45–1.51
	hh(14)	2.30–2.41	hh(2)	1.84–1.85	hh 1.40–1.44
B	ph(24)	2.36–2.51	ph(12)	1.86–1.92	ph 1.45–1.52
	hh(12)	2.27–2.40	hh(6)	1.82–1.86	hh 1.40–1.44
C	ph(24)	2.34–2.66	ph(12)	1.88–1.98	ph 1.45–1.53
	hh(15)	2.27–2.53	hh(0)		hh 1.40–1.45
D	ph(24)	2.34–2.45	ph(12)	1.88–2.02	ph 1.44–1.52
	hh(15)	2.31–2.43	hh(0)		hh 1.40–1.44
E	ph(30)	2.30–2.48	ph(0)		ph 1.45–1.49
	hh(10)	2.31–2.40	hh(10)	1.85–1.91	hh 1.40–1.44
F	ph(26)	2.31–2.48	ph(8)	1.89–1.97	ph 1.45–1.51
	hh(13)	2.32–2.48	hh(4)	1.85–1.86	hh 1.40–1.46
G	ph(28)	2.31–2.52	ph(4)	1.90–1.98	ph 1.44–1.51
	hh(12)	2.33–2.48	hh(6)	1.84–1.91	hh 1.40–1.46
H	ph(28)	2.32–2.53	ph(4)	1.90–2.00	ph 1.44–1.50
	hh(12)	2.29–2.39	hh(6)	1.86–1.92	hh 1.40–1.45
I	ph(26)	2.30–2.42	ph(8)	1.87–1.98	ph 1.45–1.51
	hh(13)	2.29–2.36	hh(4)	1.84–1.89	hh 1.40–1.45
J	ph(27)	2.29–2.53	ph(6)	1.87–1.93	ph 1.44–1.49
	hh(12)	2.30–2.45	hh(6)	1.83–1.90	hh 1.39–1.45
K	ph(26)	2.30–2.59	ph(8)	1.89–2.02	ph 1.44–1.51
	hh(13)	2.30–2.45	hh(4)	1.86–1.88	hh 1.39–1.45
L	ph(30)	2.31–2.63	ph(0)		ph 1.45–1.50
	hh(9)	2.32–2.38	hh(12)	1.87–1.90	hh 1.39–1.44
M	ph(24)	2.33–2.57	ph(12)	1.89–2.02	ph 1.44–1.51
	hh(14)	2.30–2.46	hh(2)	1.88–1.89	hh 1.40–1.44
N	ph(22)	2.34–2.51	ph(16)	1.89–2.05	ph 1.45–1.49
	hh(15)	2.31–2.56	hh(0)		hh 1.40–1.46
O	ph(22)	2.33–2.53	ph(16)	1.90–2.02	ph 1.44–1.50
	hh(14)	2.30–2.56	hh(2)	1.87	hh 1.42–1.47

on Si atoms. Overall, these results indicate two major facts. First, Si–C bonds at the Si-C border are highly coulombic due to significant charge transfer from the Si atoms to the neighboring C atoms. Second, the charges associated with the Si atoms located far from the frontier with C atoms are vanishingly small and either positive or negative. Therefore, a correspondence can be established between the location in the cage and the electronic character of the Si atoms, leading to the identification of two groups of Si atoms. The first takes advantage of a strong coulombic interaction with the C atoms to be tightly linked to C-made part of the cage. The second is formed by those Si atoms less affected by the coulombic attraction. These atoms have been found to behave as seeds of dynamical instability, since any residual repulsion occurring among a pair of them can result in rapid fragmentation [24].

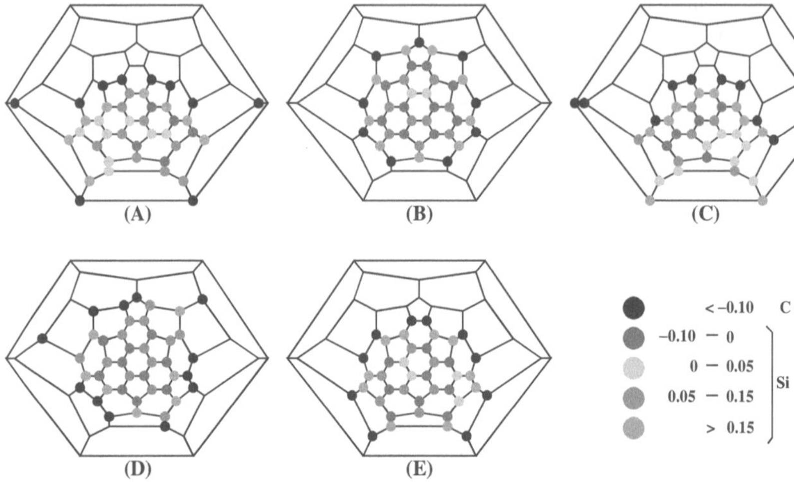

Figure 2. A flattened view of the five most stable isomers with the shading codes corresponding to values of Mulliken charges.

In Figures 3 and 4, we show a three dimensional representation of the isomers **A** and **C** (see (a)). The plots of their total electron densities ρ for the isosurfaces corresponding to $\rho = 0.07\ e/(\text{a.u.})^3$ are also given as (b) in each figure. The probability density of the KS orbitals corresponding to the isosurfaces of the highest occupied molecular orbital (HOMO) and the lowest unoccupied molecular orbital (LUMO) at $\rho = 0.005\ e/(\text{a.u.})^3$ are given as (c) and (d), respectively, in each figure. The charge transfer effect caused by doping is clearly visualized in the total electron density plot, where the electron density appears largely predominant in the carbon region. By keeping in mind that the results presented here refer to the structural optimization at zero temperature only, it is conceivable to associate the HOMO localization sites to those most reactive for these specific conditions (T = 0 K). Similarly, the LUMO corresponds to regions where enhanced chemical reactivity can manifest itself when sufficient energy is injected as thermal motion. A substantial gap reduction goes along with the introduction of temperature (going from typical values of 0.6 eV at T = 0 K down to 0.1 eV at T = 3000K). This allows to attribute to the LUMO, in terms of localization, the same significance of the HOMO but at a higher temperatures. Accordingly, looking for the LUMO localization sites is a viable approximation to identify the HOMO localization sites at high temperature. In view of these considerations, we can conclude that the inner Si atoms are the best candidates for important bonding changes at finite temperatures, as those occurring when cluster fragmentation takes place. Indeed, the LUMO is localized on these inner Si sites. This analysis has been recently substantiated by an extensive set of first-principles molecular dynamics simulations, showing that at high temperatures the HOMO shifts its region of localization from the outer to the inner part of the Si region [24].

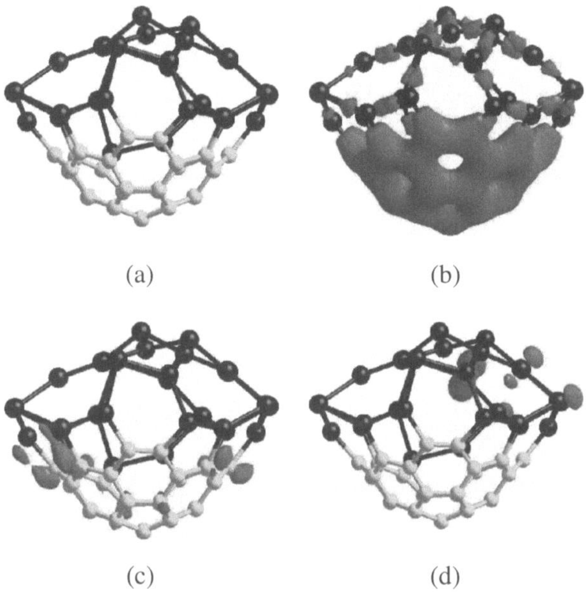

Figure 3. (a) Three-dimensional ball and stick representations of the most stable isomer **A**. Isodensity surfaces associated with (b) the total electronic density at $\rho = 0.07\ e/(\text{a.u.})^3$ and (c) the HOMO and (d) the LUMO at $\rho = 0.005\ e/(\text{a.u.})^3$.

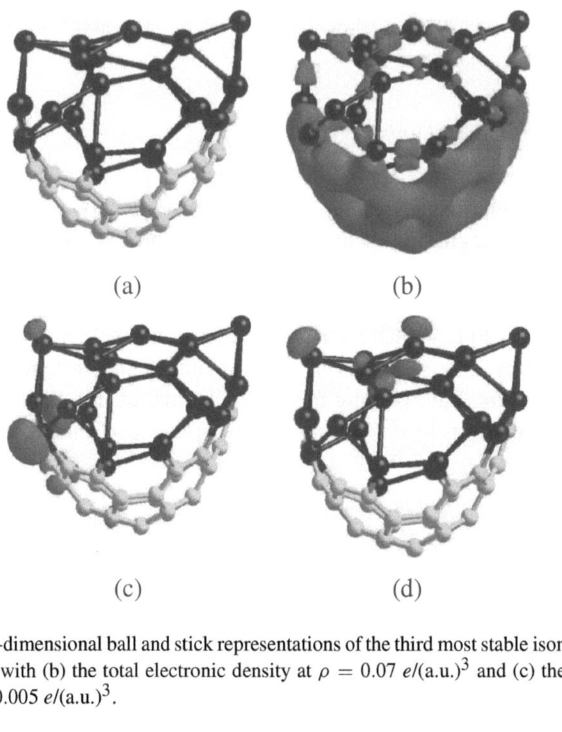

Figure 4. (a) Three-dimensional ball and stick representations of the third most stable isomer **C**. Isodensity surfaces associated with (b) the total electronic density at $\rho = 0.07\ e/(\text{a.u.})^3$ and (c) the HOMO and (d) the LUMO at $\rho = 0.005\ e/(\text{a.u.})^3$.

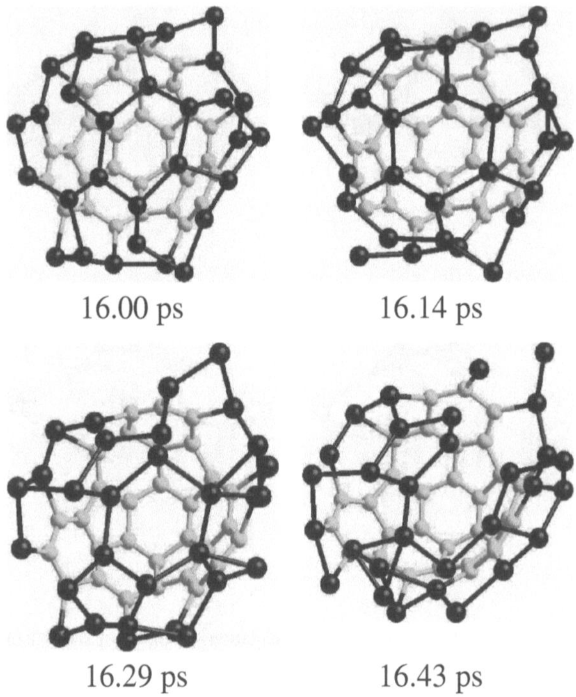

Figure 5. Snapshots of configurations taken at T = 3000K for the isomer D during the first stages of fragmentation.

3.1.2. Fragmentation mechanism

The above pieces of evidence are indicative of the role played by loosely connected Si atoms in inducing thermal instability. In Figure 5 we display a sequence of configurations taken at the beginning of the fragmentation process occurring at T = 3000 K. The Si atoms belonging to the inner regions are the first to move away from each other, even though fluctuations can even induce temporary recombination. The border region between Si and C is characterized by a remarkable stability, with Si–C pairs firmly connected by ionic interactions. On the other hand, whenever Si atoms find neighbors of the same kind bearing the same residual charge, energetically unfavorable local configurations lead to the disruption of the cage, that opens in the inner Si region.

4. CONCLUSION

We have performed extensive search of equilibrium geometries for a fairly large number of configurations of $C_{30}Si_{30}$, as much as 15 isomers, and investigated their structural and electronic properties. Whenever the cage is formed by fully segregated Si and C regions, the stability of the heterofullerene is enhanced. Visual inspection of the electronic densities associated with the total valence charge, the HOMO and the LUMO gives clear evidence for the highly polar character of these isomers. We highlighted the existence of chemically reactive sites moving from the Si–C border to the inner Si regions with increasing temperature.

Acknowledgements

Calculations were performed on the computers of the French national centers IDRIS (Orsay) and CINES (Montpellier). M.M. acknowledges the Center of Excellence Program of the Academy of Finland.

References

1. H.W. Kroto, J.R. Heath, S.C. O'Brien, R.R. Curl, and R.E. Smalley, *Nature* **318**, 162, 1985.
2. T. Guo, C. Jin, and R.E. Smalley, *J. Phys. Chem.* **95**, 4948, 1991.
3. H.-J. Muhr, R. Nesper, B. Schnyder, and R. Kötz, *Chem. Phys. Lett.* **249**, 399, 1996.
4. C.M. Brown, E. Beer, C. Bellavia, L. Cristofolini, R. Gonzárez, M. Hanfland, D. Häusermann, M. Keshavarz-K, K. Kordatos, K. Prassides, and F. Wudl, *J. Am. Chem. Soc.* **118**, 8715, 1996.
5. W. Branz, I.M.L. Billas, N. Malinowski, F. Tast, M. Heinebrodt, and T.P. Martin, *J. Chem. Phys.* **109**, 3425, 1998.
6. I.M.L. Billas, W. Branz, N. Malinowski, F. Tast, M. Heinebrodt, T.P. Martin, C. Massobrio, M. Boero, and M. Parrinello, *Nanostruct. Mater.* **12**, 1071, 1999.
7. I.M.L. Billas, C. Massobrio, M. Boero, M. Parrinello, W. Branz, F. Tast, N. Malinowski, M. Heinebrodt, and T.P. Martin, *Comp. Mat. Sci.* **17**, 191, 2000.
8. L. Hultman, S. Stafström, Z. Czigány, J. Neidhardt, N. Hellgren, I.F. Brunell, K. Suenaga, and C. Colliex, *Phys. Rev. Lett.* **87**, 225503, 2001.
9. T. Kimura, T. Sugai, and H. Shinohara, *Chem. Phys. Lett.* **256**, 269, 1996.
10. J.L. Fye, and M.F. Jarrold, *J. Phys. Chem. A* **101**, 1836, 1997.
11. M. Pellarin, C. Ray, P. Mélinon, J. Lermé, J.L. Vialle, P. Kéghélian, A. Perez, and M. Broyer, *Chem. Phys. Lett.* **277**, 96, 1997.
12. C. Ray, M. Pellarin, J.L. Lermé, J.L. Vialle, M. Broyer, X. Blase, P. Mélinon, P. Kéghélian, and A. Perez, *Phys. Rev. Lett.* **80**, 5365, 1998.
13. M. Pellarin, C. Ray, J. Lermé, J.L. Vialle, M. Broyer, X. Blase, P. Kéghélian, P. Mélinon, and A. Perez, *J. Chem. Phys.* **110**, 6927, 1999.
14. M. Pellarin, C. Ray, J. Lermé, J.L. Vialle, M. Broyer, X. Blase, P. Kéghélian, P. Mélinon, and A. Perez, *Eur. Phys. J. D* **9**, 49, 1999.
15. I.M.L. Billas, C. Massobrio, M. Boero, M. Parrinello, W. Branz, F. Tast, N. Malinowski, M. Heinebrodt, and T.P. Martin, *J. Chem. Phys.* **111**, 6787, 1999.
16. C.-C. Fu, M. Weissmann, M. Machado, and P. Ordejón, *Phys. Rev. B* **63**, 085411, 2001.
17. M. Menon, *J. Chem. Phys.* **114**, 7731, 2001.
18. P.A. Marcos, J.A. Alonso, L.M. Molina, A. Rubio, and M.J. López, *J. Chem. Phys.* **119**, 1127, 2003.
19. M. Matsubara and C. Massobrio, *J. Chem. Phys.* **122**, 084304, 2005.
20. M. Matsubara, C. Massobrio, and J.C. Parlebas, *Comp. Mat. Sci.* **33**, 237, 2005.
21. P.A. Marcos, J.A. Alonso, and M.J. López, *J. Chem. Phys.* **123**, 204323, 2005.
22. M. Matsubara and C. Massobrio, *J. Phys. Chem. A* **109**, 4415, 2005.
23. M. Matsubara and C. Massobrio, First principles study of extensive doping of C_{60} with silicon, to be published on Materials Science and Engineering C.
24. M. Matsubara, J. Kortus, J.C. Parlebas, and C. Massobrio, *Phys. Rev. Lett.* **96**, 155502, 2006.
25. R. Car and M. Parrinello, *Phys. Rev. Lett.* **55**, 2471, 1985.

26. We have used the code CPMD: CPMD Version 3.9.1, Copyright IBM Corp (1990–2004) and MPI für Festkörperforschung Stuttgart (1997–2001).
27. A.D. Becke, *Phys. Rev. A* **38**, 3098, 1988.
28. C. Lee, W. Yang, and R.G. Parr, *Phys. Rev. B* **37**, 785, 1988.
29. M.D. Segall, R. Shah, C.J. Pickard, and M.C. Payne, *Phys. Rev. B* **54**, 16317, 1996.

THEORETICAL STUDY OF THE MAGNETISM IN MOLECULAR CRYSTALS USING A FIRST-PRINCIPLES *BOTTOM-UP* METHODOLOGY

M. DEUMAL[1], M.A. ROBB[2], AND J.J. NOVOA[1]

[1] *Departament de Química Física, Facultat de Química and CERQT, Parc Científic, Universitat de Barcelona, Av. Diagonal 647, 08028-Barcelona, Spain*
[2] *Department of Chemistry, Imperial College London, South Kensington Campus, London SW7 2AZ, UK*

Abstract The study of molecular materials presenting magnetic properties is currently one of the main research areas in Materials Science. However, the progress in this field has been hampered by the lack of tools allowing a proper rationalization of the mechanism of magnetic interactions. Previous theories commonly used to explain the nature of magnetic interactions at the microscopic level (like the so-called McConnell-I and II approaches) fail in many cases, and do not have a sound theoretical basis. In the present work, we review a recent theoretical methodology developed by us, called first-principles *bottom-up* methodology, which allows an accurate and systematic study of the magnetism in molecule-based crystals. The approach is *bottom-up* because it begins by computing the microscopic magnetic interactions and, using these values, the macroscopic magnetic properties. It is also "first-principles" because the magnetic interactions are evaluated without making preliminary assumptions about the nature of these interactions. The way in which the methodology works is illustrated on a molecular crystal, the 2-hydronitronyl-nitroxide, experimentally known to present antiferromagnetic properties.

1. INTRODUCTION

The discovery of bulk ferromagnetic properties in molecular crystals of radicals containing only C, O, N, and H atoms in their structure [1] started a widespread interest, within the scientific community, towards what are now known as purely organic molecular magnets. These magnets are solids resulting from the aggregation of purely organic stable radicals. The crystal packing allows the propagation of the magnetic interactions along two or more directions within the solid (basic principles preclude the existence of magnetism in one dimension above 0 K [2]). Currently, the field of purely organic molecular magnets is a subset of the more general molecule-based

S. Lahmar et al. (eds.), Topics in the Theory of Chemical and Physical Systems, 271–289.

magnets field, which includes crystals whose radicals contain atoms other than C, O, N, and H.

Well-known pure organic molecule-based magnets are many derivatives of the nitronyl nitroxide family of neutral radicals [3] (see Figure 1 for the general structure) and some C_{60} molecular crystals (e.g. C_{60}TDAE salts [4a] and pressed / heated polymeric C_{60}-fullerene neutral crystals [4b]). Among the non-purely organic subset one can mention the Prussian Blue derivatives [5] and [Fe(Cp*)$_2$]TCNE (decamethyl ferrocinium tetracyano ethenide) salts [6], both presenting ferromagnetism above room temperature.

Progress in the field of molecule-based magnets has been fast, and a large number of compounds showing interesting macroscopic magnetic behaviors have been found (e.g. ferromagnetism, ferrimagnetism, antiferromagnetism, metamagnetism, spin-glass, spin-ladders ...). However, due to improper knowledge of the magnetic interactions at the microscopic level, and also due to a lack of rigorous procedures to compute macroscopic properties from microscopic interactions, the progress in this field has been difficult. As a consequence, there are currently no rigorous procedures to design molecule-based magnets showing a given desired macroscopic magnetic behavior.

With the aim of filling this gap, we have recently proposed a methodological procedure [7], hereafter called *first-principles bottom-up methodology*, which allows a rigorous computation of the macroscopic magnetic behavior from the only knowledge of the microscopic magnetic interactions (obtained from first-principles computations of all unique radical–radical pairs found in the crystal, making no *a priori* assumptions about the mechanism of the magnetic interaction). Therefore, this procedure allows connecting the crystal geometry with the macroscopic magnetic properties in a rigorous and unbiased manner. Consequently, the macroscopic behavior can be associated to specific interactions between pairs of radicals.

The nature and strength of all radical–radical microscopic magnetic interactions can be described in terms of the associated J_{AB} parameter, which can be computed

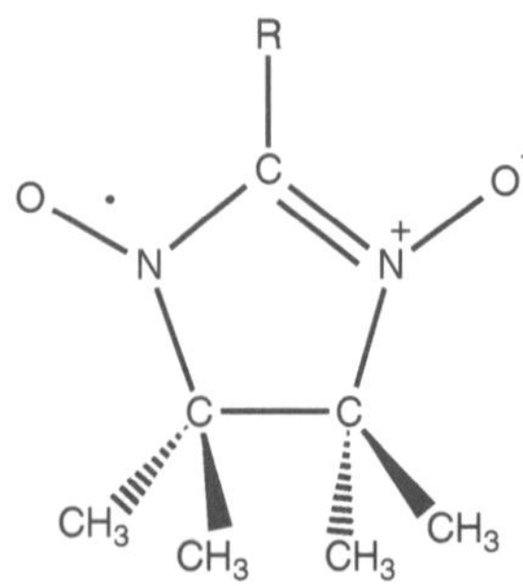

Figure 1. Chemical structure of a nitronyl nitroxide radical, where R is the substituent that differentiates among members of this family (representative cases range from a simple H atom up to substituted six-membered rings).

from the energy difference between different electronic states of the radical–radical pairs (the energies of these states are computed using first-principles methods, like density functional or *ab initio* techniques). Once all J_{AB} parameters are obtained, the connectivity they establish among the radicals within the crystal defines the *magnetic topology* of the crystal. Such a topology is necessary to find a finite model of the crystal that would properly represent its magnetic properties. The spin functions of this finite model are used as a basis to compute the matrix representation of the Heisenberg Hamiltonian (which is a function only of the J_{AB} parameters). The full energy spectrum of the Heisenberg Hamiltonian is then obtained by diagonalizing the matrix. Once these energies are inserted in the proper Statistical Mechanics expressions, one can compute the macroscopic magnetic properties of interest and also define the macroscopic magnetic behavior of the crystals.

In the following sections, our first-principles *bottom-up* methodology will be described from a physical perspective (a full description on mathematical terms can be found in Ref. [7]). Then, the use of this methodology will be illustrated on the α-2-hydro nitronyl nitroxide [8] (hereafter called α-HNN), the simplest member of the nitronyl nitroxide family (R = H in Figure 1). The macroscopic magnetic properties and crystal structure of this compound are well known (its crystal structure is stored in the Cambridge Crystallographic Database [9] with refcode name TOLKEK).

2. MAGNETISM IN MOLECULAR CRYSTALS

The presence of magnetism in a crystal requires the existence of spin-containing units capable of interacting with each other. Examples of spin-containing units are molecules having unpaired electrons, e.g. free radicals (if all electrons are coupled, no net spin is left in the molecule and its aggregates become *diamagnetic*). Experimentally, it has been observed [3] that the same spin-containing units can pack in different forms (called polymorphs) which, in some cases, present different magnetic properties. Therefore, *the type of magnetic interactions depends on the spatial arrangement of the spin-containing units in the crystal.*

The net spin of these spin-containing units in organic molecules is usually well represented by the total spin quantum number S. For non-purely organic molecules the use of the quantum number J is sometimes required, when a non-negligible coupling with the angular momentum L is present. When the spin-containing units do not interact (the spin states of the aggregate being degenerate), the aggregate behaves as a set of isolated spin units, and is usually referred to as a *paramagnet*. When the spin-containing units interact with nearby units, non-degenerate states result from that interaction. For instance, in the case of the interaction between two doublet units, one can obtain one singlet state and three energetically degenerate triplet states. When the triplet state is energetically the lowest in energy, one talks about the presence of a *ferromagnetic interaction*. When the ground state is a singlet one talks about an *antiferromagnetic interaction*. The strength of the magnetic interaction between two radicals A and B is evaluated by the parameter J_{AB}, which is proportional to the energy difference between the highest and lowest spin states. Using the Heisenberg Hamiltonian: $\hat{H} = -2\sum J_{AB}\hat{S}_A \cdot \hat{S}_B$, values of $J_{AB} > 0$ stand for ferromagnetic interactions, while $J_{AB} < 0$ stand for antiferromagnetic interactions. It is also worth pointing out the existence of a dimerization problem. Sometimes, the unpaired electrons of two adjacent radicals form a new bond between the two radicals. The result is the formation of diamagnetic dimers. Dimerization should be avoided if one wants to obtain crystals presenting magnetic properties.

When the coupled spin-containing units are structurally different and also have different total spins, S_1 and S_2, the allowed spin states for the dimer vary from $(S_1 + S_2)$ to $|S_1 - S_2|$ in -1 steps. In the high spin ferromagnetic case $(S_1 + S_2)$, one still talks about ferromagnetic interactions, but in the low spin case, necessarily different from zero, one talks about *ferrimagnetic interactions* (in a certain way, all the allowed states from $(S_1 + S_2) - 1$ to $|S_1 - S_2|$ can be considered as ferrimagnetic).

In general, any crystal can present more than one type of magnetic interactions. They can be ferromagnetic between some of the nearby units and antiferromagnetic between others. Sometimes, one or some of these interactions are much larger than the rest and are said to be the *dominant magnetic interactions*. In many of the experimental crystals, the interplay between the different classes of interactions is far from having a clear dominancy, and computations are required to establish the macroscopic magnetic behavior.

The macroscopic properties depend on the temperature (for instance, above a temperature called critical temperature all ferromagnets become paramagnets, but more complex phenomena are also possible). The dependence of the macroscopic magnetic properties with the temperature can help define the dominant magnetic character of the crystal. Consequently, one of the first objectives of any computational procedure is to study the dependence of the common macroscopic magnetic properties against temperature.

The interaction between the spin-containing units can be $0D$, $1D$, $2D$, or $3D$. In the first case (0D), the units A and B interact, but their interaction with the other surrounding spin-containing units is negligible and the crystal has to be visualized as a set of isolated pairs (the extension to isolated aggregates is obvious). For $1D$ magnetic interactions, the A–B pair interacts in such a form that the A–B interaction is extended along one crystallographic direction, i.e. there are infinite $\cdots ABABAB \cdots$ chains within the crystal that do not interact among them. One refers to $2D$ magnetic interactions, if these chains interact with nearby chains in such a way that infinite layers along two dimensions are formed. An n-leg spin ladder (i.e. n chains that interact among themselves but not with the remaining $n+1$ chains) can be considered as a $1D$ motif with partial $2D$ character (i.e., it is a $1D/2D$ magnetic motif). Finally, the chains can interact along all three directions of space. In this case one speaks of $3D$ magnetic interactions. A particular case of $3D$ magnetism is bulk ferromagnetism, where the dominant J_{AB} interactions are ferromagnetic.

From a microscopic point of view, the cause for the existence of magnetism is the energy splitting among states of different multiplicities when the radicals interact. In purely organic molecules, using a simplistic molecular orbital model, the splitting is induced by the overlap of the singly occupied orbitals (SOMO) of the interacting units. This explains why these interactions are sometimes identified as *through-space magnetism* (there is another type of magnetism, called *through-bond* magnetism, where the two interacting units are covalently bonded by a diamagnetic ligand, as in the case of the Cu(II)–OH–Cu(II) interaction).

The underlying principles behind the magnetic through-space interactions are an extrapolation to the supramolecular world of the principles governing the interaction between two H atoms to form H_2 (see Figure 2). The overlap of two 1 s orbitals, centered on atoms H_A and H_B, at very large distances gives rise to four energetically degenerate states, resulting from antisymmetrization of $1\,s_A\alpha(1)1\,s_B\alpha(2)$, $1\,s_A\alpha(1)1\,s_B\beta(2)$, $1\,s_A\beta(1)1\,s_B\alpha(2)$, and $1\,s_A\beta(1)1\,s_B\beta(2)$ configurations. Such degenerate states are equally populated at any temperature, a fact that explains the paramagnetism (lack of net spin) of a pair of non-interacting H_A and H_B atoms. As the H atoms move closer to each other, the 1 s orbitals of *A* and *B* interact to form $1\,s_A + 1\,s_B$ bonding and $1\,s_A - 1\,s_B$ antibonding molecular orbitals (MOs). Once again, there are four states clustered into one singlet (S_1) and three degenerate triplet (T_1) states. At larger distances, the bonding and antibonding MOs are nearly degenerate, and Hund's rule indicates that the ground state is the maximum occupancy and highest multiplicity MO, i.e. the triplet T_1 state. This state has one electron in the bonding MO and another in the antibonding MO. It is energetically repulsive (see Figure 2).

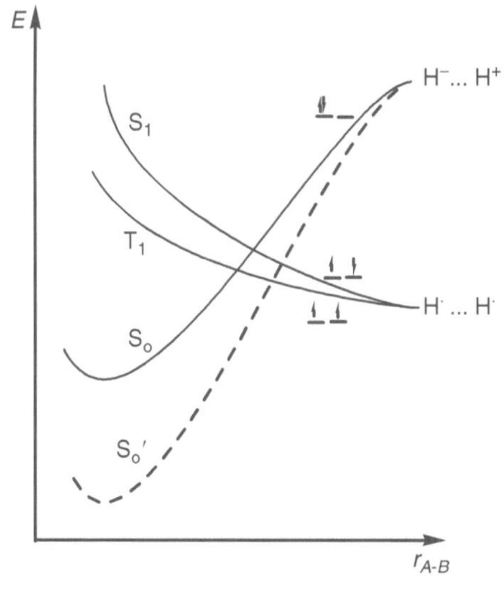

Figure 2. Scheme showing the variation of the energy of the S_o, S_1, and T_1 diabatic curves as a function of the interfragment distance $r_{A\text{-}B}$ in the case of two atoms whose electrons are placed in the corresponding 1 s orbitals. The S_o curve describes the situation in which the new bond formed is not very strong (notice the presence of a barrier in the S_o–S_1 and S_o–T_1 crossing). The $S_o{}'$ curve describes a much stronger bond (notice the decrease in $H^{\cdot}\cdots H^{\cdot}$ towards H_2 energy barriers when going from the S_o to the $S_o{}'$ curve).

The S_1 state is an open-shell singlet and, for similar reasons, is also repulsive (though less stable than T_1, according to Hund's rule). As the distance between the H atoms decreases the splitting between the bonding and antibonding MOs increases, finally leading to a lowest energy configuration with two electrons of opposite spins in the bonding MO. In valence-bond (VB) language, it is equivalent to say there is a crossing between S_1 and S_o diabatic surfaces. Notice that the S_o diabatic surface dissociates into H^- and H^+ (see Figure 2). The S_o diabatic surface is then associated to the H–H bond formation. Thus, S_o is energetically much stable than S_1, which is related to the dissociation into two H atoms, and the S_o–S_1 crossing is an early one implying a negligible barrier, if any. When designing molecule-based crystals, one is particularly interested in working with molecular fragments with S_1 and T_1 curves being the lowest in energy (see large r_{AB} region in Figure 2). In such a case, one has antiferromagnetic or ferromagnetic interactions between the pair of fragments. Instead, when the ground state is the S_o curve, the interfragment interaction is diamagnetic and thus of no magnetic relevance.

The diagram of Figure 2 can be more complex for radicals where overlapping orbitals are molecular and distributed over many atoms. They often present nodes between the atoms (some qualitative theories [10] suggest that these nodal regions are orientations where possible ferromagnetic interactions can be induced). The radical center must be protected to avoid the formation of radical dimers covalently bonded. Such a protection is provided by the presence of bulky groups or various lone-pair electrons in the vicinity of the radical center. The ONCNO radical center in the nitronyl nitroxides (see Figures 2 and 3) are of the latter class. Their electronic structure presents an open-shell CX_2H center ($X = NO$) and two open-shell NO groups. Notice that each NO group has one electron on the O atom, two lone-pair orbitals on the O atom and one on the N atom.

The electronic structure of such a group (see Figure 3a) is similar to that of the H_3 radical, with two electrons in the bonding MO (which has no nodes) and one electron in the nonbonding MO (which has a node on the central H atom). In this model, the spin is delocalized half on each lateral H atom, and so is the case in the ONCNO radical centers where the spin is delocalized over the lateral NO fragments. Accurate *ab initio* computations on a large number of nitronyl nitroxides [11] show that this is indeed the case in most of these radicals, although a small polarization is found in the central C atom of the ONCNO group, normally referred to as spin polarization, which is only reproduced [11] by methods which go beyond the restricted-open-Hartree–Fock (ROHF) formalism like the UHF method, among others. The interaction between two ONCNO groups can be analyzed for the smallest molecule containing such a group, the (HNO)–CH–(HNO) molecule, as shown in Figure 3b. As seen there, the tendency of the open-shell centers to form a new bond between the spin centers is counterbalanced by the repulsive interactions between the lone-pair electrons of the NO groups. The net effect of the interfragment lone-pair$\cdots$lone-pair interactions is a shift of the S_o diabatic towards less stable values (see Figure 2). By doing this we avoid the formation of a new bond between these fragments. A similar effect is induced in spin-containing units that have net charges

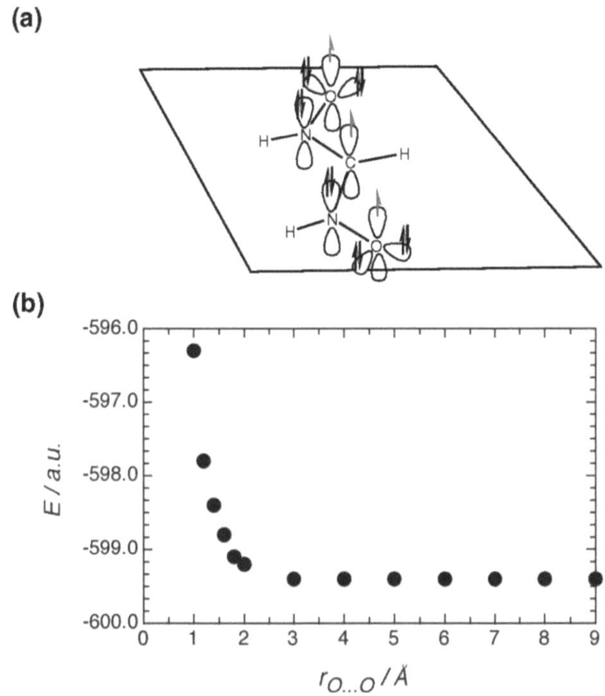

Figure 3. (a) Valence bond representation of the electronic structure of the (HON)–CH–(NOH) radical, a prototype of nitronyl nitroxide. (b) Potential energy curve for the interaction of two (HON)–CH–(NOH) radicals placed one on top of the other (the coordinates differing by a displacement along the vertical coordinate z). Each fragment is a doublet state, and the curve was computed for the triplet state at the UB3LYP/6−31+G(d) level.

of the same sign. In this case, the resulting dimer can be, at best, a metastable system (that is, the dimer minimum lies above the dissociation energy of the neutral fragments, as in $TCNE^-$ dimers [12]).

Most currently available theoretical analyses of purely organic crystals are focused only on estimating the sign and size of the J_{AB} pair interaction, using qualitative or quantitative evaluations, and then the macroscopic behavior is inferred in simple terms ("one interaction dominates", "there is a competition between two or more interactions", ...). This is clearly not a valid procedure in many cases. Our first-principles methodology supersedes these theories, since it allows a rigorous quantitative connection between the microscopic and macroscopic magnetic properties. The sign and size of the through-space J_{AB} pair interactions are normally estimated using the so-called McConnell-I [13] and McConnell-II [14] mechanisms. Both are qualitative proposals that predict the sign and order of magnitude of the J_{AB} parameters by looking at the orientations of the interacting radicals. The McConnell-I mechanism looks at the signs of the atomic spin population on the atoms of the dimer establishing the shortest contacts (it predicts that the atomic spin populations must be of opposite signs for a ferromagnetic interaction). The McConnell-II mechanism works on the basis of the stability of the charge transfer electronic configurations in the dimer.

Both McConnell mechanisms of through-space magnetism have been shown to fail [15, 16] due to the oversimplifications that they present. Their validity in other cases is associated to fortuitous error compensations. Consequently, its general use is untrustworthy. The failure of the McConnell-I mechanism is also seen when doing statistical analysis of the packing of nitronyl nitroxide crystals that present dominant ferromagnetic or antiferromagnetic interactions. According to McConnell-I these nitronyl nitroxide crystals should present specific orientations of the radicals within crystals showing an overall dominant ferromagnetic (or antiferromagnetic) behavior. However, a recent study [17] showed the absence of simple correlations between the relative orientations of the spin-containing groups (the ONCNO group of the five-membered ring, see Figure 1) and the dominant magnetic interactions.

Given the lack of solid magneto-structural correlations, a proper determination of the J_{AB} values can be obtained using quantum mechanical methods [18]. Once these pair interactions are known, one still needs a quantitative and rigorous procedure in order to connect the microscopic magnetic information contained in the J_{AB} pairs and the macroscopic magnetic properties, for instance, the magnetic susceptibility. A numerical procedure of this kind would provide the possibility of quantitatively exploring how changes in the microscopic pair interactions induce changes in the macroscopic properties, in particular magnetic susceptibility. This would allow to find what changes should be induced in the crystal packing to improve the macroscopic magnetic properties, thus providing new perspectives towards crystal engineering of purely organic molecular magnets. In the following sections, we review a quantitative numerical procedure connecting the microscopic and macroscopic data, the so-called first-principles *bottom-up* methodology [7], and we will illustrate it on the α-HNN crystal. Experimental studies indicate that this crystal presents antiferromagnetic interactions.

3. THE FIRST-PRINCIPLES *BOTTOM-UP* METHODOLOGY

The first-principles *bottom-up* methodology is a quantitative procedure to analyze rigorously the magnetism of molecular crystals in a totally unbiased form. Its only input data are the geometry of the crystal to be analyzed. The magnetic properties are computed using first-principles methods, without making any hypothesis on the nature of the magnetic interactions. More specifically, the procedure computes the J_{AB} parameters for all unique microscopic magnetic interactions present in the crystal. Then, using these J_{AB} values, it calculates in a rigorous form its macroscopic magnetic properties. These calculations are done following a series of steps designed to make the process systematic and rigorous. The validity of the procedure and the quality of its results have been proven on a series of representative crystals [19]. In the following section, we shall describe the main steps of this procedure for the α-HNN crystal in order to get a proper picture of the work strategy.

The HNN crystal presents two different polymorphic phases [8], called in the literature the α and β phases. As already mentioned, here we focus only on the α phase, sometimes called TOLKEK (Cambridge Crystallographic Database). This crystal belongs to the $P2_1/n$ space group ($a = 11.879$ Å, $b = 11.611$ Å, $c = 6.332$ Å, $\beta = 104.48°$, $Z = 4$) with four radicals per unit cell. Its crystal packing (see Figure 4) can been rationalized [20] as planes that pile up along the *a* axis (four per unit cell), stacked in an ABBA pattern (Figure 4a). Each plane is identical, and only differs from the others in its relative position within the unit cell. These planes result from the aggregation of HNN radicals (Figure 4b). *Ab initio* computations [20] indicate that the strongest radical–radical intermolecular interactions are two $C(sp^2)$–H···O–N hydrogen bonds (they form the primary structure of the crystal). Thus, they first form HNN dimers. These dimers then aggregate by means of the $C(sp^3)$–H···O–N

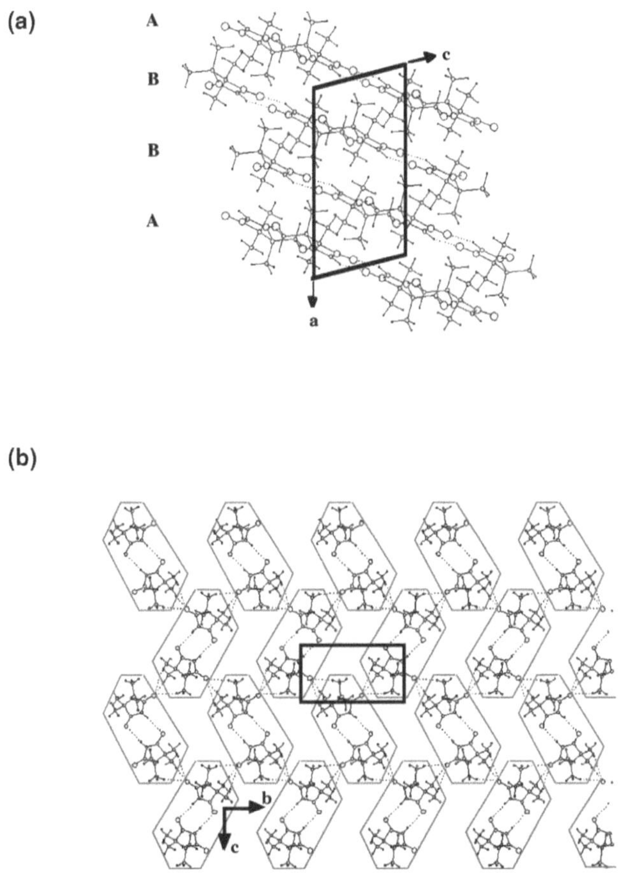

Figure 4. Crystal packing of TOLKEK showing the C–H···O contacts shorter than 3.20 Å. (a) view along the *b* axis, where one can see four layers (ABBA) within the unit cell; (b) view of one layer (along the *bc* plane), stressing the dimeric entities and their $C(sp^2)$–H···ON hydrogen bonds.

hydrogen bonds and form planes (secondary structure of the crystal). Nearby planes can aggregate using the remaining $C(sp^3)$–H···O–N hydrogen bonds (the stacks are the tertiary structure of the crystal).

As already mentioned the magnetic properties of TOLKEK are those of a crystal which presents antiferromagnetic interactions. The dependence of the experimental magnetic susceptibility on temperature has been fitted to a "pure" Bleaney–Bowers model (isolated radical dimers), the fitted values being C = 0.5 emu $K\,mol^{-1}$ and $\mathbf{J}/k_B = -11.2\,K$ (that is, $-7.8\,cm^{-1}$). Note that this **J** parameter is the result of a least-squares fitting and does not necessarily have a real physical meaning (such meaning is only possible when the model chosen properly represents the real magnetic topology of the crystal).

In order to carry out a first-principles *bottom-up* study of the TOLKEK crystal, one has to perform the following four steps in sequential order.

1. Analysis of the crystal structure in order to find all unique radical–radical pairs present in the crystal.
2. Computation of the J_{AB} microscopic magnetic interactions for all pairs.
3. Determination of the real magnetic topology of the crystal and of its minimal magnetic model space.
4. Calculation of the macroscopic magnetic properties.

We now describe each of these steps in detail for the case of the antiferromagnetic HNN crystal.

Step 1: Computation of all unique radical–radical pair interactions in the crystal.

For such a task, it is necessary to analyze in a systematic way the crystal structure (see Figure 4) and find all unique radical–radical pairs potentially relevant to magnetic interactions. As it has been shown that the strength of the through-space magnetic interactions decreases exponentially with the distance between the molecules [21], only those pairs involving first and second-nearest neighbors to a given radical will be important. In practical terms, the unique radical pairs are identified by looking at all radical–radical pairs whose distance is smaller than a given cutoff value (which is selected to include all first-nearest neighbors and the closest next-nearest neighbors). Notice that this procedure does not assume the presence of a given microscopic magnetic interaction or the size of a J_{AB} parameter. In the case of TOLKEK, a cutoff of 7.4 Å between the ONCNO groups of the five-membered ring (those found to localize 90% of the spin density of the HNN molecule) is enough to satisfy the above criteria [22].

Figure 5 shows the seven unique radical–radical pairs found in the crystal (*d1–d7*).

Step 2: Computation of the J_{AB} microscopic magnetic interactions for all pairs.

The value of the J_{AB} magnetic interaction, hereafter identified as $J_{AB}(d1)$–$J_{AB}(d7)$, is now computed for each of the radical–radical pairs selected in Step 1. The HNN radical is a doublet. Therefore, the only two possible states for the radical–radical pairs are a singlet and a triplet. The value of the J_{AB} parameter for each pair is obtained from the energy difference of the singlet and the triplet using the geometry that each pair has in the crystal.

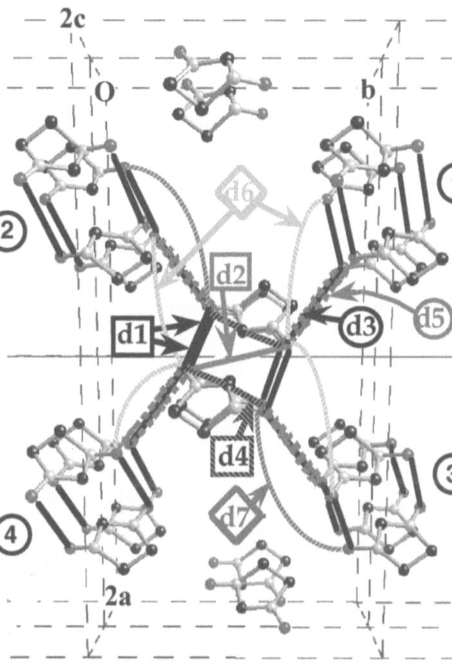

Figure 5. Unique radical–radical pairs (d1–d7) found in the α-HNN crystal. There are five identical piles of dimers, the central one is surrounded by four lateral ones identified by the numbers 1, 2, 3, and 4. Methyl groups are omitted for clarity.

The exact expression of J_{AB} depends on the Heisenberg Hamiltonian used. In the case of the most widely-used expression for this Hamiltonian:

$$(1) \qquad \hat{H} = -2\sum J_{AB}\hat{S}_A \cdot \hat{S}_B$$

one finds that $E^S - E^T = 2J_{AB} = 2(E^S_{BS} - E^T)$, where E^T is the energy of the triplet and E^S_{BS} that of the singlet, computed using the broken symmetry approach [23] (this expression derives from the original broken-symmetry equations when doing $S_{ab} = 0$ [23], the normal case for through-space interactions). According to our experience, this expression gives values closer to the experimental results, although one should mention that the use of the broken-symmetry approach within the DFT context has been controversial, particularly in relation to the use of projection in computing the values of the J_{AB} parameters (for a detailed discussion, the reader is addressed to reference [23]).

The values of the J_{AB} parameters for the seven radical–radical pairs in Figure 5 were computed using the UB3LYP functional [24] with the 6−31+G(d) basis set. The computations were performed using the Gaussian-98 program suite [25]. The results obtained are collected in Table 1. Only five of the seven $J_{AB}(di)$ parameters are non-negligible. Two of them correspond to antiferromagnetic interactions while

Table 1. Values of the J_{AB} pair interactions computed for the $d1$–$d7$ radical pairs using broken symmetry UB3LYP/6−31+G(d)

Pair	$d_i(O\cdots O)/Å$	$J_{AB}(di)/cm^{-1}$
d1	3.80	−7.26
d2	4.27	+1.54
d3	4.65	+0.22
d4	4.79	+0.24
d5	6.12	−0.13
d6	6.22	< |0.05|
d7	7.01	< |0.05|

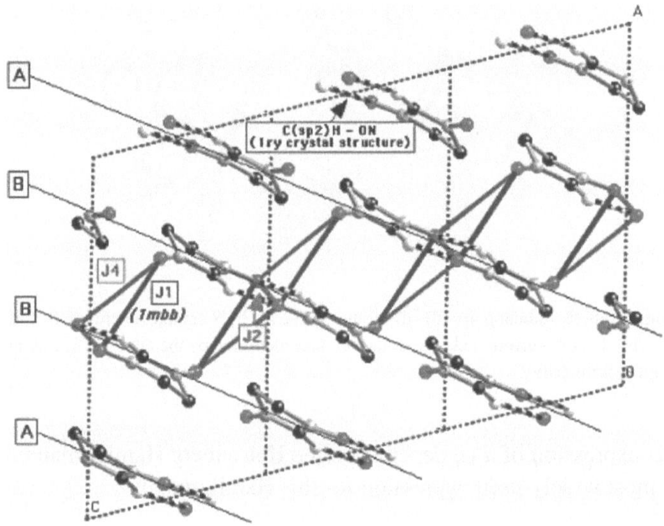

Figure 6. Lateral view of the TOLKEK crystal, showing the position of the $J_{AB}(d1)$, $J_{AB}(d2)$, and $J_{AB}(d4)$ pairs (the $J_{AB}(d3)$, and $J_{AB}(d5)$ pairs are not shown for simplicity). Note that in the Figure $J_{AB}(di) = J_i$. Methyl groups are omitted for clarity.

the remaining ones correspond to ferromagnetic interactions. One can now correlate these pair interactions with the crystal packing. The dimers involved in the primary crystal packing structure of TOLKEK (that is, the dimers encircled in Figure 4b) correspond to the $d2$ dimer, and have a $J_{AB}(d2)$ value of $+1.54\,cm^{-1}$. The strongest J_{AB} pair ($J_{AB}(d1) = -7.26\,cm^{-1}$) connects two radicals in adjacent planes placed in opposite directions, while $J_{AB}(d4)$ connects radicals pointing in the same direction (see Figure 6).

Step 3: Determination of the real magnetic topology of the crystal and of its minimal magnetic model space.

From the magnetic point of view, the magnetic interactions define the topology of a 2-leg spin-ladder. Figure 5 shows five of such spin-ladders: a central one surrounded by four others, identified as 1, 2, 3, and 4 (Figure 6 presents a lateral view). $J_{AB}(d3)$ and $J_{AB}(d5)$ link magnetically adjacent ladders. This structure of connected two-leg spin-ladders is the *magnetic topology* of the crystal. Figure 7 shows a schematic view of this magnetic topology, where each radical is replaced by a point site located in its center of mass. Notice that the magnetic topology represents the network of magnetic interactions that are found between all the radicals of the crystal, and it is a good graphical depiction of its magnetic properties.

The topology displayed in Figure 7 can be visualized as five weakly interacting two-leg spin-ladders (a central one and four surrounding ladders, numbered 1, 2, 3, and 4, according to Figure 5). Within each two-leg spin-ladder, one finds a strong $J_{AB}(d1)$ interaction along the rungs, a weaker $J_{AB}(d2)$ interaction along the diagonals of the rails, and a weak $J_{AB}(d4)$ along the rails. The ladders are interconnected by means of very weak $J_{AB}(d3)$and $J_{AB}(d5)$ interactions. Depending on the relevance of these latter two interactions with respect to the others, the magnetic topology is: (i) a weakly interacting set of spin-ladders (a 3D magnetic motif), or (ii) a set of isolated spin-ladders (a 1D/2D magnetic motif). One way of establishing which of these two descriptions is correct is to check which one reproduces best macroscopic magnetic properties (for instance, the temperature dependence curve of the magnetic susceptibility χ(T)).

The computation of the macroscopic properties requires a previous calculation of the energy spectrum (the eigenvalues for all possible states that can be found in the crystal). In principle, such a calculation can be performed by diagonalizing the matrix representation of the Heisenberg Hamiltonian for a finite crystal, using the basis of all

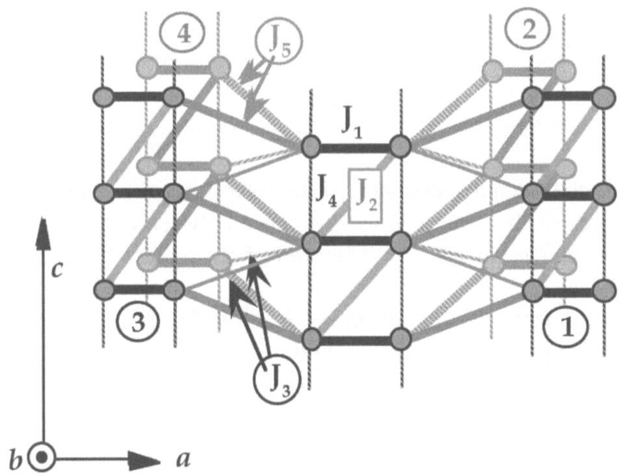

Figure 7. Magnetic topology of the TOLKEK crystal. Each radical is represented by a point site and the connection lines indicate pairs of radicals where the $J_{AB}(di)$ interaction is non-negligible. Note that in the Figure $J_{AB}(di) = J_i$. There is a central spin-ladder (their J_1, J_2, and J_4 pairs are specified) and four surrounding spin-ladders, numbered **1** to **4** as in Figure 5.

the spin functions of this finite crystal. However our crystal is infinite. How then do we represent properly the energy spectrum of the infinite crystal by a finite one?

We have done numerical simulations with models of increasing size that mimic the crystal, and also methodological studies using a *regionally reduced density matrix approach* [26]. Both show it is possible to obtain an energy spectrum that has similar macroscopic results than larger spaces by using properly selected subspaces. The key is to use subspaces that include all the physically significant J_{AB} interactions and in the same proportion found in the larger space. When these subspaces are properly selected, the macroscopic results converge towards the results obtained with the larger space when increasing the size of the subspace. The process can be extrapolated to an infinite crystal if one uses as convergence data the experimental data of the macroscopic property of interest. Obviously the full crystal must result from the extension of the smaller model subspaces. The smallest finite model capable of describing the magnetic properties of the infinite crystal is called the *minimal magnetic model space*.

Step 4: Calculation of the macroscopic magnetic properties of the crystal.

Once a reasonable minimal magnetic model space has been selected, the basis set for the representation of the Heisenberg Hamiltonian is built by adding all the spin functions for each spin-containing unit (for instance, α and β functions when the spin unit is a doublet) and selecting just those that are antisymmetric (because electrons are Fermions). Therefore, the matrix representation increases with the number of sites (N) as $N!/[(N/2)!(N/2)!]$, thus increasing the computational cost and required resources. Using this basis set, the matrix representation of the Heisenberg Hamiltonian is computed and diagonalized. In many ways, such a matrix representation is like a topological Hückel matrix, where the β parameters are substituted by J_{AB} pairs. We have used the following Heisenberg Hamiltonian, which allows the use of faster computer codes:

$$\hat{H} = -\sum_{A,B}^{N} J_{AB}\left(2\hat{S}_A \cdot \hat{S}_B + 1/2\hat{I}_{AB}\right) \tag{2}$$

where $\hat{\mathrm{I}}_{\mathrm{AB}}$ is the identity operator. The energy spectra computed using this Hamiltonian present the same energy difference between the different eigenvalues than those obtained using Eq. (1). Therefore, both predict the same macroscopic properties.

As a final comment, we should like to mention that the subspaces can be built either by imposing periodic boundary conditions or in an open way (that is, just truncating the crystal without periodic boundary conditions). Extensive numerical tests indicate that both choices provide nearly identical results. We decided to use open finite models in our implementation.

Once the energy spectrum of the finite model space is known, one can use its eigenvalues to compute the macroscopic property of interest, using its exact Statistical Mechanics expression. For instance, the magnetic susceptibility χ at a given temperature (T) is computed by using the formula:

$$(3) \qquad \chi = \frac{Ng^2\mu_B^2}{3k_BT}\mu_o \left[\frac{\sum_n S_n(S_n+1)(2S_n+1)\exp\left[-\frac{E_n-E_o}{k_BT}\right]}{\sum_n (2S_n+1)\exp\left[-\frac{E_n-E_o}{k_BT}\right]} \right]$$

where S_n and E_n are the spin quantum number and total energy for the nth state, E_o is the energy of the ground state, T is the absolute temperature, μ_B the Bohr magneton, k_B the Boltzmann factor, N is Avogadro's number, g the electron gyromagnetic factor, and μ_o the permeability of free space. This equation can be solved for a given range of temperatures (usually from room temperature to nearly 0 K). This allows a direct comparison between the theoretical and experimental $\chi(T)$ curves.

The similarity of shape of these two curves is an indication of the goodness-of-fit of the selected minimal model space. One should keep in mind that as one extends this space, the computed curves should converge towards the experimental curve.

Let us illustrate the previous steps on the α-HNN crystal. As mentioned above, a first reasonable form of visualizing the magnetic topology of this crystal is as a set of noninteracting two-leg spin-ladders (see Figure 7). A minimal magnetic model space capable of describing the isolated 2-leg spin-ladders found in α-HNN is the 4-site model defined by J_1, J_2, and J_4 (see Figure 7). Such a model includes two rungs of the ladder and can be called 2 *mbb* (each rung in one *mbb*). Extending the 2*mbb* model by adding more rungs to the ladder generates the rest of the ladder. This 2*mbb* model contains 4 spin centers and two rungs. As the spin centers are doublets, the size of the matrix representation of the Heisenberg Hamiltonian for the 2*mbb* model is 6 × 6. Thus, for the 5*mbb* model (5 connected rungs, that is 10 spin centers), the matrix representation of the Heisenberg Hamiltonian is of dimension 252 × 252.

An alternative form of visualizing the magnetic topology of the α-HNN crystal is as a set of interacting spin-ladders. Such model can be described by adding the 2*mbb* model (that describes one isolated 2-leg spin-ladder) and the four rungs of the nearby ladders (identified as **1–4** in Figure 7). This addition introduces the J_3 and J_5 interactions. The resulting extended model will be called the 2*mbb*–(*mbb*)$_4$ model. The convergence of this model can be tested by increasing the number of rungs in the central spin-ladder (as in the 3mbb, 4mbb, 5mbb, ... models) and/or the lateral ladders.

The $\chi(T)$ curves computed by using the 2*mbb* (+), 3*mbb* (□), 4*mbb* (△), and 2*mbb*–(*mbb*)$_4$ (×) models are shown in Figure 8. A comparison of the first three curves shows that convergence is achieved. The same convergence (not shown) is also found when extending the lateral ladders. The negligible difference between the 2*mbb* (+) and 2*mbb*–(*mbb*)$_4$ (×) curves also shows that the isolated spin-ladder is the proper form for describing the magnetic topology of the α-HNN crystal.

Although the convergence is achieved, at low temperatures there is a small difference between the curve calculated with the largest model and the experimental one. We have found, in this and many other cases, that such differences disappear when a small linear-scaling factor is applied to all eigenvalues of the energy spectrum (in this case, the linear factor required is 1.1). Such a scaling factor accounts for the

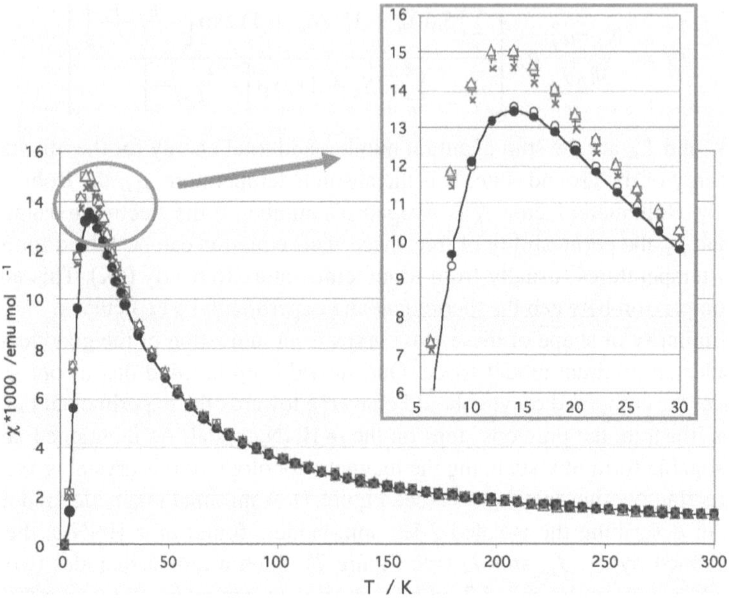

Figure 8. Experimental (*solid line with solid circles*, ●) and computed magnetic susceptibility curves (*the inset expands the low temperature region*). The computed curves are for model spaces 2*mbb* (+), 3*mbb* (□), 4*mbb* (△), 2*mbb*–(*mbb*)$_4$ (×), and model 2*mbb* with a scaling factor 1.1 (●), which reproduces best the experimental curve.

small errors introduced in the computed J_{AB} values using DFT methods instead of more accurate ones [27], the use of high-temperature crystal structures instead of low-temperature structures, and the neglect of polarization effects that could be present in the crystal but are not accounted for in our radical–radical calculations. We thus obtain a very good agreement between the computed and experimental magnetic susceptibility curves. Although not shown here, agreement is also found for other macroscopic properties, as the heat capacity curve. Such good agreement has also been found in many other magnetic crystals presenting a wide variety of macroscopic magnetic properties. Therefore, we can be confident in the quality of the predictions achieved by the first-principles *bottom-up* methodology.

We also reproduce other well-known properties of two-leg spin-ladders. For instance, it is well known that antiferromagnetic two-leg spin-ladders should present a gap between their lowest-energy singlet and triplet states. We have found such a gap (the singlet-triplet energy gap being $-12.5 \pm 3.5\,\text{cm}^{-1}$ when no scaling factor is used, and $-13.8\,\text{cm}^{-1}$ when the 1.1 linear-scaling factor is applied).

Let us summarize the main physical information obtained from this first-principles *bottom-up* study, and how it can be used to make a rational design of crystals presenting improved magnetic properties. We have found that the α-HNN crystal presents the magnetic topology of an isolated two-leg spin-ladder having ferro and antiferromagnetic interactions, these latter being the dominant ones. Such a topology is not predicted by a simple inspection of the crystal packing. We have also found that the strongest magnetic interaction is $J_{AB}(d1)$. By changing the chemical structure of the radical, one can modify the geometry of the dimer and thus tailor the strength of $J_{AB}(d1)$. Therefore, this methodology provides a tool to rationally design new molecular crystals presenting improved magnetic properties.

4. CONCLUSIONS

The results presented in this study show that the first-principles *bottom-up* methodology is a powerful tool for analyzing the mechanisms of magnetic interactions in molecule-based magnetic crystals. This approach not only accurately reproduces the macroscopic magnetic properties of these crystals but also provides a deep insight into the details of the interactions at the microscopic level. Furthermore, it connects the macroscopic properties to their microscopic origin, thus providing a tool for starting the rational design of such molecular materials.

As an illustrative example, we have applied the first-principles *bottom-up* methodology to the study of the magnetism of the α-HNN crystal. We have been able to reproduce with accuracy the macroscopic magnetic susceptibility curve of this crystal. We have also concluded that the magnetic topology of the crystal can be described as a set of isolated two-leg spin-ladders presenting ferro and antiferromagnetic interactions, the latter being the dominant ones.

Acknowledgements

The authors gratefully acknowledge the continuous support of the Spanish Science and Education Ministry (projects BQU2002-04587-C02-02 and CTQ2005-02329) and the Catalan Autonomous Government (Grants 1999SGR-00046 and 2005SGR-00036). They also thank CESCA–CEPBA for grants of computer time partly funded by the University of Barcelona. M.D. thanks to the Spanish Science and Education Ministry for the award of a "Ramón y Cajal" Fellowship.

References

1. (a) P. Turek, K. Nozawa, D. Shiomi, K. Awaga, T. Inabe, Y. Maruyama, and M. Kinoshita, *Chem. Phys. Lett.* **180**, 327, 1991. (b) M. Tamura, Y. Nakazawa, D. Shiomi, K. Nozawa, Y. Hosokoshi, M. Ishikawa, M. Takahashi, and M. Kinoshita, *Chem. Phys. Lett.* **186**, 401, 1991.
2. R.L. Carlin, *Magnetochemistry*, Springer, Berlin, 1986.
3. M. Kinoshita, *Jpn. J. Appl. Phys.* **33**, 5718, 1994; M. Kinoshita, *Handbook of Organic Conductive Molecules and Polymers* Vol. 1, Nalwa H.S. (ed.), Wiley, New York, 15, 1997.
4. (a) P.-M. Allemand, K.C. Khemani, A. Koch, F. Wudl, K. Holczer, S. Donovan, G. Grüner, J.D. Thompson, *Science* **253**, 301, 1991. (b) T.L. Makarova, B. Sundqvist, R. Höhne, P. Esquinazi, Y. Kopelevich, P. Scharff, V.A. Davydov, L.S. Kashevarova, A.V. Rakhnmanina, Nature **413**, 716, 2001.
5. S. Ferlay, T. Mallah, R. Ouahes, P. Veillet, and M. Verdaguer, *Nature* **378**, 701, 1995.
6. J.S. Miller, J.C. Callabrese, H. Rommelmann, S.R. Chittipedi, J.H. Zhang, W.M. Reiff, A.J. Epstein, *J. Am. Chem. Soc.* **109**, 769, 1987.
7. M. Deumal, M. Bearpark, J.J. Novoa, M.A. Robb, *J. Phys. Chem.* A **106**, 1299, 2002.

8. Y. Hosokoshi, M. Tamura, K. Nozawa, S. Suzuki, M. Kinoshita, H. Sawa, and R. Kato, *Synthetic Met.* **71**, 1795, 1995.
9. Refcodes are the unique codenames given by the Cambridge Crystallographic Database (CCSD) to each crystal present in it. See for instance: (a) F.H. Allen, J.E. Davies, J.J. Galloy, O. Johnson, O. Kennard, C.F. Macrae, and D.G. Watson, *J. Chem. Inf. Comput. Sci.* **31**, 204, 1991; (b) F.H. Allen, O. Kennard, *Chem. Des. Automation News* **8**, 31, 1993.
10. (a) O. Kahn, *Molecular Magnetism*, VCH, New York, 1993. (b) P.J. Hay, J.C. Thibeault, and R. Hoffmann, *J. Am. Chem. Soc.* **97**, 4884, 1975.
11. M. Deumal, P. Lafuente, F. Mota, and J.J. Novoa, *Synth. Met.* **122**, 477, 2001.
12. J.J. Novoa, P. Lafuente, R.E. Del Sesto, and J.S. Miller, *Angew. Chem. Internat. Ed.* **40**, 2540, 2001.
13. The McConnell-I magneto-structural correlation was an empirical suggestion proposed in: H.M. McConnell, *J. Chem. Phys.* **39**, 1910, 1963.
14. The McConnell-II magneto-structural correlation was proposed a few years after the McConnell-I proposal in: H.M. McConnell, Proc. R.A. Welch *Found. Conf. Chem. Res.* **11**, 144, 1967.
15. The McConnell-I proposal was found to lack of a rigorous derivation from a general Heisenberg Hamiltonian and to give results that were correct due to error compensation associated to the high symmetry of the systems where it was calibrated; see: M. Deumal, J.J. Novoa, M.J. Bearpark, P. Celani, M. Olivucci, and M.A. Robb, *J. Phys. Chem.* A **102**, 8404, 1998.
16. The McConnell-II proposal was found to fail in some well known examples; see: C. Kollmar and O. Kahn, *Acc. Chem. Res.* **26**, 259, 1993.
17. (a) M. Deumal, J. Cirujeda, J. Veciana, and J. J. Novoa, *Adv. Mater.* **10**, 1461, 1998; (b) M. Deumal, J. Cirujeda, J. Veciana, and J.J. Novoa, *Chem. Eur. J.* **5**, 1631, 1999.
18. (a) V. Barone, A. Bencini, and A. diMatteo, *J. Am. Chem. Soc.* **119**, 10831, 1997; (b) J. Cano, P. Alemany, S. Alvarez, M. Verdaguer, and E. Ruiz, *Chem. Eur. J.* **4**, 476, 1998. (c) I.P.R. Moreira, F. Illas, C.J. Calzado, J. F. Sanz, J.-P. Malrieu, N. Ben Amor, and D. Maynau, *Phys. Rev.* B **59**, R 6593, 1999. (d) M. Mitani, H. Mori, Y. Takano, D. Yamaki, Y. Yoshioka, and K. Yama-guchi, *J. Chem. Phys.* **113**, 4035, 2000. (e) C. Blanchet-Boiteux, J.M. Mouesca, *J. Phys. Chem.* A **104**, 2091, 2000. (f) A. Rodriguez-Fortea, P. Alemany, S. Alvarez, and E. Ruiz, *Chem. Eur. J.* **7**, 627, 2001.
19. (a) M. Deumal, M.A. Robb, and J.J. Novoa, *Polyhedron* **22**, 1935, 2003; (b) M. Deumal, M.J. Bearpark, M.A. Robb, Y. Pontillon, and J.J. Novoa, *Chem. Eur. J.* **10**, 6422, 2004; (c) M. Deumal, J. Ribas-Ariño, M.A. Robb, J. Ribas, and J.J. Novoa, *Molecules* **9**, 757, 2004; (d) M. Deumal, M.A. Robb, and J.J. Novoa, *Polyhedron* **24**, 2368, 2005; (e) M. Deumal, G. Giorgi, M.A. Robb, M. M. Turnbull, C.P. Landee, and J.J. Novoa, *Eur. J. Inorg. Chem.* 4697, 2005.
20. M. Deumal, J. Cirujeda, J. Veciana, M. Kinoshita, Y. Hosokoshi, and J.J. Novoa, *Chem. Phys. Lett.* **265**, 190, 1997.
21. C. Herring, Direct exchange between well-separated atoms, in *Magnetism* Vol. IIB, p. 5, G.T. Rado, H. Shul (eds.) Academic Press, New York, 1966.
22. Measured as the shortest O···O distance between the ONCNO groups. The threshold distance comes from increasing by 1.0 Å the largest ONCNO···ONCNO distance, for which a non-negligible computed magnetic J_{AB} interaction was found in another nitronyl nitroxide radical (KAXHAS).
23. (a) L. Noodlemann, *J. Chem. Phys.* **74**, 5737, 1981. (b) L. Noodlemann, E.R. Davidson, *Chem. Phys.* **109**, 131, 1986. (c) R.G. Parr and W. Yang, *Density-Functional Theory of Atoms and Molecules*, Oxford University Press, New York, 1989; (d) L. Noodlemann, D.A. Case, *Adv. Inorg. Chem.* **38**, 423, 1992; (e) L. Noodlemann, C.Y. Peng, D.A. Case, and J.M. Mouesca, *Coord. Chem. Rev.* **144**, 199, 1995. (f) E. Ruiz, P. Alemany, S. Alvarez, and J. Cano, *J. Am. Chem. Soc.* **119**, 1297, 1997. (g) R. Caballol, O. Castell, F. Illas, I.D.R. Moreira, and J.-P. Malrieu, *J. Chem. Phys.* A **101**, 7860, 1997. (h) Ref. [8d]. (i) H. Nagao, M. Nishino, Y. Shigeta, T. Soda, Y. Kitagawa, T. Onishi, Y. Yoshioka, and K. Yamaguchi, *Coord. Chem. Rev.* **198**, 265, 2000. (j) J.-M. Mouesca, *J. Chem. Phys.* **113**, 10505, 2000. (k) F. Illas, I.D.R. Moreira, C. de Graaf, and V. Barone, *Theor. Chem. Acc.* **104**, 265, 2000.
24. (a) A.D. Becke, *Phys. Rev.* A **38**, 3098, 1988. (b) C. Lee, W. Yang, and R.G. Parr, *Phys. Rev.* B **37**, 785, 1988. (c) A.D. Becke, *J. Chem. Phys.* **98**, 5648, 1993.

25. M.J. Frisch, G.W. Trucks, H.B. Schlegel, P.M.W. Gill, B.G. Johnson, M.A. Robb, J.R. Cheeseman, T. Keith, G.A. Petersson, J.A. Montgomery, K. Raghavachari, M.A. Al-Laham, V.G. Zakrzewski, J.V. Ortiz, J.B. Foresman, J. Cioslowski, B.B. Stefanov, A. Nanayakkara, M. Challacombe, C.Y. Peng, P.Y. Ayala, W. Chen, M.W. Wong, J.L. Andres, E.S. Replogle, R. Gomperts, R.L. Martin, D.J. Fox, J.S. Binkley, D.J. Defrees, J. Baker, J.J.P. Stewart, M. Head-Gordon, C. Gonzalez, and J.A. Pople, Gaussian 98, Gaussian, Pittsburgh, PA, 1999.
26. J.-P. Malrieu and N. Guihéry, *Phys. Rev.* B **63**, 5110, 2001. These authors formulate a renormalization-group procedure where the renormalized Hamiltonian is defined as a Bloch effective Hamiltonian. This procedure is based on the real-space renormalization-group (RSRG) method: (a) K. G. Wilson, *Rev. Mod. Phys.* **47**, 773, 1975. (b) S.R. White and R.M. Noack, *Phys. Rev. Lett.* **68**, 3487, 1992.
27. 10–15% systematic error compared to FCI, DDCI (J.J. Novoa, and R. Caballol, unpublished results).

INDEX

V

W

Z

Progress in Theoretical Chemistry and Physics

1. S. Durand-Vidal, J.-P. Simonin and P. Turq: *Electrolytes at Interfaces.* 2000
ISBN 0-7923-5922-4
2. A. Hernandez-Laguna, J. Maruani, R. McWeeny and S. Wilson (eds.): *Quantum Systems in Chemistry and Physics.* Volume 1: Basic Problems and Model Systems, Granada, Spain, 1997. 2000 ISBN 0-7923-5969-0; Set 0-7923-5971-2
3. A. Hernandez-Laguna, J. Maruani, R. McWeeny and S. Wilson (eds.): *Quantum Systems in Chemistry and Physics.* Volume 2: Advanced Problems and Complex Systems, Granada, Spain, 1998. 2000 ISBN 0-7923-5970-4; Set 0-7923-5971-2
4. J.S. Avery: *Hyperspherical Harmonics and Generalized Sturmians.* 2000
ISBN 0-7923-6087-7
5. S.D. Schwartz (ed.): *Theoretical Methods in Condensed Phase Chemistry.* 2000
ISBN 0-7923-6687-5
6. J. Maruani, C. Minot, R. McWeeny, Y.G. Smeyers and S. Wilson (eds.): *New Trends in Quantum Systems in Chemistry and Physics.* Volume 1: Basic Problems and Model Systems. 2001 ISBN 0-7923-6708-1; Set: 0-7923-6710-3
7. J. Maruani, C. Minot, R. McWeeny, Y.G. Smeyers and S. Wilson (eds.): *New Trends in Quantum Systems in Chemistry and Physics.* Volume 2: Advanced Problems and Complex Systems. 2001 ISBN 0-7923-6709-X; Set: 0-7923-6710-3
8. M.A. Chaer Nascimento: *Theoretical Aspects of Heterogeneous Catalysis.* 2001
ISBN 1-4020-0127-4
9. W. Schweizer: *Numerical Quantum Dynamics.* 2001 ISBN 1-4020-0215-7
10. A. Lund and M. Shiotani (eds.): *EPR of Free Radicals in Solids.* Trends in Methods and Applications. 2003 ISBN 1-4020-1249-7
11. U. Kaldor and S. Wilson (eds.): *Theoretical Chemistry and Physics of Heavy and Superheavy Elements.* 2003 ISBN 1-4020-1371-X
12. J. Maruani, R. Lefebvre and E. Brändas (eds.): *Advanced Topics in Theoretical Chemical Physics.* 2003 ISBN 1-4020-1564-X
13. J. Rychlewski (ed.): *Explicitly Correlated Wave Functions in Chemistry and Physics.* Theory and Applications. 2003 ISBN 1-4020-1674-3
14. N.I. Gidopoulos and S. Wilson (eds.): *The Fundamentals of Electron Density, Density Matrix and Density Functional Theory in Atoms, Molecules and the Solid State.* 2003
ISBN 1-4020-1793-6
15. J.-P. Julien, J. Maruani, D. Mayou, S. Wilson and G. Delgado-Barrio (eds.): *Recent Advances in the Theory of Chemical and Physical Systems.* 2006
ISBN 1-4020-4527-1
16. S. Lahmar, J. Maruani, S. Wilson and G. Delgado-Barrio (eds.): *Topics in the Theory of Chemical and Physical Systems.* 2007
ISBN 978-1-4020-5459-4

FSC
www.fsc.org
MIX
Papier aus verantwortungsvollen Quellen
Paper from responsible sources
FSC® C105338

If you have any concerns about our products,
you can contact us on
ProductSafety@springernature.com

In case Publisher is established outside the EU,
the EU authorized representative is:
Springer Nature Customer Service Center GmbH
Europaplatz 3, 69115 Heidelberg, Germany

Printed by Libri Plureos GmbH
in Hamburg, Germany